奶牛乳腺炎防治技术

杨丰利 著

科 学 出 版 社

北 京

内 容 简 介

本书主要介绍了奶牛乳腺炎概况、奶牛乳腺炎的流行病学、奶牛乳腺炎致病性金黄色葡萄球菌病原学特征、奶牛隐性乳腺炎诊断技术、奶牛乳腺炎的治疗措施研究、奶牛乳腺炎疫苗研究、奶牛乳腺炎的危害、奶牛乳腺炎的预防措施。在开展奶牛乳腺炎的研究工作中，作者深切体会到写作本书的现实迫切性和责任感，希望本书能够为奶牛养殖企业提供理论指导，有助于其采取相应的措施，降低生产过程中由于奶牛乳腺炎引起的经济损失，进而促进奶牛养殖业健康稳定发展。

本书可供国内奶牛养殖企业的技术管理人员和畜牧兽医师、研究奶牛方向的研究生以及科研工作人员学习和参考使用，也可供国外相关技术人员和研究生学习使用。

图书在版编目(CIP)数据

奶牛乳腺炎防治技术/杨丰利著. —北京：科学出版社，2016
ISBN 978-7-03-048752-0

Ⅰ. ①奶⋯ Ⅱ. ①杨⋯ Ⅲ. ①乳牛–乳房炎–防治 Ⅳ. ①S858.23

中国版本图书馆 CIP 数据核字（2016）第 131804 号

责任编辑：王玉时 / 责任校对：彭珍珍
责任印制：张 伟 / 封面设计：迷底书装

科 学 出 版 社 出版
北京东黄城根北街 16 号
邮政编码：100717
http://www.sciencep.com

北京凌奇印刷有限责任公司 印刷
科学出版社发行 各地新华书店经销

*

2016 年 6 月第 一 版　　开本：787×1092　1/16
2023 年 1 月第三次印刷　印张：16
字数：400 000

定价：88.00 元
（如有印装质量问题，我社负责调换）

前　言

近年来，我国奶牛养殖业不断快速发展，与此同时，奶牛疾病仍然是阻碍奶牛养殖业健康发展的一个不可回避的重要因素。奶牛乳腺炎是世界范围内奶牛三大主要疾病之一，每年都会导致奶牛业遭受巨大的经济损失。

据资料报道，一头奶牛发生一次临床乳腺炎后，本胎次减少产奶量 500 kg 左右，约占总产奶量的 20%，同时影响其终身产奶水平，乳汁质量也受影响，乳脂率下降 0.3%，乳体细胞数上升到 100 万～150 万/mL。另有文献报道，在所有导致奶牛养殖收入减少的因素中，14%是由乳腺炎造成的奶牛淘汰及死亡所引起的，8%是牛奶废弃所引起的，另外 8%是治疗费用开支，剩下的 70%是由产奶量下降导致的。世界各国奶牛饲养管理技术有所差异，临床乳腺炎和隐性乳腺炎发病率差异很大，临床乳腺炎发病率常为 1.8%～37.1%，隐性乳腺炎阳性率常为 22.3%～62.9%。我国奶牛临床乳腺炎发病率大部分在 20%左右，隐性乳腺炎奶牛阳性率也在 50%左右。

奶牛乳腺炎主要致病菌有金黄色葡萄球菌、大肠杆菌、乳房链球菌、停乳链球菌、无乳链球菌和酵母菌。目前国内外防治奶牛乳腺炎主要靠加强牛场管理和抗生素治疗，由于抗生素疗法在公共卫生上的局限性和耐药菌株的产生，研究者进行了大量的疫苗接种实验。结果证明，利用疫苗预防和控制奶牛乳腺炎发生有一定的效果，但是由于致病菌种类繁多，采用疫苗很难将乳腺炎控制到极低的发病率。鉴于奶牛乳腺炎的危害性及其研究现状，作者将近年来关于该领域研究工作的成果在本书中进行了总结。本书内容共八章，分别对奶牛乳腺炎概况、奶牛乳腺炎的流行病学、奶牛乳腺炎致病性金黄色葡萄球菌病原学特征、奶牛隐性乳腺炎诊断技术、奶牛乳腺炎的治疗措施研究、奶牛乳腺炎疫苗研究、奶牛乳腺炎的危害、奶牛乳腺炎的预防措施，进行了系统而全面的阐述。附录以英文列出近年较有价值的奶牛乳腺炎研究成果，以供参阅。

本书在撰写过程中，得到了广西大学动物科学技术学院何宝祥教授、郑喜邦教授、钟诚副教授、李恭贺副教授，以及长江大学动物科学学院杨玉莹教授、李鹏副教授、李助南副教授、汪招雄老师和李小杉老师的悉心指导和帮助，在此向他们表示由衷的感谢！由于作者水平有限，书中难免存在一些不足，诚恳希望同行专家和广大读者予以批评指正，以便再版时修正。

杨丰利

2015 年 12 月 12 日

目 录

前言
第一章　奶牛乳腺炎概况 ··· 1
　　第一节　奶牛乳腺炎的定义及分类 ··· 1
　　第二节　奶牛乳腺炎的病因 ··· 5
　　第三节　奶牛乳腺炎病原学特点 ·· 10
　　第四节　奶牛乳腺炎的发病规律 ·· 15
　　第五节　奶牛乳腺炎的发病机制 ·· 15
　　第六节　牛奶中的体细胞 ··· 44
　　第七节　动物乳腺炎模型的建立 ·· 49
第二章　奶牛乳腺炎的流行病学 ··· 55
　　第一节　奶牛乳腺炎的发病率和主要致病菌调查 ························· 55
　　第二节　热应激对不同泌乳阶段奶牛日产奶量的影响 ···················· 71
第三章　奶牛乳腺炎致病性金黄色葡萄球菌病原学特征 ······················ 76
　　第一节　金黄色葡萄球菌研究进展 ··· 76
　　第二节　奶牛乳腺炎致病性金黄色葡萄球菌分子特征性分析 ·········· 85
第四章　奶牛隐性乳腺炎诊断技术 ·· 93
　　第一节　奶牛隐性乳腺炎常用的诊断技术 ···································· 93
　　第二节　奶牛隐性乳腺炎乳汁中丙二醛浓度和酶类活性分析 ·········· 98
第五章　奶牛乳腺炎的治疗措施研究 ··· 104
　　第一节　奶牛乳腺炎治疗措施研究进展 ···································· 105
　　第二节　中草药制剂对奶牛乳腺炎致病菌抗菌效果分析 ·············· 116
第六章　奶牛乳腺炎疫苗研究 ·· 125
　　第一节　奶牛乳腺炎疫苗研究进展 ··· 125
　　第二节　奶牛乳腺炎金黄色葡萄球菌疫苗的初步研制 ·················· 130
第七章　奶牛乳腺炎的危害 ·· 147
　　第一节　奶牛乳腺炎对生产性能的危害 ····································· 147
　　第二节　奶牛临床乳腺炎对繁殖性能的影响 ······························· 155
第八章　奶牛乳腺炎的预防措施 ·· 162
附录（Appendix）　The research results in dairy mastitis ················ 167
　　Section 1　Role of antioxidant vitamins and trace elements in mastitis in dairy cows ···· 167
　　Section 2　Effects of vitamins and trace-elements supplementation on milk production in dairy cows ···· 174
　　Section 3　Bovine mastitis in subtropical dairy farms ················ 180

Section 4	The prevalence of heifer mastitis and its associated risk factors in Huanggang	186
Section 5	Somatic cell counts positive effects on the DNA yield extracted directly from Murrah buffalo milk	193
Section 6	Detection of virulence-associated genes in *Staphylococcus aureus* isolated from bovine clinical mastitis milk samples in Guangxi	198
Section 7	Malonaldehyde level and some enzymatic activities in subclinical mastitis milk	204
Section 8	Clinical mastitis from calving to next conception negatively affected reproductive performance of dairy cows in Nanning, China	210

参考文献 218

第一章 奶牛乳腺炎概况

随着生活水平的提高，人们的健康意识越来越强，2007年国务院总理温家宝在重庆视察奶业时曾说过："我有一个梦想，让每个中国人，首先是孩子，每天都能喝上一斤奶"。国家进一步推动了"学生奶计划"，提出了"一杯奶强壮一个民族"的口号，从而使我国的奶牛养殖业在国家和各级政府的积极引导和扶持下得以长足发展。但是人们在注重加快乳业发展步伐、追求高产的同时，却面临对奶牛疾病预防和治疗的压力，尤其是普通病的发病率一直居高不下，奶牛乳腺炎就是其中之一，已严重影响了奶牛养殖效益和产品质量，给养殖业造成了巨大的经济损失（张颖，2014）。

第一节 奶牛乳腺炎的定义及分类

一、奶牛乳腺炎的定义

奶牛乳腺炎又称乳房炎（mastitis），顾名思义是乳腺发生炎症，简单地定义为乳腺受到物理、化学、致病微生物等因素的刺激所产生的一种炎症变化，主要致病菌包括金黄色葡萄球菌（*Staphylococcus aureus*）、无乳链球菌（为传染性致病菌）、大肠杆菌（*Escherichia coli*）、链球菌和肠球菌（环境致病菌，来自垫料、粪便和土壤等）（Heringstad et al., 2000）。乳腺炎的主要特点是乳腺组织发生病理学变化，乳汁发生理化性质及细菌学变化。在炎症早期，感染的乳区血管扩张，血液增加，血流减慢，同时毛细血管的通透性增加。患病奶牛体内出现了炎症反应，血液中的炎性介质变多，肥大细胞和巨噬细胞被活化后释放了细胞因子，淋巴细胞及中性多形核白细胞等进入受损部位。此过程中的细胞因子主要有白细胞介素-1（IL-1）、白细胞介素-6（IL-6）、白细胞介素-8（IL-8）和肿瘤坏死因子-α（TNF-α）等。临床乳腺炎的奶牛全血黏度、全血还原黏度、红细胞聚积指数、红细胞变形指数、红细胞刚性指数等血液流变指数均显著高于正常奶牛和患隐性乳腺炎的奶牛，临床乳腺炎奶牛血液中 IL-1β、IL-8 和 TNF-α 显著高于正常奶牛，其中 IL-8、TNF-α 显著高于隐性乳腺炎奶牛（冯士彬等，2011）。

二、奶牛乳腺炎的分类

（一）国际乳业联盟对乳腺炎的分类

国际乳业联盟（IDF）（1986）对奶牛乳腺炎的分类如下：①正常乳房，乳汁外观和成分均正常，并且无病原微生物生长；②潜在性感染，乳汁体细胞数正常，但存在病原微生物；③亚临床乳腺炎，乳房没有临床症状，但乳汁体细胞数增加，成分改变；④临床乳腺炎，包括急性、亚急性和慢性，急性乳腺炎的乳腺炎性反应和乳汁变化明显，体温升高；亚急性乳腺炎有炎症反应，乳汁变化，如有絮块存在，但体温正常；慢性乳腺炎症状不明显，乳房逐渐萎缩硬化，病程较长；⑤非特异性乳腺炎（无细菌乳腺炎），也可分为临床型和亚临床型,临床型的症状与由细菌感染的临床乳腺炎相似;亚临床型则只有乳汁变化,微生物检查均为阴性;⑥乳头管菌落形成,即病原只在乳头管繁殖,由乳头采样,菌检阳性,但乳房穿刺取样,则菌检阴性。

(二)美国国家乳腺炎委员会对乳腺炎的分类

美国国家乳腺炎委员会（National Mastitis Council,NMC）根据乳房和乳汁有无肉眼可见的变化将奶牛乳腺炎分为:①临床乳腺炎（clinical mastitis）,用肉眼观察就可见乳汁和（或）乳房的临床变化;②隐性乳腺炎（subclinical mastitis）,用肉眼观察乳房和乳汁均正常,但乳汁的电导率、体细胞数（somatic cell counts,SCC）、pH等理化性质已发生变化;③慢性乳腺炎（chronic mastitis）,通常是由于急性乳腺炎没有及时治疗或者没有得到彻底治愈,从而持续感染,使乳腺组织渐进性发炎引起的（刘朝,2007）。

临床乳腺炎为乳房间质、实质或间质组织的炎症。其特征是乳房和乳汁都有不同程度的异常,乳房红肿、疼痛、发热,拒绝哺乳和挤乳,轻者乳汁中有絮片、凝块,有时呈水样,乳房轻度肿胀;重者乳区重度肿胀,乳汁出现渗出物,泌乳减少甚至停止,同时出现全身反应,如食欲降低、脉搏加快、体温升高、精神沉郁等（赵中利,2012）。

隐性乳腺炎又称亚临床乳腺炎,是奶牛乳腺炎中最为常见的一种乳腺内感染,它既没有乳房局部和全身性的肉眼可见临床症状,乳汁从外观上也看不到任何变化,但是乳汁在理化性质及细菌学上已经发生变化。通过乳汁检查可发现体细胞数明显增加;pH 7.0以上,呈偏碱性;氯化钠含量在0.14%以上;乳汁中的细菌数和电导率上升等许多变化。由于乳房和乳汁看上去无异常,隐性乳腺炎经常被挤奶员和牛群管理人员忽视,许多隐性乳腺炎都是转变成临床乳腺炎后才被发现。相对于临床乳腺炎,奶牛的隐性乳腺炎很少对乳房组织造成迅速的损伤,一般也不会危及奶牛的生命,但有其明显的特点:首先,隐性乳腺炎感染牛虽然没有临床症状,却可以成为牛群感染源,继续感染周围的健康奶牛;其次,隐性乳腺炎常常发生于临床症状出现之前,而且病程缓慢,经常持续很长一段时间仍未被发现,造成产奶量长期缓慢下降,牛奶质量降低;再次,隐性乳腺炎经过长时间的发生,后期可转变成临床乳腺炎,导致产奶量急剧降低;最后,应用抗生素治疗效果不佳,治愈率低。隐性乳腺炎和临床乳腺炎一样,会直接导致牛奶品质下降,产奶量降低甚至奶牛的经济价值降低,对奶牛产业的影响十分巨大。研究表明,在一个牛群中大多数乳腺炎是隐性型,由乳腺炎造成的奶牛场的经济损失,70%是由隐性乳腺炎引起的。在奶牛业发达的美国,患有隐性乳腺炎的奶牛达50%。据文献记载,我国奶牛隐性乳腺炎阳性率为6.4%~75%,每头患牛每年造成损失1526.25元。因此,从经济角度上来说,隐性乳腺炎是所有奶牛乳腺炎中危害最为严重的一种（双金等,2001;李进,2014）。

慢性乳腺炎是由乳房持续感染引起的,通常没有临床症状,多以亚临床的形式持续几个月甚至几年,偶尔可发展成临床乳腺炎,但在短时间突发以后通常又转为非临床型,每次从隐性乳腺炎变成临床乳腺炎的过程,都伴随着乳腺组织的损坏过程,慢性乳腺炎一般由急性临床乳腺炎处理不及时或治疗不完全引起,造成持续感染,使乳腺组织渐进性发炎。全身症状不明显,仅是乳房有肿块,乳汁变清,有絮状物,产乳量明显下降,长期可导致乳腺组织纤维化,乳房萎缩。慢性顽固性乳腺炎,乳房可见坚硬的、大小不一的肿块,形状呈圆形或卵圆形，且容易复发很难治愈（Fox et al.，2003；李进，2014）。慢性乳腺炎病例发展的结果是乳房萎缩，成为瞎乳头。患病奶牛乳汁有不同程度的发黄和变稠，常含有块状或絮状物，患区乳房组织弹性降低，泌乳量减少，部分病牛会出现体温升高或食欲减退，甚至乳房肿大等症状（赵中利，2012）。

（三）国内对乳腺炎的分类

国内学者主要是根据病程的长短和临床症状将奶牛乳腺炎分类如下：①最急性乳腺炎，通常是一个乳区突然发生，乳腺组织红、肿、热、痛等炎症反应明显，体温升高，乳汁稀薄或混有较多的絮状物，产奶基本停止，炎症反应可由病原微生物自身，组织或细菌产生的酶、内毒素或外毒素，白细胞产物等引起。奶牛的全身性症状（发热、厌食、抑郁，有时甚至死亡等）主要是由败血症和毒血症造成的，而且这些症状通常出现在乳汁和乳腺有明显临床表现之前。②急性乳腺炎，乳腺组织出现红、肿、热、痛等炎症反应，产奶量下降，乳汁中混有较多的絮状物和（或）凝块，全身性临床症状与最急性乳腺炎相似，但没有其严重，如体温可能略微升高。③亚急性乳腺炎，乳腺组织基本没有红、肿、热、痛等炎症反应，也没有全身性临床症状，但乳汁中有凝块或絮状物。④慢性乳腺炎，触诊乳腺组织有硬块，乳汁稀薄易分层，产奶量下降。⑤隐性乳腺炎，乳房和乳汁肉眼观察均正常，但乳汁中体细胞数增多，需借助实验室方法检查。

根据急性乳腺炎发病特点的不同，可分为浆液性乳腺炎、卡他性乳腺炎、纤维蛋白性乳腺炎、出血性乳腺炎、化脓性乳腺炎、乳房坏疽和乳头管狭窄及闭锁等类型。

1. 浆液性乳腺炎 浆液性乳腺炎是以乳房充血、皮下及乳腺小叶结缔组织内有大量浆液性渗出物和白细胞为特征的一种急性乳腺疾病。引起浆液性乳腺炎的病原微生物有链球菌、葡萄球菌、大肠杆菌等。它们在乳房发生创伤和挤乳不当造成损伤时，通过皮肤和血管侵入乳房结缔组织而引起发病。浆液性乳腺炎可由胃肠或子宫疾病继发。牛患病后病程发展快，病势重。临床表现为乳房肿胀增大，皮肤充血发红，触诊发热，有疼痛反应，乳房淋巴结肿大，乳房实质坚硬。此病多发于母牛产后最初几天，4个乳区同时发病的较为少见，往往发生于4个乳区中的一个，或是乳房的一半。轻症者初期乳汁变化不大，以后逐渐变稀薄并带有絮状物，絮状物由少而多。重症病例乳房严重肿胀，患病乳区的产乳量减少，伴随病程的发展，乳汁变稀薄，内有絮状物，严重时有明显全身症状，患牛食欲减退、体温升高、精神不振。剖检可见乳腺小叶呈灰黄色，小叶间的间质及皮下结缔组织内有大量的浆液性渗出物和血管充血现象（顾有方等，2005；李进，2014）。

2. 卡他性乳腺炎 卡他性乳腺炎是乳腺泡上皮和输乳管上皮发生变形而形成的卡他性炎症。其病理特征主要是腺泡、腺管、输乳管和乳池的腺状上皮及其他上皮细胞剥脱和变性，渗出的白细胞及脱落的上皮沉积在上皮细胞的表面，因炎症只涉及上皮表层，故称卡他性乳腺炎。链球菌、葡萄球菌及大肠杆菌是其主要病原，当乳头黏膜受损伤、乳头括约肌松弛或乳房受冻时，病原菌即可侵入乳房引起发病。发病初期乳房不表现发热、疼痛等炎症过程，也不表现全身症状。随着病程的进一步发展，发热、精神沉郁等全身症状逐渐出现。由于病变部位的不同，症状也有所不同。腺泡卡他为个别小叶或数个小叶的局限性炎症，由炎症部位挤出的乳汁呈水样，有絮状凝块。如果炎症扩大，患部温度增高，挤奶有痛感，并有体温升高、食欲减退等全身症状。输乳管和乳池卡他的局部症状表现为乳头壁变厚，输乳管扩大，患部充血，乳汁中含有絮状凝块，凝块可阻塞输乳管使管腔扩大，乳头基部触诊检查可摸到波动，其絮状物凝块可阻塞乳头，不易被挤出。剖检可见乳腺小叶肿大，浅黄色颗粒状，按压时小叶流出浑浊渗出物。随着病区乳汁挤尽，后挤出的奶转为正常（李进，2014）。

3. 纤维蛋白性乳腺炎 纤维蛋白性乳腺炎是一种非常严重的急性乳腺炎，其特征是纤维蛋白渗出到乳池、乳管的黏膜表面，或沉淀在乳腺实质深处，可继发乳腺坏死或脓性液化。通常由卡他性乳腺炎发展而来，患部乳区发热，严重肿胀、疼痛，乳房淋巴结肿大，触诊乳房基部时，可听到捻发音。全身症状明显，体温升高，饮食减退或废绝。

4. 出血性乳腺炎 出血性乳腺炎是以乳汁呈现血样变化为特征的乳腺炎。一般由卡他性或浆液性乳腺炎转变而来，多发生在半个或整个乳房，常为急性发病。这些类型的乳腺炎能使乳房血管渗透性增高，导致血细胞涌出血管壁而进入乳腺细胞。此外，乳房受钝性创伤引起乳房内血管破裂，是发生本病的次要原因。某些细菌或腐败菌感染也可引起疾病，由此而引发的乳腺炎，往往伴有严重的全身症状。出血性乳腺炎主要临床特征是乳腺深部或输乳管出血。在发病的初期只是乳汁中发现血液，呈粉红色血样或已凝结成血块、血丝等，挤奶时表现为剧烈疼痛。随着病情发展，患病部位显著肿胀。有时乳房上出现严重水肿、发硬，皮肤出现红色或紫红色斑点。各乳区可同时发现血液，有的仅出现在一两个乳区中。血量的多少不一，差异较大，病牛通常全身状况无明显变化。个别病牛可见体温升高、精神萎靡、食欲减少等症状（曹立亭等，2009；李进，2014）。

5. 化脓性乳腺炎 化脓性乳腺炎分为化脓性卡他性乳腺炎、乳房肿胀和乳房蜂窝织炎等几种。临床上均以患部初期发热、肿胀、疼痛，后期化脓，并伴有体温升高、乳房淋巴结肿大、饮食减少、精神萎靡为特征。化脓性乳腺炎可由细菌感染引起，也可由非感染性乳腺炎（如机械性损伤、创伤引起的乳腺炎）转化而来，但最终都有病原菌的参与。化脓性卡他性乳腺炎除了以上特点以外，还表现类似卡他性乳腺炎的症状。急性期过后，患部的炎症程度渐渐减轻，肿胀缩小，精神及饮食正常。但患部组织变性，乳汁稀薄，呈黄色或浅黄色。

6. 乳房坏疽 乳房坏疽是乳腺炎的并发病，由腐败菌通过乳管、血液循环系统或乳房创口而感染乳房所引起。临床表现为乳房的皮肤上出现圆形病灶，局部疼痛，呈红色或紫色，随着病程的发展，患部组织逐渐腐败分解，流出污秽分泌物，有臭味；患侧淋巴结增大，泌乳停止或可挤出少量的脓样分泌物。在患病期间，患牛行走时后肢展开，由于疼痛拒绝挤奶器或犊牛吸奶。

7. 乳头管狭窄及闭锁 乳头管狭窄表现为挤乳时只可挤出一股细的乳流，是由于平时挤乳操作不当，如人工挤乳动作粗鲁和机械挤乳间隔时间过长等造成乳头机械性损伤，使乳头管黏膜结缔组织增生，致使乳头管狭窄，排乳困难。外伤性或炎症引起的狭窄，可见到局部症状。当仔细捻转乳头时，可发现乳头管粗硬。此外，乳头管腔内生长肿瘤也可发生闭锁。乳头管闭锁的症状表现为乳房充满，但挤不出乳汁，仔细进行触诊检查，可发现乳头管局部有增厚变硬的地方。

8. 亚急性乳腺炎 亚急性乳腺炎是一种温和的炎症，乳房和乳汁肉眼观察无变化，通常奶中可见小的薄片或凝乳块，牛奶颜色变淡，一般没有全身症状。有时乳房有肿胀，产奶量下降。亚急性乳腺炎通常发生于急性乳腺炎经过抗生素治疗的牛。未经治疗的牛，尽管症状温和，往往导致乳腺炎反复发作，尤其是金黄色葡萄球菌引起的临床乳腺炎。

第二节　奶牛乳腺炎的病因

乳腺炎是奶牛最常见的疾病之一，已经报道的能够引发奶牛乳腺炎的不同致病性微生物共有137种，包括细菌、支原体、酵母菌和藻类（Bradley，2002）。国内报道奶牛乳腺炎主要致病菌有金黄色葡萄球菌、大肠杆菌、乳房链球菌、停乳链球菌、无乳链球菌和酵母菌。某些情况下其他一些疾病也会继发乳腺炎，临床上最多见的继发乳腺炎的疾病是子宫内膜炎和阴道炎，其次是中毒性疾病、高热性疾病（Nash et al.，2000；Riollet et al.，2001）。卫生条件、环境、奶牛年龄、泌乳阶段、温度等因素都会影响感染乳腺炎的概率。挤奶员不正规操作、未消毒挤奶设备、多重使用抗生素治疗、没有适当消毒的器具、不当的卫生习惯等因素，都会导致细菌在奶牛间的传播。

一、病原微生物方面的因素

在健康奶牛的乳房上或乳房内有大量的微生物寄居。据报道，可以从奶牛乳腺中分离出137种微生物。国内外众多学者调查研究发现，引起乳腺炎最主要的细菌有金黄色葡萄球菌、大肠杆菌、链球菌、酵母菌，其中链球菌有五类：无乳链球菌、停乳链球菌、乳房链球菌、化脓链球菌及粪链球菌。感染链球菌的奶牛，大多取慢性经过，多数无临床症状。无乳链球菌有高度传染性，潜伏期较长，可达数月，不易被发现，主要通过挤奶员的手和消毒不彻底的奶杯传染（马吉锋等，2010）。

病原微生物按其传染方式及生活繁殖场所大致可分为两类，一类是接触传染性病原微生物，如无乳链球菌、金黄色葡萄球菌、停乳链球菌和支原体等，这些微生物定植于乳腺内并通过挤奶员或挤奶设备传播；另一类是环境性病原菌，此类病原菌主要在粪便、垫料或褥草、污水、运动场等处生活和繁殖，奶牛因接触污染的外界环境而感染，并导致发病，主要包括致病性大肠杆菌、沙门氏菌、化脓棒状杆菌、乳房链球菌等。此外，其他的病原微生物如真菌、支原体等也可引发乳腺炎。易本驰（2007）在对河南信阳地区奶牛乳腺炎发病规律调查研究中发现，该地区奶牛隐性乳腺炎发病率为51.81%，其中，葡萄球菌导致发病的占35.36%，链球菌占14.49%，革兰氏阴性杆菌引发的占49.43%。

二、管理因素

管理不当也是引发乳腺炎的主要原因之一。管理不妥，如对临床乳腺炎未能采取针对性治疗，不能及时淘汰久治不愈病牛，致使传染源一直存在于牛群中，给健康奶牛造成极大威胁。

挤奶操作不当造成的乳房损伤也是引起乳腺炎最常见或最主要的原因之一，主要有手工挤奶时没有使用拳握式挤奶法或机器挤奶时真空压过高、频率过快或过慢、挤奶杯有损伤等。

另外，挤奶时间过长、挤奶工具未定期消毒或挤奶后不用消毒液浸泡、乳腺炎病牛与健康奶牛共用一套挤奶设备、牛舍和运动场有尖锐异物、牛只自己踩伤或压伤等也可引起乳房损伤，继而发生乳腺炎（肖定汉，2002）。干奶不合理，如对已达到干奶期的奶牛不能及时科学地进行干奶或是干奶时间掌握不好；干奶前最后一次挤奶未能挤净或是挤净了但没用抗菌药物封闭乳头，以及对高产奶牛自然干奶时不能减少精料和限制饮水次数，这些都可导致干奶期或是下一个泌乳期奶牛乳腺炎发生。

三、营养因素

机体抵抗力与营养紧密相关，特别是产犊前后的营养水平对母牛产后疾病的易感性具有重要影响。饲料营养不全，如饲料中维生素或微量元素的缺乏会降低机体免疫力，进而促进乳腺炎的发生和发展。饲喂过量的氮或蛋白质也是促发乳腺炎发病的因素之一，饲料中脲的含量增多可使感染乳房的数量和敏感性增加16%。干奶期日粮中的矿物质不平衡也会增加乳腺炎的风险，如产褥热或低血钙症的母牛，产犊后3天内乳腺炎风险增加5.4倍，乳房水肿的风险增加约3倍（丁丹丹，2014）。

（一）氮和蛋白质

日粮中含有过多的氮和蛋白质是引发乳腺炎的原因之一。很多人认为，奶牛日粮中的蛋白质含量与乳腺炎的发生率没有明确的联系。有研究报道，氮不是以蛋白质的形式影响乳腺炎的发生率，而是以尿素、氨气等形式来影响的。

（二）钙磷比例

日粮中钙磷比例失调会导致奶牛产后发生产褥热。在大型奶牛场中，如果日粮中缺乏钙元素，奶牛将会在产后几小时发生由大肠菌群引起的乳腺炎。在干乳期，低血钙症通常是由日粮中钙磷比例不协调引起的。

（三）硒和维生素E

有研究人员对硒和维生素E在乳腺炎防治中的作用进行调查发现，日粮中保持合理的硒含量有助于预防乳腺炎，能够降低感染的严重程度以及缩短感染的持续时间（Mukherjee，2008）。另外，硒还能够通过增加白细胞的释放和吞噬细胞的功能来加强机体的免疫功能。日粮中同时添加适量的硒和维生素E才能够充分发挥其作用，仅添加维生素E能够降低乳汁中的体细胞数，但是不能改变乳腺炎的发生率。仅添加大剂量的硒对奶牛不但无益，而且还有毒性作用。奶牛日粮中每天添加16 mg的硒而不添加维生素E可增加乳腺炎的发生率（付云贺，2015）。

大量的研究表明，硒和维生素E增强中性粒细胞吞噬活性，降低氧化应激水平，提高机体免疫调控能力，并能降低乳腺炎发生风险（张雯，2014）。有研究证明，维生素E在奶牛围产期对其他肠道疾病以及临床乳腺炎仅有有限的治疗作用，对于不清楚奶牛状况的情况下使用维生素E反倒有不良影响（Bouwstra et al.，2010），这表明奶牛应该在有效监测或者基于硒和维生素E缺乏的情况下使用。此外，在补充维生素、矿物质和微量元素时，应该考虑到不同的国家和地区的差异（土壤、谷物、青贮等方面的差异）。奶牛产犊前1个月应该补硒以增加血液谷胱甘肽过氧化物酶（GPX）活性，降低体细胞数（SCC）在奶牛泌乳时的含量，各种来源的硒，如注射用亚硒酸钠、钡硒或酵母硒均具有较好的预防和治疗作用（Ceballos-Marquez et al., 2010）。矿物质和维生素的补充与奶牛产犊前后血液及乳汁中中性粒细胞的活性、数量有关，而其与维持乳腺健康有关（Green and Bradley, 2013）。

反刍动物严重缺硒可导致营养性肌病，又称白肌病（white muscle disease, WMD）。临界水平的硒缺乏在成年奶牛中常见，长期处于临界水平的缺硒，增加了奶牛罹患乳腺炎、子宫内膜炎和胎衣不下的风险（Bourne et al., 2008）。多项研究表明，添加硒能够增强细胞的吞噬活性并降低临床乳腺炎风险，硒缺乏与牛奶中高体细胞数和泌乳早期易患临床乳腺炎有直接关系。有趣的是，饲喂硒浓度在16 mg/天时，反而提高了奶牛临床乳腺炎的发病率，因此硒浓度超过临界阈值时对机体有损伤作用（张雯，2014）。奶牛乳汁中硒含量水平与奶牛罹患金黄色葡萄球菌型乳腺炎风险呈负相关，饲粮中添加硒降低了放牧奶牛乳腺内感染病原菌的概率，同时降低了泌乳早期较高的奶牛乳汁中体细胞数（Ceballos et al., 2010）。每千克饲料干物质添加亚硒酸钠0.05~0.35 mg饲喂奶牛，在泌乳14~16周通过乳头管人工感染大肠杆菌，加硒组未发生临床乳腺炎，而对照组发生了较为严重的乳腺炎症状。硒的添加浓度，美国研究委员会（National Research Council, NRC）建议为0.3 mg/kg干物质，英国推荐使用量是0.1 mg/kg干物质。多项研究均已表明，饲料中添加适量的硒，能够最大限度地降低乳腺炎、白肌病和呼吸系统疾病的发病率。由于奶牛机体硒浓度的提高，人类可以通过牛奶和肉类摄取足够的营养硒，而预防和治疗某些疾病（Sheppard and Sanipelli, 2012）。

研究较多的与奶牛乳腺炎相关的硒蛋白是谷胱甘肽过氧化物酶（GPX），其具有保护吞噬细胞免受氧化损伤的作用，且其血液浓度与SCC呈负相关，因此，GPX的抗氧化作用对亚临床乳腺炎的发病率和临床乳腺炎的疾病严重程度均具有一定的影响（Spears and Weiss, 2008）。研究表明，饲粮加硒可以显著提高超氧化物歧化酶的表达，其血清活性（16.01 IU/mL）与乳腺炎组（12.85 IU/mL）和正常组（14.78 IU/mL）相比，具有显著提高。临床添加营养硒，特别是与维生素E合用时，提高了GPX的表达，同时增强了血液和乳汁中多形核白细胞的吞噬活性和吞噬指数，与单独的恩诺沙星治疗组相比，加硒组显著提高了乳腺中中性粒细胞的抗炎作用，降低了乳汁中的体细胞数，提高了奶牛对乳腺内感染的抗病能力（Ugall et al., 2008）。研究表明，牛外周血淋巴细胞硫氧还蛋白氧化还原酶（thioredoxin reductase, TrxR）的活性降低导致了围产期奶牛发生氧化应激和乳腺促炎因子的表达升高，添加亚硒酸钠显著提高了牛内皮细胞的GPX和TRxR活性，减弱了氧化应激诱导的炎症反应和免疫亢进。围产期和泌乳中前期奶牛乳腺组织中GPX、TrxR与促炎细胞因子TNF-α、IL-2、IFN-γ等基因表达呈显著的负相关，因此，硒蛋白在调节免疫功能和炎症反应方面起关键作用（Klein, 2009）。

（四）营养均衡

一般情况下，奶牛乳腺炎的易感性与饲粮密切相关。奶牛过多的身体能量损耗和血液高β-羟丁酸浓度会增加牛群水肿的风险，而这又与临床乳腺炎的发生风险较高相关（Compton et al., 2007）。因此，减少体质损耗和过度动员个体的能量损失似乎是降低奶牛乳腺炎易感性的一种合理策略。治疗用的离子载体（最初被开发为抗球虫药的家禽抗生素，用于牛生长促进饲料添加剂，如莫能菌素）可以降低奶牛的乳腺炎发生风险（Parker et al.,

2008），而离子载体在奶牛中的使用，促使奶牛体况评分升高及 β-羟丁酸和游离脂肪酸浓度降低，并进一步导致奶牛隐性乳腺炎感染率的降低（Ugall et al.，2009）。虽然产犊前饲喂干草、青贮饲料、甜菜浆或玉米青贮可能会增加乳腺炎的风险，但对其相关性阐述并不明显，可能与特定的饲养方式或牛场管理水平抑或环境卫生有关，而不是营养和乳腺炎之间的实际关联（Vliegher et al.，2012）。

铜具有抗氧化功能，能减少奶牛临床乳腺炎发病；锌在防止上皮细胞感染方面有重要作用，但其在保护乳腺免受病原感染方面的作用还有待研究。维生素 A 和 β-胡萝卜素具有保护黏膜表面免受氧化应激损伤的作用（Heinrichs et al.，2009）。哺乳奶牛对维生素和矿物质的需求一般是由生长速率和身体相对成熟时的体重决定的，但很少有具体的数据。从管理的角度来看，奶牛日粮维生素和矿物质的补充应因地/因群而异，但围产期奶牛和泌乳奶牛的饮食结构务必最优化。

日粮中营养不均衡，如蛋白质、豆科植物（青苜蓿等）含量过多，胡萝卜素、维生素 A、维生素 E 缺乏都会提高乳腺炎的发病率，因此科学合理的管理与饲养是降低乳腺炎发病率的必要措施（李诗莹，2015）。

四、环境性因素

气温突变、气候寒冷或过热、空气过湿或曝晒等都会使奶牛处于应激状态，使机体免疫力下降，直接或间接影响乳腺炎的发生率。牛舍采光不好、通风不良、空气质量差、尘埃多等是导致乳腺炎发生率上升的因素；运动场牛粪堆积、未定期消毒、垫料不干净等在乳腺炎的发生过程中起重要作用，不用垫料的奶牛比用垫料的奶牛乳腺炎的感染水平高一倍（周成青，2013）。

（一）气候

气候对乳腺炎的发生有着直接或间接的影响，严寒、干燥、湿度过高、热应激等均有助于乳腺炎的发生。有研究报道，气温的剧烈变化常会导致乳腺炎发生，但是需要与其他因素相互作用才能引发乳腺炎。夏季乳腺炎是一种特殊的乳腺炎类型，主要是由昆虫叮咬使乳腺感染化脓棒状杆菌和其他厌氧菌而引发的，该类乳腺炎的发生率和奶牛场所处的地理位置的气候有关（付云贺，2015）。

（二）舍饲

如果一直在室内饲养奶牛，乳腺炎的发生率会增加，因为滞留在圈舍内乳腺损伤的风险较大，而且室内的病原微生物数量多于室外。在草场上自由活动的奶牛群，只有挤奶时奶牛才会进入室内，这种情况下很少有奶牛发生由大肠菌群引起的乳腺炎。

（三）圈舍空气质量

圈舍内过于干燥、湿度过高、温度频变等都能使乳腺炎的发病率增加。这些因素对圈舍中病原菌的密度有较大影响，如当相对湿度较低时，奶牛会发生肺炎杆菌感染，而由大肠杆菌引起的感染则不会随着湿度有太大变化（付云贺，2015）。

（四）垫草

感染乳腺炎的奶牛所产的乳汁，其湿度有利于微生物的生长。奶牛一天中约有 14 h 与垫草接触，没有垫草的牛场乳腺炎的发生率是有垫草牛场的两倍。另外，不同材质的垫草可能对不同病原微生物的生长有不同的影响。最常用的是稻草，切碎的稻草和锯末比报纸好，但是稻草比锯末更有利于克雷伯菌的生长（付云贺，2015）。

五、乳房形状的影响因素

从解剖学角度来看，如果乳腺位置浅及向内、4 个乳房附着牢固、乳头管括约肌紧张，奶牛在活动、趴卧时，乳头不易拖地，在地上、垫草及土壤上的病原菌不易进入乳房，因此可以减少乳腺炎的发生。而且形状良好、匀称的乳房产奶量偏高。所以在选择高产奶牛时要考虑乳房的形状，避免选择乳房位置深及向外、乳房松弛，以及悬池、乳头管括约肌松弛的奶牛（李诗莹，2015）。

奶牛的乳房位于耻骨部的腹下壁，两股根部之间。其外形各异，有盆形、山羊形、发育不平衡形和扁平型等，以盆形为好（郭梦尧，2014）。乳房的外形与品种、年龄、泌乳期、护理及组织发育程度等因素有关。乳房分 4 个区，称乳区，左右乳区间有浅的乳房间沟，前后乳区间则无可见界限。每个乳区之下各有一个乳头。乳头大体呈圆锥形，长 5～8 cm，5 cm 以下为短型，10 cm 以上为长型，其长度受挤奶方法影响很大。乳头下端中央，有一乳头孔，为乳头管的开口。20%～40%的牛有副乳头，形小，多位于后乳头后方或前后乳头之间，有一定的遗传性。个别的副乳头有小的乳腺及乳头管，但与泌乳量无关。乳头损伤引起疼痛，同时破坏乳头管内的黏膜组织，为病原的入侵提供了门户。病原微生物通过乳头管进入乳腺，通过内部各个乳导管感染乳腺小叶，继而引发感染。引起乳头损伤的因素主要包括：①挤奶的方式不正确，如挤奶手法不当、榨奶器不合格等；②乳头内陷或先天生长过小，导致幼畜吸奶用力过度，引起乳头内压增大造成损伤；③频繁且长时间的挤奶增加乳头受损的机会等（郭梦尧，2014）。

另有报道，前乳房附着松弛、乳房悬垂不良、乳房位置较深、后乳房附着松弛、乳头位置向外、长乳头的奶牛易患乳腺炎，而盆碗形乳房的发病率低于圆筒形乳房，乳头管紧的奶牛发病率低（Sabini et al.，2001；丁丹丹，2014）。

六、遗传因素

目前，外国有报道说，乳腺炎一定程度上与遗传因素有关，发病率较高的母牛，其后代往往也具有较高的发病率。不同品种奶牛对乳腺炎的易感性也不相同，产奶量高与乳腺炎发生率有一定的相关性，高产奶牛一般情况下更易受到感染。从遗传学角度来看，乳脂率和临床乳腺炎的发病率有着一定的关系。奶牛产奶量越高，对乳腺炎的敏感性越高（Sordillo，2005）。

荷斯坦奶牛乳腺炎的发生率比其他品种奶牛相对较高（Compton et al.，2007）。先天免疫水平也影响着动物乳腺炎的发生，研究显示防御素分泌多的奶牛不容易被病原菌感染，Toll 样受体-2（Toll-like receptor-2，TLR-2）等病原体相关分子模式（PAMP）与乳腺炎的发生也具有密切的联系（Compton et al.，2009）。

研究表明，种公牛的差异对后代患乳腺炎的概率具有重要的影响（Heringstad et al.，2001；郭梦尧，2014），高产奶牛乳腺炎的发病率一般较高。不同品种的奶牛乳腺炎的发病情况也存在差异，最直观的证据是，黄牛、牦牛、水牛等品种不易患乳腺炎（高树新，2005）。

七、年龄因素

乳腺炎发生率及乳汁中体细胞数会随着奶牛年龄的增大而增大。这主要是由于随着泌乳期的增多，乳导管不断扩大，使得更多的病原微生物能进入乳房，导致乳腺组织损伤和体细胞数大幅上升。也有报道称，这可能是由于它们机体的代谢功能减退或是被致病菌感染导致的。

八、其他因素

产后奶牛对疾病抵抗力降低，大量的泌乳造成营养不良、贫血、应激或疲劳，也使动物机体对病原菌的抵抗力降低。此时犊牛口、鼻、喉处的细菌，以及工作人员带入的细菌更容易通过乳导管感染乳腺。此外，乳腺炎病史及其他乳腺疾病也是导致乳腺炎发生的因素。

有研究表明，产科疾病与奶牛乳腺炎的发生具有明显的相关性。乳腺水肿、乳头水肿、乳汁中带血等显著地促进产后 1 周内乳腺炎发生及后期金黄色葡萄球菌性奶牛乳腺炎的发生。产前挤奶会减少乳腺水肿及促进致病菌的排出，降低乳腺炎的发生（Bowers et al.，2006；Daniels et al.，2007）。瑞典的奶牛场里，围产期产科疾病，如胎盘滞留、子宫内膜炎、子宫积脓、难产等会诱发乳腺炎（Svensson et al.，2006）。

第三节　奶牛乳腺炎病原学特点

已经报道的能够引发奶牛乳腺炎的不同致病性微生物共有137种，包括细菌、支原体、酵母菌和藻类（Bradley，2002）。在过去很长一段时间内，世界各国的研究人员也把主要精力投入到对于奶牛乳腺炎致病微生物的分析中，并且对各自的研究进行了报道，如来自北美地区（Middleton et al.，2005）、欧洲（Piepers et al，2010）、新西兰（Compton et al.，2007；Parker et al.，2008）和南美地区等的报道。国内报道奶牛乳腺炎主要致病菌有金黄色葡萄球菌、大肠杆菌、乳房链球菌、停乳链球菌、无乳链球菌和酵母菌。总体上可以分为两大类：第一类是接触传染性病原微生物，它们通过挤奶员或挤奶机在奶牛个体间和牛群间进行传播，是引起奶牛发生乳腺炎的最主要的致病菌，这类致病微生物包括金黄葡萄球菌、无乳链球菌和停乳链球菌等；第二类是环境性病原微生物，包括大肠杆菌、肺炎克雷伯菌、沙门氏菌、凝固酶阴性葡萄球菌、乳房链球菌、支原体和真菌等，它们通常存在于奶牛的皮肤、乳导管、垫料、粪便和土壤中，甚至是来自于犊牛的口腔。在一些乳汁样品中，可分离出两种或者两种以上病原菌混合感染。

（一）金黄色葡萄球菌

金黄色葡萄球菌引起的乳房感染主要为慢性感染，并且难以诊断，主要感染源为感染乳区的分泌物，即经过挤奶员的手和挤奶设备进行传播。当挤奶设备不良或奶牛乳头有损伤、挤奶程序不合理时，金黄色葡萄球菌就会趁机侵入乳腺组织，且金黄色葡萄球菌容易对抗生素产生耐药性，因此，金黄色葡萄球菌一旦在牛群中定植下来就很难被清除（丁丹丹，2014）。有研究认为，金黄色葡萄球菌是奶牛乳腺炎致病菌里面最难治疗和防控的病原（Barkema et al.，2006）。感染的奶牛不仅自身发病，同时还会随乳汁将金黄色葡萄球菌分泌，致使其他奶牛被感染，对饲养人员及消费者身体健康造成危害。虽然目前的研究认为金黄色葡萄球菌主要引发亚临床型奶牛乳腺炎（Piepers et al.，2010），但也不可忽视其还可以引发临床乳腺炎。金黄色葡萄球菌的传播很容易，饲养场里的苍蝇就可以做到，而且该菌的变异及流行速度很快。金黄色葡萄球菌应该作为奶牛乳腺炎的主要病原进行更为深入的研究（Waller et al.，2009）。

金黄色葡萄球菌性乳腺炎通常表现为慢性或者呈最急性。急性患牛可致其死亡，慢性患牛其乳房被侵害。金黄色葡萄球菌主要感染源来自乳区的分泌物，经挤奶员的手臂、挤奶设备传播，引起慢性感染，对抗生素易产生抗药性。由金黄色葡萄球菌引起的慢性病例极少自愈，除了降低产量之外，病变乳区还会纤维化，造成永久实质性破坏。

金黄色葡萄球菌的感染率为5%～50%（Leitner et al.，2003），其主要损害乳房上皮细胞和腺泡的功能，使牛奶的产量和质量下降，并且它引起的奶牛乳腺炎很难治愈和根除（Talbot and Lacasse，2005）。金黄色葡萄球菌入侵乳腺组织的分子机制目前还不很清楚，推测主要有以下因素引起：①定植（colonization）；②局部感染（local infection）；③进入血液（systemic dissemination）；④转移性感染（metastatic infection）；⑤中毒症状（toxinosis）。此外，金黄色葡萄球菌产生的超抗原形成免疫抑制，以及它的内化作用可逃避免疫系统的监视，也是它能引起疾病发生的重要因素（Ferens and Bohach，2000）。

由于抗生素的持续选择和滥用、细菌基因突变，以及耐药质粒、转座子或整合子在细菌间的传递，使得金黄色葡萄球菌对多种药物产生抗性，如β-内酰胺类、青霉素类、头孢类、氨基糖苷类、四环素类、氟喹诺若酮类等。耐药机制主要是：①产生钝化酶，如β-内酰胺酶使β-内酰胺环裂解而使该类抗生素药物失效；②抗菌药物作用靶点的改变，如耐甲氧西林金黄色葡萄球菌（methicillin-resistant *Staphylococcus aureus*，MRSA）的青霉素结合蛋白（PBP）要比甲氧西林敏感金黄色葡萄球菌（MSSA）的 PBP 蛋白多个 PBP2a 蛋白，PBP2a 蛋白的转肽酶区与现有抗菌药物的亲和力极低，当抗菌药物抑制正常 PBP 蛋白活性时，它便提供转肽酶活性，继续合成肽聚糖，呈现高度耐药性；③外排泵系统，某些细菌能将进入菌体的药物泵出体外，如金黄色葡萄球菌的 MepA 泵出系统能将四环素、环丙沙星、诺氟沙星等药物泵出菌体；④细菌外膜通透性的改变，主要是细菌外膜上通透性蛋白的性质和数量改变，减少药物的摄入量而导致耐药，如大肠杆菌的 OmpF 蛋白多次接触抗生素后，其结构基因发生突变，引起 OmpF 通道蛋白丢失，而导致β-内酰胺类、喹诺酮类等药物进入菌体的量减少（苏洋，2013）。邓海平等（2010）报道甘肃地区乳腺炎金黄色葡萄球菌的感染率为 23.08%，菌株对复方新诺明、阿莫西林、青霉素、庆大霉素及四环素等药物具有较强抗性；贵州地区乳腺炎金黄色葡萄球菌的感染率为 58.78%，菌株对复方新诺明、四环素等药物产生较强的抗性。

20 世纪 40 年代以前，金黄色葡萄球菌引起感染性疾病的死亡率很高。青霉素在 1941 年被发现后，使得金黄色葡萄球菌病例的治愈率大有改观，但仅在一年后便发现耐青霉素的金黄色葡萄球菌。到20世纪50年代末，半合成青霉素——甲氧西林在1959年首次用于治疗因PBP2a蛋白异常表达引起对青霉素产生抗性的金黄色葡萄球菌后，1961年英国便报道了MRSA的出现。在随后的几十年内该菌株迅速向全世界扩散，耐药性也越来越强，耐药变种菌株也在不断地出现，由于其高度耐药性和异质性给临床治疗带来了极大的困难。MRSA的感染已经成为各国学者研究的热点，许多学者将MRSA的感染同艾滋病、乙型肝炎列为世界范围内最难解决的三大感染性疾病。美国每年因感染MRSA而死亡的人数已超过艾滋病，我国MRSA的感染率高达33.3%~80.4%（苏洋，2013）。

1972年，首次从乳腺炎奶样中分离到动物源性MRSA，但当时并没有检测它的 *mecA* 耐药基因。直到2003年，一篇韩国的研究才首次报道了 *mecA* 基因呈阳性的乳腺炎金黄色葡萄球菌（Lee，2003）。后来，在欧洲、美洲和亚洲等地都有零星的、低频率的MRSA存在于临床型和亚临床型奶牛乳腺炎中（Sabour et al.，2004；Moon et al.，2007；Vanderhaeghen and Cerpentier，2010）。

MRSA产生耐药性的主要原因是：甲氧西林敏感金黄色葡萄球菌（methicillin-susceptible *Staphylococcus aureus*，MSSA）获得并整合一个可移动基因原件SCCmec染色体盒到自身基因组，它携带甲氧西林、重金属或其他药物的抗性基因；通过 *mecA* 基因编码产生额外的、经修饰的PBP2a蛋白，它与β-内酰胺类抗生素的亲和力极低，使得金黄色葡萄球菌在β-内酰胺类抗生素存在的条件下能继续合成细胞壁，呈现高耐药性。*mecA* 基因的表达受其他基因的调节，包括 *mecRI-mecI* 基因、*blaRI-blaI* 基因及 *orf1* 和 *orf2*。其中由 *mecI* 基因编码的MecI蛋白（抑制因子）和 *mecRI* 基因编码的MecRI蛋白（诱导因子）在调控 *mecA* 基因表达时起重要作用，MecI蛋白与 *mecA* 基因的启动子区域结合而使PBP2a蛋白的表达量很少或不表达，但当诱导因子MecRI蛋白存在时可除去抑制因子MecI蛋白的阻碍效应而使PBP2a蛋白大量表达（苏洋，2013）。

奶牛场员工和临床兽医师是与奶牛接触最为密切的MRSA携带者，因此奶牛中MRSA的流行情况也备受重视，但目前的报道表明，奶牛乳腺炎MRSA的分离率还是比较低的。在比利时，从118份临床型奶牛乳腺炎奶样中分离到11株（9.3%）MRSA，分型结果表明，这些菌株都属于ST398、spatypes T011或t567株，这些菌株携带 *SCCmec-type IVa* 或 *SCCmec-type V* 基因证实它们属于新出现的LA-MRSA CC398菌株；在德国，从17家奶牛养殖场临床型奶牛乳腺炎奶样中分离到25株MRSA，牛场员工身上分离到2株MRSA，研究表明，23株为spa type011、3株为spa type034、1株为spatype2576，MLST分型发现所有菌株都归属于ST398，SCCmec-type V占主要比例。各地区的研究报道表明，ST398 MRSA为欧洲国家奶牛乳腺炎MRSA的一优势菌系，同样，ST8、ST1、ST239、ST329、ST8、ST425、ST130和ST1245在土耳其、匈牙利等欧洲国家的奶牛乳腺炎奶样中零星发现。Laura等（2011）第一次从奶牛乳腺炎奶样中检出SCCmecⅪ型MRSA。亚洲地区关于奶牛乳腺炎MRSA的研究报道较少，来自日本和韩国地区的研究报道表明，ST5、ST580、ST1和ST72 MRSA也可从牛奶样中检出。以上研究表明，ST398和其他STs MRSA都能引起奶牛乳腺炎，且存在地区性差异，虽然感染率不高，但也引起广大兽医界同行们和相关部门的高度重视。就像目前所了解的，像ST398这样物种特异性不强且具有传播性的菌株，一旦在牛群中传播开来将会给奶牛养殖业带来严重的威胁（苏洋，2013）。

(二) 凝固酶阴性葡萄球菌

凝固酶阴性葡萄球菌致病性较低，但大量的研究表明，有10余种该类细菌普遍在患病牛乳汁中被发现（Piessens et al.，2011；Supré et al.，2011）。虽然对于此类细菌与奶牛乳腺炎的关系已经有来自世界各地的大量报道（Makovec and Ruegg，2003；Pitkälä et al.，2004；Piepers et al.，2007），但其致病机制及对奶牛乳腺的影响还有待深入研究（Taponen et al.，2007；Schukken et al.，2009）。此外，仅利用生化反应鉴定和研究该类细菌已经不适用于如今的科研工作（Sampimon and Sampimon，2009），我们应当更多地探讨其分子生物学致病机制（Zadoks and Watts，2009），深入地分析其种属特异性效应、毒力特征及抗生素耐药性的产生模式（Supré and Vliegher，2009；Piessens et al.，2010；Braem et al.，2011）。

(三) 大肠杆菌性乳腺炎

大肠杆菌是革兰氏阴性杆菌致病菌，为环境性致病菌，在环境中到处存在，侵入乳房的机会极多，因此不清洁的垫料往往是革兰氏阴性杆菌性乳腺炎发病的主要原因。细菌很可能是从损伤的生殖道进入血液或淋巴而引起感染，炎症大多呈急性或最急性，甚至引起乳房坏疽。

大肠杆菌性乳腺炎，多见于高产牛分娩后不久的泌乳高峰时期，主要是由于产后初期和泌乳期的前30天内体细胞数低，奶牛抵抗力弱，易感染大肠杆菌性乳腺炎（丁丹丹，2014）。临床上通常呈最急性，是由于病菌细胞膜所产生的内毒素引起的毒血症，多为单乳区发病，整个病乳区形成肿块，坚硬，可挤出灰黄色液体。其病情急，病程短，可于数日内死亡，主要症状是乳房肿胀、高热、精神不振、食欲废绝、腹泻、乳汁水样黄色、迅速停止泌乳。溶血性大肠杆菌性乳腺炎，愈后不良，甚至死亡，一般使用抗生素进行治疗即可收到良好效果。

(四) 链球菌属

链球菌属主要包括无乳链球菌、停乳链球菌、乳房链球菌、兽疫链球菌、兰氏D类肠球菌、兰氏G类链球菌、兰氏L类链球菌等（李进，2014）。感染本属菌时，多无临床症状或无明显的临床症状。本属菌中的无乳链球菌几乎是专门寄生在乳腺之中，其他组织极少感染。其传播途径是挤奶员的手或消毒不完全（甚至不消毒）的挤奶杯，有时是以牛舍中的蝇类为媒介进行传播。感染牛大多取慢性经过，有时取急性-亚急性经过（Compton et al.，2007）。有的病例出现严重症状，但持续时间短。化脓链球菌主要是人的化脓菌（如扁桃腺炎、猩红热、化脓性咽喉疼痛的病原菌）常经挤奶员的手传播给奶牛。常发生在分娩之后，表现为急性-最急性乳腺炎。兽疫链球菌是各种家畜均可感染的化脓菌，呈散发性，此菌毒性很大，牛感染后，可引起败血症而死亡，多因乳房的病灶转移到全身所致。

(五) 化脓棒状杆菌

化脓棒状杆菌能引起急性-亚急性化脓性乳腺炎，具有地方性，发病率仅1%～2%，一般呈散发性。但有的地方呈季节性流行，如在英国多于6～9月雨季发病，故有"夏季乳腺炎"之称。本菌除了由洗涤乳牛乳房用的水和擦布传播外，苍蝇也起着重要的传播作用。

（六）绿脓杆菌

绿脓杆菌性乳腺炎，呈散发性，多为急性局限性病例，发病率1%以下。该病菌通常存在于水和土壤中，饮水也可感染。患病乳区肿胀，脓液呈蓝绿色，乳汁水样有凝块，患牛高热，可因败血症而死。也有呈慢性、亚急性的，治疗困难。诊断时必须先将乳样置温箱内使细菌繁殖，再接种和培养，否则不易生长。

（七）产气荚膜杆菌

产气荚膜杆菌感染乳腺主要是由于乳房注入污染了的抗生素制剂所致，将发生急性坏疽性乳腺炎，常为暴发性。

（八）支原体

支原体也是引发奶牛乳腺炎的病原（Barkema et al.，2009；Passchyn et al.，2012），支原体很难在体外培养，而且乳汁中检测也较为困难（Biddle et al.，2004）。因此对于该类病原的流行情况报道相对较少。有关研究认为，这些支原体感染病原应该是来自于牛的呼吸道（Barkema et al.，2009）。已确定致病性牛乳腺炎支原体至少有12种，较常分离到的有牛支原体等6种。牛支原体性乳腺炎是传染性的，临床特征是：患乳区肿胀，但无热、痛反应，泌乳异常，且常伴有关节炎性跛行和呼吸道症状。病原体能通过挤奶、呼吸和配种过程等传播。多数支原体不仅能引起乳腺炎，还能引起其他器官组织疾病。病原体存在于牛的乳腺、呼吸道、生殖道及关节滑液等处，可随挤奶、咳嗽或喷嚏排出体外，在冰箱中可存活数天，在粪便中可存活数周。自然感染时，通常由一种支原体为主，另外几种支原体为辅，共同引起牛的乳腺炎。有的病例还伴有金黄色葡萄球菌、无乳链球菌等其他细菌。感染后常呈地方性流行，干乳期敏感性较高。乳汁呈黄褐色，水样，有颗粒状或絮状凝乳块，轻症乳汁似正常，但静置后底层出现粉状或条状沉淀，上层变清亮。泌乳量明显下降，通常不表现明显的全身症状，有时乳汁细胞数也不增加，常伴有关节炎性跛行，有时跛行比乳腺炎更严重。呼吸道症状多见于青年病牛。本病尚无有效疗法。

（九）真菌

早在1901年就已从乳汁中分离出酵母菌样真菌，以后不断发现，主要由念珠菌属、隐球菌属、毛孢子菌属和曲霉菌属等引起真菌性乳腺炎，但不多，呈散发性，多发生于应用抗生素治疗之后，或由于药品、器械被真菌污染。

真菌感染引起严重的临床乳腺炎。其临床特点是：乳房肿大、发热、不食、泌乳量不断下降、乳腺组织为肉芽肿样组织所代替，很难治疗。念珠菌属和毛孢子菌属感染可引起局限性严重炎症，但大多数患牛都能在两周到一月内转为正常且无后遗症。真菌感染常因使用污染的抗生素制剂或乳房注入时操作不小心所致。

第四节 奶牛乳腺炎的发病规律

(一) 不同季节奶牛乳腺炎发病情况

隐性乳腺炎在不同季节的发病率不同，夏季气温高不仅有利于病原微生物的生长繁殖，而且热应激还会导致奶牛的抵抗力下降，因此夏季奶牛隐性乳腺炎发病率高；而冬季由于气温低不利于微生物的生长繁殖，隐性乳腺炎发病率相对较低。研究发现，每年隐性乳腺炎的发病率都是从 3 月份开始明显增多，7~9 月达到高峰，进入 11 月份后乳腺炎发病率又开始回落。

(二) 不同乳区乳腺炎发病情况

杨章平和贝水荣研究发现，奶牛左侧乳区乳腺炎的发病率比右侧高 4.5%，且两侧的发病率差异极显著。这是由于在人工挤奶时，挤奶人员往往习惯于用右手，致使左侧乳区挤奶操作不方便，而造成左侧乳区的损伤，导致左侧乳区发病率远高于右侧（苏俊强等，2004）。利用管道式机器挤奶则能避免人工挤奶的这一弊端，使左右乳区的隐性乳腺炎发病率无明显差异（谢怀根和许世勇，2003）。

(三) 不同年龄、胎次奶牛乳腺炎发病情况

奶牛隐性乳腺炎发病率会随着奶牛年龄、胎次的增加而增加。奶牛随着年龄增加，体质减弱，自身免疫功能也随之下降。另外，乳房在挤奶过程中长期受挤压，造成乳头的机械损伤增多，乳头括约肌机能减退，出现松弛与闭合不严的问题，为病原微生物入侵创造了条件。

因而会出现隐性乳腺炎的发病率随着年龄、胎次的增加而增高的现象。李庆国（2006）在研究中发现，7~8 岁奶牛隐性乳腺炎发病率与 3~4 岁比较差异显著，3~4 胎、5~6 胎与 1~2 胎比较差异显著。杨克礼等（2005）在研究中发现，不同胎次的奶牛隐性乳腺炎的发病率差异极显著，其中以 2~3 胎奶牛发病率最高，而头胎牛发病率最低。

(四) 不同泌乳阶段奶牛乳腺炎发病情况

杨克礼等（2005）在对合肥某奶牛场调查研究时发现，当奶牛泌乳至第 5、第 6 泌乳月时，高产奶牛将获得较高的日产奶量，此时奶牛的乳房负担较重，这种应激会加剧隐性乳腺炎发病率的增高。另外，随着挤奶时间的加长，外界环境中致病菌的入侵机会就会增多。此外，奶牛在泌乳高峰期后淋巴免疫细胞数量下降，这也会导致隐性乳腺炎发病率增高。

第五节 奶牛乳腺炎的发病机制

一、奶牛乳腺的防御机制

对乳腺免疫防御机制的充分认识有助于制定乳腺炎预防和控制措施，从而控制奶牛等反刍动物的主要疾病——乳腺炎。尽管已有 10 多年的研究，但由于先天免疫领域的广泛性，目前对乳房先天防御系统的认识仍然是不完整的。然而，这方面的资料正随着乳腺病

原和诱导防御试验的深入而在完善，乳腺上皮细胞在局部的防御和白细胞的动员作用也正在深入的研究中。在上皮部位研究天然免疫最合适的模型莫过于乳腺上皮细胞与 *E. coli* 和 *S. aureus* 等引起乳腺炎的病原微生物之间的相互作用的研究模型。免疫反应遗传分析、微阵列基因技术、转录技术及 RNA 干扰等强大的新研究工具从分子和基因水平对奶牛控制易感/耐受乳腺炎的关键机制的研究成为可能，也使乳腺先天防御和乳腺炎致病相互作用的研究前景更为光明，从而通过免疫调节和遗传改进提高奶牛等产乳反刍动物抗乳腺炎的能力（敬晓棋，2013）。

天然免疫不等同于非特异性免疫，尽管所涉及的是非抗原特异性的防御机制，但天然免疫却具有分子特异性。与适应性免疫不同，天然免疫不需要与提前存在的抗原相遇，即不需要特异性抗原刺激。反之，天然免疫反应不会因为同一抗原的重复刺激而放大，不像适应性免疫可以依赖免疫记忆，也就是无记忆性。然而，天然免疫遇到感染性抗原时可在数小时内被诱导活化。天然免疫与组织急、慢性炎症和脓毒症等其他反应紧密相关。

天然免疫可被认为是由感应和效应两个部分组成，前者是机体如何知道受到感染，后者则是机体如何抵抗感染，二者分别可继续划分为体液和细胞两个部分。也可分为固有（直接存在于病原入侵的部位）和可移动（可移动到感染发生的部位）两个部分，二者可通过炎症的发生与否加以区别。当固有防御不足以清除病原微生物时，则系统性的可动防御被调动进行补救。

（一）固有防御

1. 乳头管屏障 乳腺具有天然的防御能力，病原微生物侵入乳腺后能否发病，不仅取决于病原的特性，也取决于机体的防御能力，乳头管是乳腺防御的第一道防线，乳头管周围有一层平滑肌构成的括约肌，可以使乳头管处于关闭状态，防止乳汁溢出，阻止病原微生物进入乳腺内。由于乳头管是病原进入乳腺的途径，在挤奶间期及干乳期，乳头管的多层上皮细胞分泌的角蛋白形成角质栓将其封闭，这种蜡样的栓塞可物理性地阻止病原微生物进入乳腺。乳头管上皮分泌的角蛋白（keratin），是一种含有脂质的纤维性蛋白，具有抑菌特性。已经证实乳头角蛋白中含有高浓度的具有抑菌性的脂肪酸，如豆蔻酸、棕榈酸、亚油酸等，角蛋白内层中所含的阳离子蛋白能结合到链球菌和葡萄球菌胞壁上，破坏它们的细胞壁结构，从而起到杀菌作用（Paulrud，2005）。乳头管上皮还可以分泌抗菌性脂类和蛋白质组成防御体系，抑制细菌的生长。乳头管黏膜和其上的角化上皮细胞产生的溶菌霉素，具有抑菌和杀菌的作用。乳头管的近端可产生局部抗体和淋巴因子。挤奶与乳头管屏障作用关系密切，挤奶时真空泵的抽吸和乳汁的流出使乳头管扩大，并冲出角质栓，挤奶后乳头管括约肌收缩并重新关闭乳头管。

挤奶时，细菌可能会通过受损乳头、乳头末端杂乱潮湿的环境、不卫生的挤奶设备等进入乳头管内。乳头创伤有可能会使乳导管处于开放状态，而且还会损害乳头窦内的角蛋白或黏膜，从而使细菌等病原微生物更加容易侵入乳腺内。挤奶后，乳头管仍会处于开放状态 1~2 h，在这期间，病原微生物可以侵入乳头管。

机器挤奶对乳头管有明显的影响，可引起乳头组织机械的和循环性的障碍，长期不合适的挤奶器使用可改变乳头管括约肌功能,同时也会改变乳头管的免疫防御。干净卫生的乳头皮肤能有效地减少其表面 *S. aureus* 对乳腺内部的感染。许多研究也表明,即使进行定期的乳头浸泡消毒,在发生乳腺感染时,细菌已经在乳头管部位存在至少数周,病原菌在乳头管的长期存在表明其可以适应乳头管的微环境,更进一步表明,乳头管区域在乳腺免疫防御方面具有特殊的作用。因此,乳头管可能的作用是感知和防止病原微生物的侵入,当然也需要大量存在于上皮内的白细胞的作用。

乳头管屏障功能的完善可有效地保持乳腺内部的无菌微环境。与肠道、口腔、上呼吸道等部位的上皮细胞不同,乳腺上皮细胞很少受到细菌成分刺激,任何细菌都可看作入侵者。乳腺上皮细胞内的白细胞的特殊性和乳腺上皮细胞表面的感应受体的种类和分布受乳腺内的感染状况所调节。

2. 非特异性抗菌分子 补体以低但恰当的浓度存在于健康且未发炎乳房的乳汁中,由于缺乏补体 C1q 而通过替代途径激活,结果细菌表面调理素 C3b 和 C3bi 积累,并产生前炎症分子 C5a。血清 C3 含量(2.5%)比所期望的血清渗出物高,但 C3b/C3b1 的积累则相当。乳汁中的 C5 因个体而差异较大(0.2%~1.9%),结果产生 C5a 的能力差异也较大。C5a 与乳腺炎症反应的起始有关,作为趋化因子能诱导中性粒细胞穿过乳腺上皮细胞,结果导致乳腺炎症反应的差异,但这种起始炎症反应的作用还有待进一步研究。

乳铁蛋白(lactoferrin,Lf)的多种功能与机体天然免疫有关。Lf 具有两个级别功能,即螯铁作用及抑菌和防止由铁离子催化的氧自由基损伤。由乳腺上皮细胞分泌的柠檬酸是泌乳期即将开始的标志,牛奶或初乳中柠檬酸盐物质的量比较高时会抑制 Lf 的抑菌作用。在复杂的乳腺中降低柠檬酸浓度和提高碳酸浓度对 Lf 的螯铁作用是有益的。

转铁蛋白是乳汁中发现的另一种铁离子结合蛋白。与啮齿动物和兔子等不同,反刍动物乳腺不能合成转铁蛋白,但可通过乳腺上皮的转胞作用和乳腺炎发生时的血浆渗出进入乳汁中,因此反刍动物乳汁中只含有很低的转铁蛋白(初乳中约 1 mg/mL,常乳中 0.02~0.04 mg/mL,而血清中可达到 4~5 mg/mL)。发生急性的 *E. coli* 乳腺炎时乳汁中转铁蛋白可达到血清白蛋白的水平(1 mg/mL)。在 Lf 含量升高之前,转铁蛋白就已经在乳汁中形成了具有抑菌作用的铁螯合物。

溶菌酶(*N*-acetylmuramyl hydrolase,*N*-乙酰胞壁酰水解酶)是一种杀菌蛋白,能裂解 G^+ 和 G^- 菌细胞壁的肽聚糖。虽然这种效应只对少数细菌有效,但溶菌酶能与抗体、补体、乳铁蛋白等协同作用。例如,乳铁蛋白与结合细菌表面的脂胞壁酸质能使葡萄球菌对溶菌酶更敏感。然而与人乳的溶菌酶浓度(10 mg/100mL)相比,100 mL 牛乳只含有 13 μg 溶菌酶。因此,溶菌酶不是乳腺防御的关键。

(二)乳腺的体液免疫

乳腺内体液免疫主要由 4 种免疫球蛋白来完成,分别为 IgG1、IgG2、IgA、IgM。IgG1 是最主要的,它的主要作用是抗毒和高度的凝集,并能在整个泌乳期持续存在;在急性感染时,免疫球蛋白的含量在几小时内就会显著升高,且 IgG2 的含量会高过 IgG1(正常时

IgG1 的含量高），IgA 和 IgG2 具有调理素的活性而 IgG1 没有。IgG 与病原体结合，通过中性粒细胞的吞噬，增强对病原体的杀伤作用。IgA 可以阻止病原体与乳腺上皮的黏附，抑制细菌的增殖。IgA 在局部免疫当中具有重要作用。IgM 只存在于免疫后的一段时间，且作用效果不明显。

（三）乳腺的细胞免疫

健康乳腺组织仅有少量白细胞，乳汁中发现的细胞也被认为与乳腺组织浸润有关，其中主要是 T 淋巴细胞（$CD4^+$ T 在腺泡间组织、$CD8^+$ T 在腺泡周围）。同时还有少量的中性粒细胞、巨噬细胞、自然杀伤细胞（NK 细胞）、B 细胞及少量的树突状细胞（Sordillo，2005），这些免疫细胞与乳腺上皮细胞构成了乳腺组织的防御细胞体系。乳腺炎发生时乳汁中的白细胞数量显著升高，这也是目前多种乳腺炎检测方法的根据所在（Roy et al.，2008）。乳腺免疫包括天然免疫和获得性免疫共同的保护（Sordillo and Streicher，2002），但在乳腺感染的最初阶段以天然免疫为主，在白细胞介素-1β（IL-1β）、肿瘤坏死因子-α（TNF-α）、IL-6 等炎性细胞因子的作用下免疫细胞趋化到炎症部位（Bannerman et al.，2004），这也是乳腺炎的重要特征之一。另外，上皮细胞、内皮细胞及乳腺免疫细胞与病原菌之间复杂的相互作用，对识别和促进炎性细胞因子分泌和调控乳腺天然免疫反应也具有重要作用。

微生物感染引起乳腺炎有三个阶段：①病原通过乳头管侵入；②在乳池中存活并增殖，向乳腺其他部位扩散的感染阶段；③以 SCC 升高为主要特征的炎症反应阶段，此时乳腺炎的临床症状逐渐显现。细菌穿过乳头括约肌和乳头管后，黏附于乳腺上皮细胞上，并释放细菌毒素刺激，使上皮细胞分泌 TNF-α、IL-6、IL-8（Rainard and Riollet，2003）。随后巨噬细胞分泌的 TNF-α 和 IL-1β 等炎性因子募大量的中性粒细胞进入乳腺组织（Rainard，2003），此时中性粒细胞替代健康乳腺组织中巨噬细胞成为主要的细胞类型。在细胞因子的刺激下，中性粒细胞的杀菌活性进一步增强，并产生能够增强局部炎症反应的前列腺素和白三烯等活性分子（Bannerman et al.，2004）。

1. 巨噬细胞 巨噬细胞是常乳、停乳期乳腺组织中主要的细胞类型之一，一般可消化常见的乳腺炎病原，但其吞噬能力弱于中性粒细胞，同时二者的消化能力也弱于血中的同类细胞。巨噬细胞和上皮细胞分泌的炎性因子及化学因子能刺激内皮细胞分泌细胞黏附分子 E-selectin、胞间黏附分子 1（intercellular adhesion molecule 1，ICAM-1）及血管黏附分子 1（vascular cellular adhesion molecule 1），使血液中性粒细胞与内皮细胞结合并穿过上皮迁移至感染部位（敬晓棋，2013）。

虽然乳汁中的巨噬细胞并不起主要的防御作用，但却是潜在的抗原递呈细胞，能提示抗原的入侵和启动炎症反应，这是其在天然免疫中的主要功能。围产期乳腺巨噬细胞功能会显著下降，这种变化正好与围产期乳腺炎发生率的升高相一致（Sordillo and Streicher，2002）。巨噬细胞具有吞噬、杀伤病原微生物的作用，还可以在与组织相容性复合体（major histocompatibility complex，MHC）结合的过程中对抗原进行提呈和加工。巨噬细胞还可产生细胞因子和少量的炎性介质。因此，巨噬细胞也是乳腺免疫防御体系中的重要部分（王华等，2010；冯超，2013）。

2. 中性粒细胞　中性粒细胞（poly-moiphonuclearneutrophil，PMN）、巨噬细胞（macrophage）、淋巴细胞（lymphocyte）、脱落的乳腺上皮细胞（bovine mammary epithelial cell，BMEC）等是乳汁中主要的免疫细胞，具有吞噬病原微生物和发挥免疫防御的作用，当乳腺受到病原微生物感染时，这些免疫细胞发挥作用，将病原微生物杀灭或清除。乳腺一旦受到病原微生物的入侵，在细胞因子的作用下中性粒细胞会最早从血液中游走出来，进入到乳汁当中，通过吞噬作用和分泌溶菌酶等杀灭病原微生物。

中性粒细胞核呈杆状或 2～5 分叶状，叶与叶间有细丝相连。其颗粒表面有一层膜包裹，可分 1 型至 4 型，颗粒中含髓过氧化物酶（myeloperoxidase）、酸性磷酸酶、吞噬素（phagocytin）、溶菌酶、β-葡糖苷酸酶、碱性磷酸酶等。中性粒细胞具趋化作用、吞噬作用和杀菌作用。中性粒细胞是乳腺组织和乳汁中最主要的细胞类型。正常乳汁中的中性粒细胞对乳腺炎发生的防御机制还不清楚，而且在乳池乳汁中的相当一部分中性粒细胞是死亡的或正在发生凋亡的，或者不在活化状态。另外，健康乳腺中中性粒细胞的存在似乎与乳腺内感染呈负相关（Burton and Erskine，2003），感染前乳汁中的中性粒细胞的活力与肠杆菌性乳腺炎的严重程度密切相关（Mehrzad et al.，2004）。

乳腺组织受到感染时，巨噬细胞和上皮细胞等分泌的炎性细胞因子及化学因子，募集中性粒细胞到达乳腺炎症部位。除了趋化因子外，补体分子 C5a、C3a、IL-8、IL-12、脂多糖（LPS）等都有募集中性粒细胞的作用。中性粒细胞在感染部位吞噬细菌，产生活性氧、低分子质量抗菌肽及防御素，从而清除能引起乳腺炎的各种病原微生物。围产期乳腺炎的高发生率可能与该时期中性粒细胞的下降有关。如果细菌侵入乳腺组织并存活时，乳腺组织的中性粒细胞浸润将很快转变为 T 淋巴细胞、B 淋巴细胞和单核细胞，但中性粒细胞仍然是慢性乳腺炎的主要浸润细胞类型（Rainard，2003；敬晓棋，2013）。

3. NK 细胞　NK 细胞是一类具有大颗粒、依赖于主要 MHC 发挥 ADCC 的淋巴细胞，虽然中性粒细胞和巨噬细胞也都具有发现和清除病原菌的能力，但 NK 细胞是清除胞内菌的关键细胞类型。此外，在受到刺激后，NK 细胞能释放皂化蛋白（saposin）家族的杀菌蛋白杀伤革兰氏阳性和阴性细菌，说明细胞因子刺激牛乳腺淋巴细胞所拥有的体外杀菌能力是不受 MHC 限制的。有研究发现，牛 NK 样细胞（$CD2^+$/$CD3^-$ T 淋巴细胞）在 IL-2 刺激下能产生抗 *S. aureus* 的杀菌活性，也编码与 saposin-like 蛋白家族同源的细胞裂解酶基因（Sordillo，2005），表明 NK 细胞在乳腺天然防御中具有一定的作用。因此 NK 细胞通过细胞毒作用和分泌细胞因子，在乳腺炎发生时，能发挥抗感染、免疫调节等重要作用。

4. 淋巴细胞　淋巴细胞有 T 淋巴细胞（简称 T 细胞）和 B 淋巴细胞（简称 B 细胞），T 淋巴细胞负责识别抗原、激活 B 淋巴细胞；B 淋巴细胞负责产生抗体，两者共同作用发挥免疫应答（immune response）作用。淋巴细胞能通过膜受体特异性识别各种抗原，即具有特异性、多样性和记忆性的免疫特征。B 淋巴细胞利用自身膜表面的受体特异性，识别树突状细胞和巨噬细胞等抗原递呈细胞加工处理的抗原，并产生相应抗体以抵御侵入的病原微生物。T 淋巴细胞分为 αβ T 细胞和 γδ T 细胞两类，二者又可分为多个亚群，因此 T 淋巴细胞在机体免疫中具有多样性的功能（敬晓棋，2013）。此外，淋巴细胞还可产生大量的细胞因子，这些细胞因子在维持免疫平衡和抵御外界病原微生物的入侵方面有重要作用。

(1) αβ T 细胞　αβ T 细胞包括 CD4$^+$ T 细胞和 CD8$^+$ T 细胞。CD8$^+$ T 细胞是正常乳腺组织重要的淋巴细胞类型，主要清除自身细胞或在细菌感染时控制免疫反应，由于其清除衰老和受损伤的细胞，因此提高了乳腺组织对细菌感染的耐受力（Sordillo，2005）。

CD4$^+$ T 细胞受到抗原与 MHC Ⅱ、抗原递呈细胞、B 淋巴细胞或巨噬细胞形成的复合物刺激活化后，成为炎症部位的主要细胞类型。目前，T 淋巴细胞在乳腺炎中的变化和作用仅有少量的研究报道。在 BALB/c 小鼠上建立的 *S. agalactiae* 乳腺炎模型研究显示，感染 5 天时巨噬细胞及乳腺淋巴结 CD4$^+$ T 细胞和 CD8$^+$ T 细胞数量增加，与奶牛 *S. agalactiae* 乳腺炎中的检测结果一致。乳房注射 *S. aureus* α-毒素的研究显示，注射毒素后 12 h 时乳汁中的 CD8$^+$ T 细胞数量升高，96 h 时 CD4$^+$ T 细胞数量升高，表明 CD8$^+$ T 细胞和 CD4$^+$ T 细胞在乳腺免疫防御中具有重要作用，可对将来提高乳腺防御能力和乳腺炎疫苗研究提供参考（Riollet et al.，2001）。然而 *E. coli* 内毒素诱导的奶牛乳腺炎结果显示，在内毒素注射后 4 h 时淋巴结输出液中 CD8$^+$ T 细胞和 CD4$^+$ T 细胞均升高，但以 CD4$^+$ T 细胞为主，CD4$^+$：CD8$^+$ 比例显著高于淋巴结输入液和对照组淋巴结，这可能是由于不同的病原微生物引起的乳腺免疫反应存在一定的差异。这些研究共同的特点在于，研究是在 CD4$^+$ T 细胞和 CD8$^+$ T 细胞两个没有进行细分的 T 细胞亚群上进行的，特别是 CD4$^+$ T 细胞包含多个不同功能的亚群（敬晓棋，2013），因此要阐明淋巴细胞在乳腺炎发生过程中的免疫机制，还有待进一步更为细致的研究。

CD4$^+$ T 细胞被分为 Th1 和 Th2 两个细胞亚群。Th1 细胞通过分泌细胞因子 IFN-γ 来清除胞内菌、病毒及癌变细胞，介导细胞免疫；Th2 细胞辅助 B 细胞增殖和产生抗体，与体液免疫有关。然而，近年来，在 Th1 和 Th2 的基础上，又相继确定了调节性 Th 细胞（regulatory Th cell，Treg）和 IL-17 分泌性 Th 细胞（IL-17 producing Th cell，Th17）两个 CD4$^+$ T 细胞亚群，特别是 Th17 的确定，加深了对黏膜炎症和免疫机制的认识。Th17 细胞多见于发生炎症的黏膜部位，以分泌 IL-17A、IL-17F 和 IL-22 等效应细胞因子为特征，在黏膜部位清除感染的胞外菌，能诱导机体强烈的炎症反应（Zheng et al.，2008）。Treg 是与免疫耐受有关的一类 CD4$^+$ T 细胞亚群，以 CD4$^+$CD25$^+$ 为特异的表面标志，并特异地表达 FoxP3（Zhou et al.，2009），在维持机体中枢和外周免疫稳定、调控炎症反应强度方面具有重要作用。乳腺炎是一个典型的局部炎症反应，其发生是否是 Th17 介导的结果，目前还没有相关的研究报道。奶牛乳腺炎造成巨大经济损失的根本原因在于，强烈的炎症反应损伤了乳腺上皮细胞，从而导致泌乳能力迅速下降，Treg 作为调控免疫反应强度的细胞，在乳腺炎上的调控作用也需要做深入的研究。

(2) γδ T 细胞　γδ T 细胞只是 T 细胞中的一个小的细胞亚群，但却起着不同于其他细胞的独特的免疫保护功能。γδ T 细胞来源于胸腺，需要与 αβ T 细胞相似的 TCR 重组，其特有的 TCR γ 和 δ 可分为多个亚群，不同亚群的 γδ T 细胞的功能和存在的位点是不相同的，因此，γδ T 细胞在抗原识别方面具有巨大变化潜能，能够在不同的特定部位起到独特的免疫保护功能。与 CD8$^+$ 或 CD4$^+$αβ 细胞不同，γδ T 细胞并不局限于 MHC Ⅰ 类或 Ⅱ 类的识别，小鼠的部分 γδ T 细胞能够识别 T10 和 T22 等 MHC Ⅰ B 类抗原。因此，γδ T 细胞不需要外源抗原的诱导活化来促进炎症反应，而是对进入保护位点病原的内源性信号作出反应，这种非常规的活化机制可能是免疫系统内微环境选择的结果。尽管 γδ T 细胞只是人、小鼠等大多数动物外周血淋巴细胞的一小部分，然而牛的γδ T 细胞却大约占到外周循环性淋巴细胞总数的 60%，特别是青年牛，这表明 γδ T 细胞在牛免疫中比其他动物更为关键（敬晓棋，2013）。

γδ T 细胞的免疫反应存在多种机制。γδ T 细胞在炎症的天然反应阶段就能够分泌 IFN-γ、IL-17 等常规 T 细胞分泌的细胞因子，且在一定时间段内是 IL-17 的主要分泌细胞。γδ T 细胞分泌 IL-17 的独特性还在于受到刺激是即刻就可合成的，而不像其他细胞需要一个启动过程而造成 IL-17 分泌的延迟，这种特性是胸腺选择的结果，如小鼠 Vγ4 胸腺细胞能分泌 IL-17，而 Vγ1 细胞只有极其微弱的分泌能力（Roark et al.，2008）。γδ T 细胞是 IL-17 的一个重要来源，能够在黏膜部位快速地对抗原作出反应以保护机体免受侵害。

（四）细胞因子在乳腺免疫中的作用

细胞因子（cytokine，CK）是细胞与细胞交流的关键信使分子，在乳腺免疫应答过程中具有重要的作用，并与维持机体免疫自稳及病理过程的调节有关，其中，白细胞介素（interleukin，IL）、肿瘤坏死因子（tumor necrosis factor，TNF）的作用尤为明显。乳腺炎病理过程中免疫细胞的活性主要受到促炎因子的调控，促炎因子的作用包括：①增强巨噬细胞和中性粒细胞的杀菌活性；②促进中性粒细胞向感染部位迁移；③刺激树突状细胞成熟；④调控获得性免疫反应。目前，在健康乳腺和受感染的乳腺已发现 IL-1β、IL-2、IL-6、IL-8、IL-12、IL-17、集落刺激因子（colony-stimulating factors，CSF）、IFN-γ 及 TNF-α 等多种细胞因子（Sordillo and Streicher，2002；敬晓棋，2013）。其中，白细胞介素是由白细胞中的粒细胞产生的一类细胞因子，可以激活 T 淋巴细胞，使其产生免疫记忆。IL-1 能够激活中性粒细胞，使中性粒细胞的数量增加，增强中性粒细胞的活性，介导急性炎症反应。IL-2 可以促进 B 淋巴细胞的生长和分化，使其产生抗体，增强免疫应答，增强 NK 细胞杀伤肿瘤细胞的能力，IL-2 还可以提高奶牛自身的免疫能力。肿瘤坏死因子主要由单核-巨噬细胞分泌，通过刺激单核巨噬细胞，使其产生白细胞介素等炎性介质，肿瘤坏死因子还可以增强中性粒细胞的吞噬能力。TNF 还具有杀瘤抑菌、免疫调节和抗感染的作用。

1. IFN-γ   IFN-γ 是 NK 细胞、$CD4^+/CD8^+$ T 细胞及 γδ T 细胞在丝裂原或抗原刺激下所分泌的，IFN-γ 与获得性免疫和 T 淋巴细胞活化及 IL-12 分泌相关，能够增强乳腺中中性粒细胞的吞噬能力，此外，对病毒和胞内菌感染性疾病具有重要作用（Nonnecke et al.，2003）。然而，Satorres 等（2009）的研究认为，IFN-γ 在抗 *S. aureus* 感染中的作用是有害的，其能够促进 *S. aureus* 在黏膜部位集落的形成。

2. IL-1β   IL-1β 由单核-巨噬细胞及上皮细胞分泌，能够增强和维持中性粒细胞的抗微生物能力，诱导 Th1、Th2、Th17 等 $CD4^+$ T 细胞及 γδ T 细胞活化（Hebel et al.，2010；Lalor et al.，2011）。在 *E. coli* 感染的炎症反应中，IL-1β 调节内皮细胞表达黏附分子及中性粒细胞的趋化作用，而在 *S. aureus* 感染时，其仅在感染早期具有重要作用。IL-1β 也是人体微生物特异性 Th17 细胞活化所必需的细胞因子，在 IL-1β 的协同作用下可将 Treg 细胞转化为 Th17 细胞，从而促进 IL-17 分泌和增强机体炎症反应。

3. IL-6   IL-6 是由树突状细胞、巨噬细胞、B 细胞等造血系细胞，以及角质细胞、成纤维细胞、上皮细胞、星形细胞等非造血系细胞分泌的一种多效性细胞因子，可抑制Th1细胞分化，通过上调IL-4分泌促进Th2细胞分化和抗体分泌。有研究表明，大量的IL-6表达与Th17细胞的分化有关，IL-6与TGF-β协同是IL-17分泌细胞必需的条件（IL-6与杆菌或 S. aureus 引起的机体败血性休克有关），可通过对乳腺中中性粒细胞向单核细胞的转换而降低中性粒细胞对乳腺的损伤作用，还可调控肝细胞对急性期蛋白的合成。

4. IL-10　　IL-10是一种调控免疫系统炎症反应强度的细胞因子，可防止过度的炎症反应造成组织损伤，在免疫系统自稳中发挥着关键的作用。Th2细胞、Treg细胞、Tr1细胞（IL-10产生的Foxp 3-CD4$^+$ T细胞）、Th3细胞（TGF-β和IL-10产生的CD4$^+$ T细胞）、NK T细胞、B细胞、巨噬细胞及树突状细胞等都能够分泌IL-10。IL-10受体（IL-10 receptor，IL-10R）是由IL-10Rα和IL-10Rβ构成的异源二聚体，其中，IL-10Rα主要存在于造血细胞系表面，IL-10Rβ则广泛表达于各种组织细胞，因此，IL-10可作用于多种细胞类型，具有复杂而重要的调控作用。研究表明，IL-10缺陷小鼠对大肠的共生细菌高度易感，极易发生结肠炎，如果同时缺陷TGF-β，结肠炎炎症则更为严重。急性感染时阻断IL-10信号会导致严重的病理过程，甚至是致死性的。相反，如果IL-10大量分泌，一般能促进病原的清除，其大量分泌常见于持续的慢性感染。

5. IL-12　　IL-12由巨噬细胞、树突状细胞分泌，具有抗胞内菌感染、抗肿瘤等作用，能够诱导Th1细胞的分化，是天然免疫与获得性免疫之间的桥梁。

6. IL-17　　1993年首次在小鼠中鉴定IL-17基因的存在，被命名为细胞毒性T淋巴细胞相关抗原 8（cytotoxic T-lymphocyteassociated antigen 8，CTLA-8），而后定名为IL-17。之后相继发现了另外 5 种不同的 IL-17，根据发现先后将 IL-17 家族成员依次命名为IL-17A～IL-17F，其中，IL-17A和IL-17F同源性最高，约为50%，其次是IL-17B、IL-17D、IL-17C、IL-17E（Kolls and Linden，2004）。IL-17与其他家族成员氨基酸一致性介于20%～50%。IL-17家族共同的特征包括C端高度的保守，以及5个空间上保守的半胱氨酸残基。IL-17A与IL-17F的同源二聚体的5个保守半胱氨酸残基中的4个形成了一个"半胱氨酸结"的结构，TGF-β及神经生长因子也具有类似的结构（Gerhardt et al.，2009）。

最初认为IL-17来源于活化的T细胞，现在的研究表明，Th17是IL-17的重要分泌细胞。此外，NK细胞、γδ T细胞、CD8$^+$记忆T细胞、中性粒细胞及单核细胞都能够分泌IL-17（敬晓棋，2013）。

IL-17作用于造血细胞、成纤维细胞、平滑肌细胞、上皮细胞，并诱导分泌 IL-6、G-CSF、GM-CSF、IL-1、TGF-β、TNF-α 等多种炎性细胞因子分泌，诱导气管、肺、肠道、皮肤等黏膜部位中性粒细胞浸润，清除细菌、病毒、真菌等感染的病原微生物，促进哮喘、银屑病、多系统硬化症等的炎症反应及抗肿瘤。

IL-17 的重要性在于炎症前期对胞外菌的清除，但也与多种自身免疫病有关。对 IL-17信号途径缺失和抗体封闭小鼠的研究表明，无 IL-17 时小鼠对克雷伯肺炎菌（Klebsiella pneumonia）和支原体肺炎菌（Mycoplasma pneumoniae）的易感性增高，也不能够清除白色念珠菌（Candida albicans）和大肠杆菌（E. coli），这可能与 IL-17 介导的中性粒细胞反应和抗微生物蛋白作用有关。IL-17能够刺激宿主细胞产生颗粒细胞集落刺激因子（granulocyte colony-stimulating factor，G-CSF）、巨噬细胞炎性蛋白（macrophage inflammatory protein，

MIP）-2、IL-8、单核细胞趋化蛋白（monocyte chemotactic protein，MCP）-1、CXCL-8、CXCL-1、CXCL-10（Kolls and Linden，2004）等各种炎症细胞因子和化学因子，以及前列腺素E2、一氧化氮、基质金属蛋白酶、急性期蛋白、IL-6等其他炎症介导因子。同样，由于这些细胞因子的作用，IL-17也促进了类风湿关节炎、银屑病、炎性大肠病、哮喘及多系统硬化症等多种自身免疫病的发生。

IL-17通过大量的细胞因子介导了强大的免疫反应，因此这些细胞因子受体也是高度多样性的。mRNA表达研究表明，受体存在于肺、肝、脾、肾的造血细胞、成骨细胞、成纤维细胞、内皮细胞、上皮细胞上，且与IL-17一样复杂。序列同源性研究表明，存在IL-17R A～IL-17R E五种受体（Kolls and Linden，2004），它们具有以前未见过的共同结构域，其结构为一个具有胞外结构域和长的胞内尾的单过性转膜蛋白。进一步的分析认为，除IL-17R A外其他受体都存在提前终止子的替代性剪切子，这可能有助于减弱免疫反应中IL-17的作用。

7. IL-22 无论慢性炎性疾病还是感染性疾病，IL-22都是调控组织炎症反应的关键细胞因子之一。通过活化Stat3信号通路，IL-22促进组织细胞增殖、抗凋亡、分泌抗微生物分子，阻止组织损伤并协助组织修复。Th17细胞、NK细胞、γδ T细胞等多种天然和适应性免疫细胞都可分泌IL-22。IL-22作用的靶细胞包括角质细胞、造血细胞、上皮细胞等多种细胞类型。

IL-22是IL-10家族细胞因子之一，也是重要的Th17相关细胞因子之一，通过TGF-β和IL-6等Th17细胞分化条件活化的未成熟T细胞能够分泌IL-22（Zheng et al.，2007）。受体由IL-10受体（IL-10Rβ）和IL-22受体（IL-22R）组成，然而与IL-10Rβ的广泛表达不同，IL-22R仅在皮肤、肝、肺和胰腺表达，而且淋巴细胞中未检测到IL-22R的存在。因此，IL-22信号仅作用于外周器官，对T细胞没有直接的影响。

IL-22在炎症中具有重要的作用，可与IL-17协同诱导抗菌肽的合成并增强清除微生物的能力。基因敲出小鼠研究表明，IL-22在银屑病和肝炎中具有重要作用，对急性肝炎具有保护作用和降低肝脏酶的作用，然而在皮肤炎症中IL-22则起到病理性作用，如使皮肤增厚（Zheng et al.，2007）。

8. IL-23 IL-23是一种新的促炎性细胞因子，IL-12及IL-27共同组成IL-12分子家族，具有十分复杂的生物学功能，在抗肿瘤免疫、抗感染免疫及抗自身免疫病中具有重要的作用。IL-23是p40和p19两个小亚基组成的异源二聚体，由活化的树突状细胞和巨噬细胞分泌。

9. TGF-β 人和小鼠几乎所有细胞都能够分泌转化生长因子（transforming growth factor，TGF-β），并与之相互作用。TGF-β最初是作为非免疫细胞的生长因子被发现的，随后人们逐渐认识到其不仅是胸腺T细胞发育的关键调控因子，而且在外周T细胞自稳、对自身抗原耐受、免疫应答中T细胞的分化等免疫应答调控中起关键作用。早期的研究认为，TGF-β的主要功能是抑制生长和免疫细胞的活化，如对抗巨噬细胞活化、阻止树突状细胞成熟、抑制B细胞抗体产生及抑制T细胞活化和增殖。然而研究显示，TGF-β与IL-6协同作用可促进Th17细胞的分化，从而促进机体的抗感染免疫和自身免疫病。

二、奶牛乳腺炎的发生过程

由于奶牛乳腺生理结构的特殊性，外界病原微生物容易入侵而发生炎症。这种炎症的发生过程主要经历三个阶段：病原微生物入侵乳房、乳腺内建立感染、乳房组织的炎症反应等。

（一）病原微生物入侵乳腺

牛的乳房由4个乳腺体构成，每个乳腺体包括一个乳腺乳池、一个乳头乳池、无数个乳导管和具有乳汁分泌功能的乳腺泡。病原微生物侵入乳房是从乳头表面或乳头管开始的，病原微生物通过乳头管屏障，进入乳池，感染乳导管，最后到达乳腺泡。一般情况下，病原微生物突破乳头管括约肌的防御作用有以下3个方法（Kerro et al.，2002）：①在挤奶过程中侵入。在挤奶过程中，乳头括约肌舒张与收缩交替发生，由于乳头口真空波动，乳头口会产生乳汁的微型小滴，病原微生物进入这些微型小滴，就有可能随真空吸进乳头管进入乳池。②在挤奶间隙期间。病原微生物在乳头管壁内增殖并侵入乳池。③在对奶牛乳头进行药浴时。如果对乳头口消毒不严格，就有可能将病原微生物带入乳腺内（王建军和税丽，2009）。

（二）乳腺内建立感染

乳腺内的免疫细胞能杀死侵入乳腺的病原微生物，防止乳腺炎的发生。病原微生物突破乳头管的屏障后，侵入乳腺内，其生长和繁殖首先需要避开奶牛免疫系统的监视。如果病原微生物存在于细胞质中，宿主的绝大多数免疫机制将失去作用。一旦病原微生物入侵到乳腺上皮细胞内，就可以为它的存活和繁殖提供安全的环境。如果病原微生物不能及时杀灭，就会在接近乳头的乳区峡部的乳管壁或分泌组织中生长和繁殖，建立感染区，然后向上扩散到乳区的内部组织。扩散的速度和范围，与入侵病原微生物的数量、繁殖速度、毒力强弱和黏附乳腺上皮的能力等因素有关（于恩琪，2014）。

（三）乳房组织的炎症反应

病原微生物感染乳腺，并成功躲避乳房的细胞和体液防御机制后，它们释放毒素、诱导内皮细胞和上皮细胞释放趋化因子，细胞因子有：肿瘤坏死因子-α（TNF-α）、白细胞介素-8（IL-8）、IL-1、类花生酸类物质［如前列腺素 F2α(PGF2α)］等，这些细胞因子能够吸引大量的白细胞进入乳腺形成炎症反应，主要是中性粒细胞（Viguier et al.，2009），虽然白细胞会吞噬细菌，但若病原微生物未被完全消灭，可能会持续性增殖进而进入小乳导管和乳腺泡区域，造成感染部分乳区血管扩张，导致更多的白细胞在驱动作用下迁移到感染部位，最终导致乳汁中的 SCC 增加。进入乳腺的中性粒细胞发挥两大用途：①释放出一种能改变血管渗透性的酶，使体液自由进入乳腺组织，稀释中和有毒产物。②聚集在乳腺泡、乳导管和乳池等感染部位周围，进入乳汁，吞噬和消化病原微生物，减少病原微生物数量。由于白细胞在正常的泌乳细胞间穿行，并且不断地释放炎性介质，泌乳细胞会被大量破坏甚至消失，最终导致泌乳组织损伤，乳汁的分泌下降甚至停止。这时会出现三种情况：①若感染轻微，病原微生物很快被杀灭，这种炎症反应表现为隐性乳腺炎，特征是乳汁中的 SCC 数轻微或中度上升，牛奶成分发生改变。②若感染持续存在，泌乳细胞大量破坏，泌乳停止，直到下次产犊才能恢复，这种炎症反应表现为临床乳腺炎，会出现乳房的红、肿、热、痛等临床症状，SCC 数明显升高。③若损伤特别严重，大量泌乳细胞被破坏，乳腺组织开始萎缩，腺泡被损坏开始失去解剖完整性（图 1-1），被破坏的腺泡组织将由结缔组织及瘢痕组织完全代替。

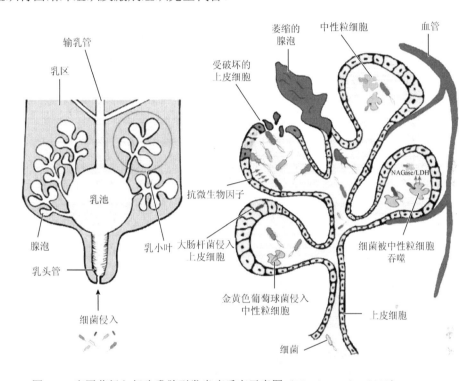

图 1-1　病原菌侵入奶牛乳腺引发炎症反应示意图（Viguier et al.，2009）

Figure 1-1　The schematic diagram of the pathogens invasived into bovine mammary and caused inflammation reaction

三、各种致病菌引起的乳腺炎的发病机制

（一）大肠杆菌性乳腺炎的致病机制

1. 大肠杆菌的毒力因子

（1）脂多糖（LPS）　　大肠杆菌是一种革兰氏阴性菌。细菌 LPS，又称为内毒素，是革兰氏阴性菌细胞外膜上的一种含有脂质 A 的重要成分，也是大肠杆菌的一种主要的毒力因子（Hogan and Larry，2003）。LPS 是细菌死亡之后才被释放出来的，可诱导机体局部和全身出现症状（Burvenich et al.，2003）。

LPS 的细胞受体 CD14 主要存在于单核细胞和巨噬细胞上，也有很少一部分存在于中性粒细胞上，其能够促进 TLR4 对 LPS 的识别。TLR4 主要由乳腺中的巨噬细胞、树突状细胞和上皮细胞表达（Werling et al.，2006）。LPS 可以刺激单核细胞和巨噬细胞产生炎性细胞因子，TNF-α 启动炎症反应，通过激活急性期蛋白（LPS 结合蛋白、结合珠蛋白、血清淀粉样蛋白 A 等）诱导急性期反应（Bannerman et al.，2004；Hiss et al.，2004）。据报道，LPS 可以引发乳腺炎，但并不是必需的因子，说明还有其他的细菌毒力因子参与乳腺炎的发生发展（Shpigel et al.，2008）。

（2）其他毒力因子　　在特定的疾病中，大肠杆菌有不同毒力因子对机体进行感染（在特殊的环境中定居、生存、繁殖，如尿道、肠）（Kaper et al.，2004）。大肠杆菌主要的毒力因子是黏附素，它可以使细菌附着并定居于黏膜表面。大肠杆菌产生的毒素能够扰乱宿主细胞的正常功能，有助于细菌穿过上皮细胞屏障，侵入组织。

2. 大肠杆菌的耐药性　　在芬兰，从乳腺炎中分离出的大肠杆菌有 9%对链霉素耐受，7%对氨苄西林耐受，5%对四环素耐受（Lehtolainen et al.，2004）。许多编码细菌耐药性的基因和一些毒力因子的基因一样，都位于质粒中，两者可能有关系（Martínez and Fernando，2002）。

3. 大肠杆菌引起的急性乳腺炎的发病机制　　LPS 通过与一些蛋白相互作用刺激细胞进行信号转导，如 LPS 结合蛋白、CD14、MD-2 和 TLR4 等。LPS 结合蛋白是一种可溶性穿梭蛋白，可直接与 LPS 结合，促使 LPS 与 CD14 结合。CD14 有利于 LPS 识别 TLR4/MD-2 受体复合物，调节机体内相关细胞对 LPS 的识别。MD-2 是一种可溶性蛋白，以非共价键的方式与 TLR4 结合在一起，也能单独与 LPS 结合成复合物。目前，还没有证据说明 TLR4 能直接与 LPS 结合，但是它能增强 MD-2 与 LPS 的结合（Hiroaki et al.，2007；付云贺，2015）。

LPS 一旦被识别，TLR4 的 TIR 结构域与下游的接头蛋白相互作用。TIR 结构域含有 3 个高度保守的区域，它们可以介导 TLR 与信号转导接头蛋白之间的相互作用。TIR 结构对于信号转导至关重要，其突变可使机体对 LPS 无反应。据报道，TLR4 可调节 5 种接头蛋白，分别为 MyD88、TIRAP、TRIF、TRAM、SARM（O'Neill and Bowie，2007）。

MyD88 在白介素 1 受体和 Toll 样受体信号通路中是至关重要的接头蛋白，具有重要的作用。有研究对 MyD88 是否参与 TLR 介导的信号通路进行研究发现，在 LPS 刺激之后，MyD88 基因缺失的巨噬细胞不能产生促炎细胞因子。

TIRAP 促使 MyD88 与 TLR4 的胞内结构域结合，启动 MyD88 依赖性信号通路（Kagan and Ruslan，2006）。TRAM 与质膜通过十四烷基化结合，在 LPS-TLR4 信号通路中同样发挥重要的作用，它主要调节下游的 MyD88 非依赖性信号通路（Rowe et al.，2006）。

（1）MyD88 依赖性信号通路　　除了 TIR 结构域外，MyD88 还含有一个死亡结构域，可以通过同型相互作用募集其他有死亡结构域的分子。机体一旦受到 LPS 刺激，MyD88 募集并激活含有死亡结构域的白介素受体相关激酶 4（IRAK-4）。肿瘤坏死因子受体相关因子 6（TRAF-6）是 IRAK-4 下游的另一个重要的接头蛋白，它与 UBC13 和 UEV1A 形成复合物，激活转化生长因子 β 活化激酶 1（TAK1）（Jin et al.，2004），然后 TAK1 激活下游的 IKK（IκB 激酶）和 MAPK 通路。IκB 磷酸化并从 NF-κB 上脱落下来，使 NF-κB 入核，进而调控各种促炎细胞因子的表达。MAPK 通路的激活同样可使 AP-1 入核并对促炎细胞因子进行调控（Chang and Karin，2001）。

（2）MyD88 非依赖性信号通路　　TRIF 是一种重要的含有 TIR 结构域的接头蛋白，

它可以介导 MyD88 非依赖性信号通路。TRIF 的 C 端区域含有一个 Rip 同型结构域，能够与 RIP1 相互作用。最初 RIP1 被当作是 TNF-α 介导的 NF-κB 激活的重要成分，RIP1 的缺失会导致 TRIF 依赖性的 NF-κB 无法激活（Meylan et al., 2004）。

（3）TLR4 信号通路的负调节　　TLR4 信号通路可以被多种负调节因子调节，缺少这些因子将会增强 TLR4 引起的反应（Liew et al., 2005）。RP105、ST2L 和 SIGIRR 等调节因子表达于细胞表面，在 TLR 信号转导的起始阶段发挥其调节作用。RP105 是 TLR4 的同系物，能够与 MD-2 的同系物 MD-1 结合，该复合物能够直接作用于 TLR4/MD-2 复合物，抑制 LPS 与其结合。ST2L 是白介素 1 受体的同系物，含有 3 个胞外免疫球蛋白结构域和 1 个胞内 TIR 结构域，它可以作用于 MyD88 和 TIRAP，抑制其功能，阻止 TLR4 对其募集。缺少 ST2 的巨噬细胞在受到 LPS 刺激之后，促炎细胞因子急剧增多（Brint et al., 2004）。

TRIAD3A 和 SOCS-1 是两个 E3 泛素蛋白连接酶，均参与 LPS/TLR4 信号转导。TRIAD3A 可以作用于某些含 TIR 结构域的蛋白，如 TIRAP、TRIF、RIP1 等，过量的 TRIAD3A 使其降解，抑制 NF-κB 激活（Colleen et al., 2006）。SOCS-1 是一种细胞因子调节子，能够通过泛素化作用降解 TIRAP，抑制 JAK-STAT 信号转导。

（二）金黄色葡萄球菌性乳腺炎的致病机制

金黄色葡萄球菌是一种广泛存在于人类生活环境和自然环境中的细菌，在人类的上呼吸道和皮肤上经常可以发现，很多人或动物都携带有这种细菌。尽管很多情况下金黄色葡萄球菌并不致病，但是金黄色葡萄球菌依然是引起人类和动物感染最多的重要病原体之一。金黄色葡萄球菌也是一种致命的病原菌，在医院中经常感染住院患者并引起严重后果，美国医院感染监控系统认为，金黄色葡萄球菌引发了绝大部分的医院和社区获得性感染，是引起人类和动物创伤化脓及感染性疾病的一种主要致病菌（王天成，2015）。

1. 金黄色葡萄球菌感染的特点和危害　　金黄色葡萄球菌可产生金黄色的类胡萝卜素，因而得名。这种细菌是由 Alexander Ogston 爵士于 1880 年首次发现。金黄色葡萄球菌为典型的革兰氏阳性菌，需氧或兼性厌氧，菌体呈圆形，直径 0.5~1.5 μm，葡萄串状、双球或短链状均常见；无芽孢，无鞭毛，运动性不强，有的菌株有荚膜或黏液层。

金黄色葡萄球菌附着在人或动物上是一种正常现象，是一种常见的共生菌，呼吸道和多个部位的皮肤表面都有金黄色葡萄球菌的存在，身体健康状态良好时不会发生感染。总体而言，金黄色葡萄球菌在人身体上存在 3 种携带形式，约有 20%的个体总是携带一种菌株，被称为持续携带者，大多数是孩子和成人；大约 60%的人间断性地携带金黄色葡萄球菌，并且菌株品系多样，被称为间歇性携带者；大约 20%的人群从来没有携带过金黄色葡萄球菌。持续携带者可以对获得其他型金黄色葡萄球菌产生一些免疫效果（王天成，2015）。但在一定条件下，当宿主的免疫状态低下时，如外伤、手术、营养不良、应激等情况，感染就会发生。手术创伤感染和乳腺炎、肺炎等疾病的罪魁祸首常常是金黄色葡萄球菌。金黄色葡萄球菌还会引发骨髓炎、心内膜炎、脓毒性关节炎、败血症和毒血症等疾病。金黄色葡萄球菌是医院获得性菌血症的主要引发原因，即便是在医疗卫生条件较好的北美地区和欧洲国家，金黄色葡萄球菌依然是不可忽视的致病菌（Fluit et al., 2001）。

近年来，金黄色葡萄球菌对抗生素产生了越来越强的耐药性，这预示着其在未来可能会大规模流行，金黄色葡萄球菌仍然是引起群体感染的一种重要人兽共患病原菌。

金黄色葡萄球菌感染机体是多种因素相互作用的结果。在感染宿主之前需要首先完成在宿主身上某个部位的定植，定植的部位呈多样性，各个部位的皮肤和鼻腔都是可定植处。金黄色葡萄球菌在鼻腔的存在和定植对于其流行病学和发病机制是一个关键因素。以人类为例，大多数情况下都可以在人前鼻腔分离出金黄色葡萄球菌，金黄色葡萄球菌菌株在鼻孔前部定植并形成了小生境。研究表明，鼻腔是金黄色葡萄球菌的最主要存在区域，此外，值得注意的是，如果鼻腔里的金黄色葡萄球菌被清除，那么全身其他区域定植的金黄色葡萄球菌也会消失（王天成，2015）。金黄色葡萄球菌定植之后才有感染宿主的可能，发生感染需要经历4个阶段，依次是局部感染、系统/全身感染或败血症、转移性感染和细菌毒素中毒症。首先需要经过皮肤或黏膜的局部感染，形成脓肿，才有可能进入血液向全身扩散。

金黄色葡萄球菌之所以能够成功感染宿主，是因为金黄色葡萄球菌能够对宿主组织和细胞进行附着，更重要的是，部分菌体成功避免了宿主免疫系统的杀伤（刘博，2013）。黏附素是金黄色葡萄球菌黏附细胞的重要因子，溶血素等毒力因子则能够裂解细胞，有很多菌株的荚膜多糖能使其抵抗吞噬和杀灭（Katherine and Lee，2004）。金黄色葡萄球菌侵入体内感染时，其本身特殊的毒力因子和相关蛋白可以让病原菌挫败最初的防御吞噬系统，突破起到隔离病原作用的脓肿边界，深入其他组织或入侵内皮细胞从而进入血液。一旦病原体进入血液，就可以扩散至机体全身，进而发生感染性休克。没有特殊的处理治疗，全身感染引起的死亡率很高。由于金黄色葡萄球菌通过血液扩散感染至全身，可以引发感染性心膜炎、骨髓炎、肾皮质脓肿、化脓性关节炎等。即使是金黄色葡萄球菌本身并没有进入血液而被限制在局部，其毒素也可能会引发各种综合征，如中毒性休克综合征、烫伤样皮肤综合征和经食物传染的胃肠炎等。显然，某些特定的条件下或者特定医疗的措施可能会使机体容易被金黄色葡萄球菌感染，如糖尿病、静脉留置针、静脉药物注射等情况都会引起金黄色葡萄球菌感染。一种病原体能够引发如此大规模的感染，是因为金黄色葡萄球菌具有多种毒力因子。这些毒力因子甚至比病原菌本身更有杀伤力。病原微生物本身还有特定的调节系统，根据具体环境因地制宜地调控毒力因子的产生和释放（王天成，2015）。

在金黄色葡萄球菌的感染过程中，中性粒细胞会受到募集，向感染组织部位集结。奶牛乳腺炎发生时，中性粒细胞聚集在乳腺组织，介导促炎性细胞因子的分泌产生，引发剧烈的炎症反应，导致宿主炎症相关免疫应答的发生，中性粒细胞在级联信号放大和各种酶的作用下导致炎症扩大，再加上金黄色葡萄球菌的外毒素和酶类的作用，损伤组织的同时降低宿主免疫系统对金黄色葡萄球菌的细胞吞噬作用，并可以干扰抗葡萄球菌抗体和细菌本身的结合，使宿主产生明显的临床症状（Piette and Verschraegen，2009）。正是金黄色葡萄球菌逃避中性粒细胞的吞噬作用，使宿主免疫系统无法有效对其清除，金黄色葡萄球菌大量繁殖，产生大量毒素，这些特点最终能导致宿主的败血性休克甚至死亡。

金黄色葡萄球菌全身性感染导致的人和动物死亡率高于肠道菌感染（Alberti et al.，2003），并且金黄色葡萄球菌感染引起人和动物败血症的发生频率呈上升趋势。随着抗生素的大量使用，耐甲氧西林金黄色葡萄球菌等耐药菌株的出现引起的感染愈发难以控制，引起了医学和兽医学领域的高度关注（Weese et al., 2006）。机体对外界其他型金黄色葡萄球菌定植的抵抗力会因为抗生素治疗而减弱。这种情况说明，耐药金黄色葡萄球菌的传播主要受抗生素治疗影响。抗生素治疗可能加剧耐药菌的产生和传播（王天成，2015）。

在兽医领域，不仅家庭伴侣动物如猫、犬等，猪、牛等经济动物也常常受到金黄色葡萄球菌感染，并造成严重损失。金黄色葡萄球菌感染引发奶牛的多种疾病，奶牛乳腺感染金黄色葡萄球菌后，感染乳区可能在 2~6 h 就会引起急性坏疽，若没有紧急治疗，严重情况下毒素被机体吸收后奶牛将在 1~2 天内死亡，死亡率很高，威胁很大，引起了人们高度重视。奶牛乳腺炎的发病和传播，与金黄色葡萄球菌和牛乳腺上皮细胞的相互作用有直接关系。金黄色葡萄球菌附着或入侵细胞具有选择性，更容易针对上皮细胞，尤其是乳腺上皮细胞，大多数被入侵的细胞都是无绒毛细胞、圆顶形的分泌型细胞（Téllez-Pérez et al., 2012）。金黄色葡萄球菌对乳导管角质化细胞进行黏附，也取决于这些细胞的分化来源。有研究表明，某些金黄色葡萄球菌的细胞壁上有一种分子质量为 145 kDa 的蛋白质，可以在金黄色葡萄球菌附着在乳腺上皮细胞表面时起重要作用。在金黄色葡萄球菌感染宿主细胞和传播过程中，其产生的黏附相关因子和多种毒力因子起到了重要作用。另外，金黄色葡萄球菌入侵机体后能逃避免疫反应和抗生素杀灭，因此危害巨大，免疫逃避等能力与细菌致病性密切相关（王天成，2015）。

尽管随着科学的发展，可以抵抗整个菌体的疫苗已经出现，能够免疫金黄色葡萄球菌对于高度分化的乳腺上皮细胞的附着，但是参与附着的抗原表位仍不清楚，效果也不够理想，并且，至今没有对人和动物宿主保护作用十分理想的疫苗出现。近年来，金黄色葡萄球菌对抗生素产生了越来越强的耐药性，这预示着其在未来可能会大规模流行，金黄色葡萄球菌仍然是引起群体感染的一种重要人兽共患病原菌（王天成，2015）。

2. 金黄色葡萄球菌的毒力因子 金黄色葡萄球菌感染人或动物后，除了细菌本身大量繁殖对机体产生作用外，其分泌的各种毒力因子也是引发各种临床症状的重要原因。细菌能分泌多种毒力因子，如细菌细胞壁相关蛋白、分泌性细菌蛋白等。该菌对人畜的感染是多种毒力因子和蛋白协同作用的结果。金黄色葡萄球菌的外毒素种类多样，主要有溶血素、肠毒素、脱皮毒素、杀白细胞素、毒性休克综合征毒素-1，还有许多与毒力相关的蛋白和多糖类物质等，如黏附素、耐热核酸酶、凝固酶、肽酶、细菌蛋白水解酶、荚膜多糖、金黄色色素等。金黄色色素可以提高细菌对免疫系统的抵抗力和细菌的毒力（Liu et al., 2005）。

金黄色葡萄球菌溶血素主要有 4 种，分别为 α、β、γ、δ 溶血素。α 溶血素是致病作用最广泛的一种。α 溶血素能够溶解红细胞，具有细胞毒性，能导致细胞死亡，α 溶血素能够对多种细胞产生裂解作用，包括红细胞、上皮细胞和各类免疫细胞等，主要通过破坏细胞钙离子通道致使细胞死亡，还能够诱导巨噬细胞激活 IL-8 等炎性因子，导致严重的炎症反应。研究表明，α 溶血素是伤害奶牛乳腺的重要毒素，敲除了表达 α 溶血素基因的金黄色葡萄球菌感染乳腺的能力显著降低，另外也减少了腹膜感染和脓毒性关节炎的发病，α 溶血素还是金黄色葡萄球菌性肺炎中起核心作用的毒素（Juliane et al., 2007）。β 溶血素于 19 世纪 30 年代被发现，也对红细胞具有溶解作用，但仅限于某些物种如山羊等。尽管 β 溶血素在金黄色葡萄球菌感染过程中的作用还不清楚，但是研究表明，敲除了 β

溶血素表达基因的菌株不易感染小鼠的乳腺，并且，β溶血素在体外实验中能够针对性地溶解牛的乳腺上皮细胞，因此β溶血素也与金黄色葡萄球菌引发奶牛乳腺炎有密切关系。γ溶血素主要影响中性粒细胞等免疫细胞，与α溶血素有协同作用。δ溶血素的作用机制不明，能裂解红细胞和其他多种组织细胞，导致组织坏死，研究表明，高浓度的δ溶血素会致命（王天成，2015）。

杀白细胞素（PVL）是细胞壁表面蛋白，归类于膜钻孔毒素家族，是一种外毒素，可以诱导炎性因子释放，裂解细胞，造成细胞坏死和凋亡，据统计，表达PVL的菌株只占总数的2%～3%。杀白细胞素同样能够在白细胞和多种细胞膜上打孔，使细胞裂解，还能经线粒体途径诱导多形核中性粒细胞和巨噬细胞等发生坏死或凋亡。低浓度的杀白细胞素诱导吞噬细胞凋亡，而高浓度则使细胞坏死（Wu et al.，2010）。PVL还可以诱导多形核白细胞释放颗粒蛋白和炎性介质，从而加剧炎症反应，引发全身症状。

致热毒素超抗原（pyrogenic toxin superantigen，PTSAg）是一个蛋白家族，具有共同特点，即致热源性和超抗原性。例如，肠毒素（enterotoxin）和毒性休克综合征毒素-1（toxin shock syndrome toxin-1，TSST-1）都属于该家族。肠毒素可导致胃肠炎，引发葡萄球菌性食物中毒（SFP），宿主发生腹泻和呕吐，并可能伴随有发热症状，白三烯可能是肠毒素诱导产生的主要炎性介质。毒性休克综合征毒素-1能直接穿透宿主黏膜，与风湿性关节炎和肾炎有密切关系（王天成，2015）。

另外，该毒素可以非特异性地刺激免疫细胞增长，激活宿主自体反应性T细胞，诱导自身免疫性疾病的发生。更危险的是，毒性休克综合征毒素激活T细胞，引起肿瘤坏死因子-α（TNF-α）和白细胞介素的大量释放，还能增加宿主对金黄色葡萄球菌内毒素的敏感性，引起中毒性休克综合征，症状为低血压、白蛋白减少和水肿，死亡率很高（McNamara et al.，2009）。致热毒素超抗原蛋白性质稳定，能够有效地抵抗酸碱性变化和蛋白酶水解所带来的失活，另外还有耐高温的特性，肠毒素经100℃煮沸30 min后仍然具有致病性。脱皮毒素（exfoliative toxin，ET）目前已知4种，与葡萄球菌烫伤样皮肤综合征和大疱性脓疱病相关（Li et al.，2011）。

葡萄金黄色素（staphyloxanthin，STX）是一种类胡萝卜素色素，能够清除自由基，抗氧化，淬灭单线氧，与金黄色葡萄球菌的免疫逃避有关，使菌体逃避免疫细胞产生的以氧化剂为基础的杀菌作用，还可以提高细菌本身的毒力（Lan et al.，2010）。

葡萄球菌蛋白A（SPA）和**脂磷壁酸蛋白**（LTA）是金黄色葡萄球菌细胞壁众多表面蛋白之一，这类蛋白能够帮助细菌抵抗吞噬作用，诱导免疫细胞和其他体细胞凋亡，促进炎症反应，使细胞变性坏死等。宿主的免疫细胞，如NK细胞、巨噬细胞等，在脂磷壁酸的刺激下会大量分泌γ-干扰素（INF-γ）和TNF-α等促炎性细胞因子。巨噬细胞产生的一氧化氮合酶和TNF-α、IL-1、IL-6等分泌量升高，加剧了炎症反应的程度。葡萄球菌蛋白A能刺激炎性因子TNF-α的分泌，与细菌细胞膜外蛋白协同增加细菌的黏附能力，帮助杀白细胞素裂解细胞，加剧炎症反应。葡萄球菌蛋白A和脂磷壁酸还能通过打乱细胞骨架结构使之发生重排，而引发细胞凋亡。

金黄色葡萄球菌毒力因子的表达主要受**群体感应系统附属基因调节子Agr**（accessory gene regulator）的调节，Agr是金黄色葡萄球菌毒力因子和毒素基因调控网络中的重要一员。Agr 通过分泌自诱导肽，调节基因的表达，影响毒力因子蛋白的合成；SarA 家族蛋白的转录调节因子（Cheung et al., 2008）、双组分调节系统 SaeRS 等也在调节毒力因子方面起重要作用。各种毒力因子及其调节系统关键蛋白，成为人们关注的焦点，这些蛋白可能是新型药物作用的靶点（王天成，2015）。

3. 金黄色葡萄球菌感染后宿主的免疫应答反应 免疫系统的调节是复杂的相互作用的平衡状态，任何导致失衡的因素都可能引起疾病的发生。机体在外界病原体入侵时，有两种防御反应方式，一种是获得性免疫反应，一种是天然免疫应答。首先，机体免疫系统会对进入体内的病原进行识别。获得性免疫系统是脊椎动物所特有的体系，在识别过程中起重要作用，主要是通过 B 淋巴细胞和 T 淋巴细胞进行识别，这两种细胞表面具有病原的抗原受体，这些抗原受体负责识别病原。B 淋巴细胞和 T 淋巴细胞能够重组免疫球蛋白和 T 细胞受体的基因，以产生多种类型的抗体，对抗各种病原。免疫球蛋白可以针对特异性抗原进行大量克隆扩增。获得性免疫是高度复杂精准的防御系统，但是需要多个步骤和较长的反应时间。因此，在获得性免疫系统发挥作用之前的时间里，天然免疫系统会发挥其快速免疫作用以对抗病原体。天然免疫系统是一种保守的存在于几乎所有动物机体内的免疫系统，是机体抵抗病原入侵的第一线，它能够识别病原体并且将机体自身成分加以区别。而 TLR 在侦测识别病原体入侵的过程中起到了核心作用（Kiyoshi and Shizuo, 2003）。机体的天然免疫系统主要由中性粒细胞和巨噬细胞等白细胞介导，吞噬处理和杀伤病原菌，并且介导机体产生广泛的炎性介质和细胞因子，激活淋巴器官中的 T 细胞，进而进一步启动后继获得性免疫。动物机体有多种天然免疫相关的受体，负责识别不同种类和不同来源的病原和抗原。这些抗原成分对病原微生物入侵机体的过程是必需的，通常，天然免疫系统所识别的物质是病原微生物结构的一部分，并且相对稳定，人们称之为病原相关分子模式（pathogen associated molecular pattern，PAMP），相应地，机体天然免疫系统中直接识别 PAMP 的物质称作模式识别受体（pattern-recognition receptor，PRR）。PAMP 都由病原所产生，具有明显的特点，可以被 PRR 识别；PAMP 对病原的致病性和存活都是不可或缺的。天然免疫系统和获得性免疫系统共同构筑了机体对抗外界病原入侵的防线。巨噬细胞和树突状细胞（DC）的细胞膜上具有 PRR，对不同的 PAMP 进行识别，进而对体内的病原微生物进行杀伤和吞噬处理，同时激活炎性信号通路，产生抗炎性和促炎性的细胞因子（Iwasaki and Medzhitov, 2004；王天成，2015）。TLR 存在于细胞膜表面，也存在于细胞质中，属于 PRR 的一种，树突状细胞和巨噬细胞及一些上皮细胞都可以分泌 TLR（Nakamura et al., 2008；Mancuso et al., 2009）。TLR 可以激活 MyD88 和 TRIF 依赖的信号转导通路，进而激活 NF-κB、MAPK 及其他转录因子的表达（Medzhitov, 2007）。TLR2 是 TLR 家族的重要成员，可以识别肽聚糖（PGN）、脂磷壁酸（LTA）、脂蛋白（lipoproteins）等物质，这些物质都是金黄色葡萄球菌细胞壁的组成成分，是金黄色葡萄球菌保持完整性和致病性所不可或缺的部分。金黄色葡萄球菌入侵机体后，TLR2 识别这些菌体蛋白，激活相关的免疫信号通路，转导级联信号，介导细胞因子的合成释放，启动免疫应答（Schmaler et al., 2011）。TLR2 所介导的信号通路对于天然免疫系统在金黄色葡萄球菌感染前期快速作出应答是十分必要的，TLR2 识别金黄色葡萄球菌后，依靠 MyD88 进行信号转导。MyD88 缺陷株细胞在金黄色葡萄球菌刺激后，细胞因子的分泌量显著下降,并且会导致机体对于金黄色葡萄球菌感染的防御反应降低,易感性升高(Tammy et al.,2007;王天成,2015)。

宿主机体的免疫系统在抵抗病原体入侵时,会与病原体相互作用,在此过程中,会产生多种细胞因子和免疫调节因子。炎症反应是机体对于不良刺激的一种防御反应,是针对机体稳态被一些极端因素打破以后的一种保护性措施,主要表现为红、肿、热、痛和机能障碍;全身反应表现为发热、白细胞增多、单核巨噬细胞系统细胞增生、实质器官病变。正常情况下,适当的炎症反应对于机体是有益的,是机体对不良刺激因素的有效防御措施。但是炎症反应过程中的多种因子对于本身机体稳态来说是敌对并且占据优势的,因此,炎症反应一旦过于剧烈,就会对机体产生伤害。炎症发生的过程,心血管系统起着中枢作用,损伤因子破坏组织和细胞,而炎症机制通过充血和渗出,以稀释、限制和杀伤损伤因子。机体免疫系统在病原体入侵时会产生免疫调节因子和促炎性细胞因子。细胞因子主要由血液循环系统中的免疫细胞产生,上皮细胞等其他类型细胞也能产生。炎性信号通路和细胞因子在体内相互协调,相互影响,广泛调节多种生理功能,对于机体免疫反应来说十分重要。有研究表明,细胞因子基因敲除的基因缺陷小鼠与正常小鼠相比,机体免疫水平的各项指标显著降低(Cavaillon et al.,2003)。促炎性细胞因子的大量产生常常会对宿主造成负面影响,如器官功能障碍,严重时甚至引起死亡。革兰氏阳性菌感染宿主后,能导致机体分泌大量的细胞因子,金黄色葡萄球菌是革兰氏阳性菌的典型代表,与革兰氏阴性菌相比,金黄色葡萄球菌等革兰氏阳性菌所引起机体释放炎性因子的时间比较滞后,峰值主要发生在感染发生后 50~70 h,产生的量也相对较低,但是同样可以引起严重的临床反应。

宿主免疫系统对抗病原微生物过程中起重要作用的细胞包括单核细胞、巨噬细胞和中性粒细胞等。这类免疫细胞可以分泌大量的促炎性细胞因子如 TNF-α、IL-1β、IL-6 和 IL-12 等,产生炎症以防御病原体,但是败血性休克与之有直接关系;另外,还有很多种上皮细胞和内皮细胞也能分泌释放细胞因子。IL-10 在炎症过程中属于抗炎性细胞因子,对 TNF-α、IL-1β、IL-6、IL-8 等产生负反馈调节。IL-8 可以诱导中性粒细胞的游出,而 MCP-1 和 MIP-1 则可以募集巨噬细胞等到达感染部位(Téllez-Pérez et al.,2012)。血液中的单核细胞和组织渗出的巨噬细胞是分泌细胞因子的主要细胞,其产生的主要细胞因子包括 TNF-α、IL-1β、IL-6 和 IL-12 等。当金黄色葡萄球菌感染发生时,常常可以发生败血性休克,TNF-α 可以介导 IL-8 的分泌,IL-12 和 IL-8 生成释放又可以介导 IFN-γ 的分泌(Hultgren et al.,2001)。机体感染金黄色葡萄球菌后,其发生的炎症反应程度与感染是否发展为败血症是密切联系的,在进行针对金黄色葡萄球菌感染的治疗时,对炎症反应程度上的把握是很重要的,合理调节炎症反应强度,对其进行纠正和适当的抑制是治疗的重要手段。金黄色葡萄球菌入侵机体后能够激活宿主的巨噬细胞和树突状细胞,Th 细胞也有可能参与其中,TLR2 通过识别 B 型葡萄球菌肠毒素,以此介导 Th2 细胞产生免疫应答(Mandron et al.,2006)。PGN 激活宿主天然免疫系统后,Th 细胞会合成释放 IL-17,而 TLR2 诱导产生的 IL-23 和 NOD2 诱导产生的 IL-1 会促进 IL-17 的合成分泌(van Beelen et al.,2007)。IL-17 可以帮助机体杀伤并清除入侵的金黄色葡萄球菌,提高机体对金黄色葡萄球菌感染的抵抗力,IL-17 的分泌不足可以导致对金黄色葡萄球菌的易感性升高,并且对病原体的免疫清除能力显著降低。另外，IL-17 的大量释放可以向感染部位和器官募集大量的中性粒细胞，可能会引起宿主的自身免疫性损伤，并对获得性免疫应答和 IFN-γ 的合成释放产生负面影响（Cho et al., 2010）。

除了淋巴细胞的识别和信号通路的激活，免疫系统各种细胞对细菌的直接杀灭和吞噬处理作用也是重要的防御反应，如巨噬细胞和树突状细胞等，中性粒细胞是直接杀灭和处理细菌的主要细胞。中性粒细胞是天然免疫应答反应中最重要的细胞，是三种吞噬细胞之一，血液中中性粒细胞所占的白细胞比例为 50%~70%。中性粒细胞内含有颗粒，里面存有杀灭病原微生物及消化组织细胞和碎片的活性物质。中性粒细胞被募集至炎症发生部位，形成抵抗病原体入侵的一道防线，其具有多重的杀菌效果。当中性粒细胞通过表面识别受体从内环境中识别了病原后，就可以对病原进行直接吞噬，吞噬后由吞噬小体进行直接杀灭（王天成，2015）。

另外，中性粒细胞还能发挥胞外杀菌作用。在中性粒细胞的胞内颗粒中储存有过氧化氢、超氧阴离子和一氧化氮等活性氧物质，由 NADPH 氧化酶等物质联合作用产生，这些超氧化物是直接的杀菌剂。中性粒细胞颗粒内物质在能够杀菌的同时，对自身细胞也有伤害。因此，为保护自身组织免受伤害，抗菌物质储存在特定的颗粒中，当吞噬发生时，形成吞噬小体，颗粒中的杀菌剂释放到吞噬溶酶体中发挥作用。

另外，Brinkmann 等（2004）发现了一种新的中性粒细胞杀菌途径，即中性粒细胞胞外捕网（neutrophil extracellular trap，NET），是天然免疫体系强有力的武器，中性粒细胞必须死亡后才能产生 NET，因此这是以自身生命为代价的，是除了细胞凋亡和坏死之外的一种独特的细胞死亡方式。中性粒细胞受到刺激因子的作用后，启动一种新的不同于凋亡的死亡过程，核膜溶解，细胞核内容物如 DNA 等释放到细胞质中，染色质 DNA 与胞质内的物质进行结合，胞内颗粒和其内容物、多种酶类和杀菌物质附着在染色质 DNA 的网格结构上，最后细胞膜崩解，组合起来的 DNA 网状结构释放出来（Fuchs et al., 2007）。通常，NET 网直径为 15~17 nm。因此，NET 网是以 DNA 为骨架，附有多种中性粒细胞颗粒酶和多肽的混合网络结构。NET 的生成也是宿主天然免疫反应的一种表现形式，也是机体天然免疫功能的组成部分。

NET 形成的网络结构起到物理屏障作用，将病原体"困在"网中，将其控制在特定区域，阻止其入侵其他细胞，进入血液循环，阻止和延迟病原的移动，以限制其扩散的速度，抑制感染。这种物理阻隔能力具有重要意义，不但自身限制、杀灭病原，还能协助其他免疫细胞的吞噬作用。吞噬细菌是消耗能量的过程，NET 固定并杀死或处理细菌，为吞噬细胞吞噬节约能量，减少了杀菌时间，使免疫系统清除病原的效率提高。NET 网还使其上附着的抗菌物质集中在一个区域内，起到相对浓度提高的作用；将病原体和颗粒蛋白聚集在一起，经不同杀菌剂的协同作用提高了杀菌的效能，同时也限制了这些抗菌蛋白和水解酶类扩散到组织中去，减少损伤和消化健康组织。NET 网上来自中性粒细胞各级颗粒的杀菌蛋白，如杀菌/通透性增强蛋白（bactericidal/permeability-increasing protein，BPI）、组蛋白和弹性蛋白酶等本身就具有杀菌能力，它们在一个区域内能形成相互的协同作用。因此，NET 网可避免免疫系统对自体的损伤作用（Brinkmann et al., 2004），通过上述的限制、集中和协同，高效消灭病原，提高免疫力。NET 能够直接抑制和杀灭多种细菌，包括金黄色葡萄球菌、沙门氏菌和志贺杆菌等，还能通过其颗粒酶降解某些病原体的毒力因子。NET 网对革兰氏阳性菌、革兰氏阴性菌、真菌和寄生虫都有杀灭作用。NET 网是机体消灭真菌和寄生虫等的重要手段。NET 降解毒素和杀菌作用是实现中性粒细胞免疫功能的重要因素。研究发现，DNase 处理过 NET 网后，就降解了 DNA 网状结构，这使得中性粒细胞的胞外杀菌能力显著下降（Brinkmann et al.，2004；Urban et al.，2006；Wartha et al.，2007）。研究表明，NET 的形成受到多种因素的影响，活性氧 ROS 的产生是一个重要因素。例如，金黄色葡萄球菌和 PMA 诱导的中性粒细胞 NET 可以被 NADPH 氧化酶抑制剂（diphenyleneiodonium，DPI）彻底阻断；过氧化氢酶能减少 PMA 诱导的 NET 的形成。NET 网通过 NADPH 蛋白酶及 RAF-MEk-ERK 途径使细胞核核膜及颗粒膜溶解，DNA 与胞质内颗粒蛋白结合，排至细胞外发挥作用；TLR4 对 NET 网的产生也有重要影响（Seeley et al.，2012），因此，NET 网与炎症反应密切相关，NET 对炎症反应的发生和发展其重要作用。NET 能提高对其他效应细胞募集的效率，具有促炎症作用（Lin et al.，2011）。除致病微生物以外，其他多种因素包括脂多糖、佛波酯、干扰素、激活的血小板和内皮细胞，以及 TNF-α、IL-1β、IL-8 等炎性因子也能刺激中性粒细胞产生 NET 网（Keshari et al.，2012）。另外，中性粒细胞吞噬病原体或组织细胞碎片后可能直接死亡，也可能形成 NET。迄今为止，调节 NET 网形成的细胞内信号途径还不清楚。NET 网的防御作用对于抵抗乳腺感染有重要意义，可以很好地协助宿主免疫系统，据报道，中性粒细胞的吞噬功能被奶中的多种蛋白类物质所影响，效能减弱，导致乳腺炎容易发病，但是中性粒细胞形成的 NET 网所具有的杀菌功能，并没有受到奶类物质的抑制（Lippolis et al.，2006）。因此，NET 网的形成和杀菌作用与乳腺炎的发病密切相关。

NET 既能提高机体炎症保护作用，也减少炎症对宿主的损伤，但是也引起一些免疫性疾病的发生。NET 与自身免疫性疾病如系统性红斑狼疮、类风湿性关节炎、小血管炎等疾病的发生有密切联系，另外，其在痛风、动脉粥样硬化和深静脉血栓的发病过程中也起到了促进作用。

4. 金黄色葡萄球菌免疫逃避的特点　　在机体健康状态良好的情况下，金黄色葡萄球菌虽然会附着在机体上，但不会引起感染。一旦宿主免疫力降低，金黄色葡萄球菌就会感染宿主，大量繁殖的细菌引起炎症、细胞死亡、脓肿等。金黄色葡萄球菌侵入机体后，会激活宿主的免疫应答反应，以消灭病原体。机体的免疫系统是强大有效的，有多种途径可以消灭病原，恢复内稳态的平衡。但是，依然有很多的病原体能成功感染，引发疾病，这与其本身具备逃避免疫的能力是分不开的（王天成，2015）。

金黄色葡萄球菌必须突破宿主天然免疫系统多种识别受体的侦测和形式多样的杀菌物质的处理和杀灭，才能顺利感染，并且能够在机体内环境内生存下来。金黄色葡萄球菌可以存活于身体各个部位和各个组织中，甚至免疫系统的巨噬细胞和中性粒细胞中，这是因为金黄色葡萄球菌可以抵抗和逃避宿主免疫系统的杀伤和清除。这种免疫逃避机制，是细菌维持自身致病性的重要因素。金黄色葡萄球菌致病是与机体产生复杂的相互作用的结果，从对宿主细胞的黏附到最终逃避免疫应答，导致细胞死亡和组织损伤，整个致病过程是细菌多种活性物质与机体共同作用所导致的。金黄色葡萄球菌本身具有金黄色葡萄球菌 A 蛋白、肽聚糖、脂多糖、磷壁酸、磷脂和荚膜等活性物质和特定结构，不仅可使金黄色葡萄球菌逃避宿主免疫系统对其的识别，还能提高自身对巨噬细胞吞噬处理和杀灭的抵抗力（Daum and Spellberg，2012）。金黄色葡萄球菌通过多种途径逃避免疫系统的杀灭。

首先是金黄色葡萄球菌本身的自我保护能力。金黄色葡萄球菌可以产生荚膜，这是一种以多糖为主要成分的薄膜，荚膜占据了菌体外层的一定空间，起到屏障作用，具有亲水的特性。荚膜使细菌能够对吞噬作用产生有效的抵抗，并且阻隔免疫活性物质如溶菌酶、补体等杀菌、抑菌物质对菌体的直接损伤。有研究表明，金黄色葡萄球菌细胞壁表面成分具有保护性抗原蛋白，有特异性调理作用，抵抗抗菌肽的杀伤作用。还有一种对胰酶敏感的细胞蛋白抗原，使宿主产生非特异性保护作用（Li et al.，2007）。荚膜多糖本身具有亲水性，并带负电荷，能降低细胞的吞噬活性，阻碍激活补体，并且产生物理屏蔽作用，阻挡抗体的 Fc 片段与吞噬细胞膜上的 Fc 受体结合，使吞噬细胞无法吞噬处理金黄色葡萄球菌。

金黄色葡萄球菌 A 蛋白可以与宿主 IgG 蛋白的 Fc 片段特异性结合，对中性粒细胞 IgGFc 受体发生竞争性抑制，从而抵抗吞噬。细菌感染后引发炎症时，其产生的超抗原类物质能够阻碍中性粒细胞向炎症部位移动，并通过化学趋化抑制蛋白阻断中性粒细胞激动剂的结合，达到逃避吞噬的目的。即使金黄色葡萄球菌被免疫细胞吞噬后，菌体也能够分泌过氧化氢酶，降低髓过氧化物酶（MPO）系统的作用。

金黄色葡萄球菌的各种毒力因子中，有一部分可以直接杀伤免疫细胞，避免自身被直接杀灭。例如，杀白细胞素对白细胞具有直接的穿孔作用，并能直接诱导白细胞的坏死或凋亡，导致脓肿形成。金黄色葡萄球菌还可以通过磷壁酸抵抗血小板分泌的抗菌肽（AMP）的杀灭作用，这也是金黄色葡萄球菌感染后不易清除并持续性感染的重要原因（Li et al.，2009）。

正常情况下，补体系统由抗原抗体复合物激活，产生多种补体受体，而金黄色葡萄球菌蛋白 A 可以与 C3b 竞争性结合 IgG，以此在空间上直接阻断与细菌表面的结合。金黄色葡萄球菌可分泌葡萄球菌补体抑制因子，这种物质能竞争性地与 C3bBb 和 C4BC2a 结合，致使 C3b 生成受到阻断，从而逃避补体的杀菌作用。IgA 对防御细菌的入侵特别重要，分泌型 IgA 能抢先与抗原结合，抢占抗原的结合位点，使细菌黏附素失去对宿主细胞的黏附能力，但是金黄色葡萄球菌还能分泌 IgA 水解酶，使其失去抑制细菌黏附的作用。另外，金黄色葡萄球菌还能够对抗抗体的调理吞噬作用（王天成，2015）。

研究表明，中性粒细胞是吞噬和杀灭金黄色葡萄球菌的主力之一，中性粒细胞受到趋化，游出并集结在炎症发生部位，不仅可以直接吞噬病原，其各种颗粒中储存的多种杀菌物质，可以直接杀灭细菌。中性粒细胞还能通过牺牲自身来形成 NET 网，这是一种高效、环保、低组织损伤的杀菌结构。但是，金黄色葡萄球菌能表达 DNase，DNA 是组成 NET 网的结构骨架，DNA 被 DNase 降解，网格结构解体，失去物理屏障作用，网格上具有杀菌作用的蛋白质和各种酶也失去了依附的基础环境，NET 也就失去了杀灭金黄色葡萄球菌的能力。研究发现，细菌毒力的大小部分取决于 DNase，DNase 是金黄色葡萄球菌逃避 NET 攻击的有效武器。

通过种种复杂的作用机制，金黄色葡萄球菌可以抵抗活性氮中介物的杀伤作用；抑制免疫系统的信号级联放大，还能存在于内皮细胞和上皮细胞以逃避免疫系统的吞噬和清除，使得金黄色葡萄球菌在宿主体内存活，感染持续存在并引起疾病。

5. 金黄色葡萄球菌对宿主炎症和凋亡的影响　　金黄色葡萄球菌感染机体后，宿主免疫系统通过一系列受体进行识别，并活化相关信号，进行识别信号的级联放大，最终启动免疫应答，消灭病原体，调控炎症和细胞死亡。病原的感染过程是复杂的相互作用的结果，主要通过细胞间的信号传递系统完成。与炎症和细胞死亡相关的信号转导通路系统极其复杂，目前研究较多的、具有代表性的有 NF-κB 和 MAPK 信号通路及几个凋亡相关蛋白通路，它们与炎症发展和细胞死亡密切联系，在此以 NF-κB 和 MAPK 信号通路和 caspase-3 等凋亡蛋白为切入点，讨论发生金黄色葡萄球菌性乳腺炎时组织细胞炎症和死亡信号的变化。

（1）NF-κB 信号转导通路　　NF-κB 信号转导通路是调节炎症反应的重要通路（图 1-2），首先在 B 细胞中发现，并且几乎所有的细胞都存在 NF-κB 信号通路。NF-κB

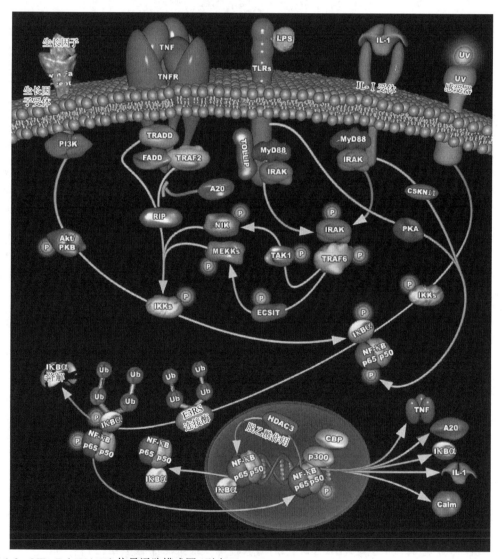

图 1-2　NF-κB（p50/p65）信号通路模式图（引自 http://www.biolegend.com/NF-kappaB-p50p65-Pathway/pathway/57）

Figure 1-2　The pattern diagram of NF-κB（p50/p65）signaling pathway

是一个蛋白家族，有 5 种结构相似的蛋白：p65、RelB、REL、p50 和 p52，它们的 N 端蛋白质同源结构域（RHD）是共享的。NF-κB 在静息状态下存在于细胞质中，并没有活性，与 IκB 家族的抑制蛋白结合在一起，IκB 能够屏蔽 NF-κB 异源二聚体中亚基的核定位序列（NLS），使 NF-κB 进入细胞核（Ghosh and Karin，2002）。当病原微生物入侵引起炎症反应时，NF-κB 信号通路可被病原微生物活性蛋白、机体先天免疫系统、获得性免疫应答和细胞凋亡的信号诱导激活。例如，LTA、内毒素、外毒素和 TLR 受体家族成员、白介素和肿瘤坏死因子（TNF）受体家族成员等促炎细胞因子，促进促炎细胞因子产生，参与炎症反应（Akira et al.，2006）。NF-κB 信号在调节炎症反应的程度上起着重要作用，另外还与细胞凋亡有关。炎症反应起始和放大都受到 NF-κB 信号的影响，NF-κB 通路激活了促炎性因子和某些凋亡抑制基因的表达，促进了 TNF-α 和 IL-1β 等炎性因子的分泌，提高了炎症反应程度。NF-κB 信号通路被干扰或阻断是引发慢性炎症和肿瘤的原因之一（Karin and Greten，2005）。NF-κB 通路通过协调多种炎症介质和自身免疫基因的表达使炎症反应与免疫细胞的生存之间相互和谐。

NF-κB 的激活通过 IKKb、IKKa 和 IKKg 发挥作用，IκB 蛋白在 IκB 激酶的作用下发生磷酸化，与 NF-κB 分离。IKKa 和 IKKb 是两个催化亚基，IKKg 是结构和调控亚基（也称 NEMO）（Israel，2010）。NF-κB 活化后，其 p65 亚基移位到细胞核内，调节有关炎症相关基因的表达。当细菌毒素、炎性细胞因子等蛋白信号被相应的膜受体所识别结合后，TLR 样受体或 TLR 样受体/CD 复合物、TNFR、IL-1R 等信号连接蛋白传递信息，激活诱导 NF-κB 激酶和其他通路的激酶。IκB 随即从 NF-κB/IκB 异源三聚体中解离出来，NF-κB 露出核心定位序列，从细胞质位移入细胞核内。NF-κB 与核内 DNA 上目的基因序列结合，从而启动或加强细胞因子基因的转录表达。

（2）MAPK 信号通路　MAPK 全称为丝裂原活化蛋白激酶（mitogen-activated protein kinase，MAPK），首先发现于真菌，存在于多种生物体内，是哺乳动物细胞一条主要炎症调节信号转导通路，参与细胞的炎症发生后的增殖、分化、凋亡及相应的免疫调节等过程。MAPK 信号转导通路的研究越来越深入。

MAPK 最为主要的三种蛋白途径分别是：细胞外信号调节激酶（extracellular-signal regulated protein kinase，ERK）、c-Jun 氨基端激酶（c-Jun amino-terminal kinase，JNK）和 p38 蛋白（Johnson and Razvan，2002）。MAPK 家族蛋白序列之间具同源性。该信号通路的激活是高度保守方式：三级激酶级联激活体系。首先，刺激因子激活 MKKK（MAP kinase kinase kinase），再激活 MKK（MAP kinase kinase），然后通过苏氨酸和酪氨酸两个作用位点的磷酸化而激活 MAPK。MAPK 信号通路活化后，参与调节广泛的生理过程。MAPK 在调控炎症反应和细胞凋亡、细胞因子生成等方面起着十分重要的作用（王天成，2015）。

ERK 分为 ERK1 和 ERK2 两种亚型蛋白，ERK 广泛参与到应激、细菌感染等反应引起的炎症和细胞死亡调控过程中。多种刺激因子，如可溶性葡萄球菌肽聚糖（solublpeptidoglycan，sPGN）就能够激活 ERK1 和 ERK2。JNK 可以被多种因子激活，LPS、肿瘤坏死因子、白介素、生长因子等生物活性蛋白和物理因素都能激活 JNK。白介素和肿瘤坏死因子激活 JNK 可活化 AP-1，促进炎症反应的程度，并且可能引起细死亡（Verma et al.，2012）。p38是一种酪氨酸磷酸化蛋白激酶，由360个氨基酸组成，几乎所有刺激因子都能激活p38。p38还能被NF-κB活化，调节转录活性，因此，MAPK与NF-κB通路是相互关联的，它们能够相互协调，相互促进，提高免疫应答水平，也促使了炎性细胞因子的分泌。模式图见图1-3。

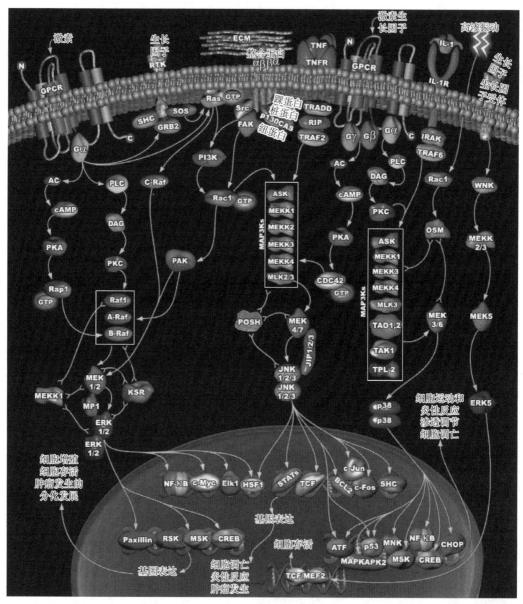

图1-3　MAPK信号通路模式图（引自http://www.biolegend.com/MAPK-Signaling/pathway/52）
Figure 1-3　The pattern diagram of MAPK signaling pathway

（3）金黄色葡萄球菌感染后引起宿主细胞的凋亡　金黄色葡萄球菌感染是病原与机体相互作用的结果，金黄色葡萄球菌对细胞进行黏附，感染宿主后，剥夺宿主营养进行繁殖，并逃避宿主免疫系统的杀灭作用，同时释放细菌毒素，既引发炎症，也造成细胞死亡。细胞死亡可根据其表观形态学分为：坏死（necrosis）、细胞凋亡（apoptosis）、细胞自噬（autophagic cell death）、细胞焦亡（pyroptosis）等。金黄色葡萄球菌的多种毒素和细胞成分都是引起细胞死亡的关键因子。金黄色葡萄球菌的 α 溶血素和 PVL^+ 毒素等都可以通过坏死的方式造成细胞死亡（Essmann et al., 2003），其产生的多种毒素和机体免疫系统相互作用产生的多种物质如 TNF-α 等，均能使细胞发生炎症进而坏死。炎症反应所产生的多种肿瘤坏死因子既能诱导细胞凋亡，也可引起细胞坏死。溶血素 α 除了能诱导 IL-1β、IL-6 等细胞因子合成释放引起炎症，还能诱导细胞凋亡，并且能够直接作用于多种体细胞并造成细胞裂解死亡。金黄色葡萄球菌感染引发的细胞死亡形式还包括细胞焦亡，其通过引发炎症反应，经炎性小体介导 IL-1β 产生细胞焦亡，尤其是巨噬细胞的焦亡（Accarias et al., 2015）。TNF-α、IL-6、CXCL1、CXCL2 和 NF-κB-依赖性的细胞因子，都可能参与到细胞焦亡过程中。

细胞凋亡是一个极为重要的生物学过程，指为维持机体内环境稳定，由基因控制的细胞自主的程序性死亡，是机体清除受损、异常和不需要的细胞的必要方法和自我保护的方式。

caspase 依赖的细胞凋亡与细胞坏死之间存在着相互联系，细胞凋亡与细胞坏死还常常发生在同一生理学或病理学过程中。金黄色葡萄球菌能够诱导各类上皮细胞、内皮细胞和免疫细胞凋亡，多种炎性因子的释放和细胞凋亡的发生，都经由 TLR 信号通路的介导。caspase-3 和 caspase-7 蛋白均为细胞凋亡调节的执行蛋白，caspase-3 和 caspase-7 具有相似的结构性、相近的底物和抑制剂特异性，它们降解 PARP、DFF-45（DNA fragmentation factor-45），导致 DNA 修复的抑制并启动 DNA 的降解。caspase-3 又称死亡蛋白，是细胞凋亡下游信号的直接执行蛋白，线粒体途径和非线粒体途径的凋亡都有 caspase-3 的参与，是凋亡调节的核心蛋白之一，一旦该蛋白被激活，其信号必然引起凋亡。caspase-3 在细胞凋亡中起非常重要的作用，是凋亡启动的先决条件之一（He et al., 2013）。

Bcl-2 和 Bax 属于 Bcl-2 蛋白家族，分别是抑制凋亡和促进凋亡的代表。Bcl-2/Bax 主要通过线粒体途径影响细胞凋亡的发生，Bcl-2 能够抗氧化损伤、调节钙离子通道和线粒体膜通透性等。这两个蛋白形成异源二聚体，也各自形成同源二聚体，凋亡的结果是由二者的比例高低决定的。细胞凋亡启动内源性途径是通过调节 Bcl-2 蛋白实现的。Bcl-2 存在于细胞内多个位置，多见于线粒体、内质网及核膜上，可以阻断线粒体内的 Cytc 释放到胞质内，抑制凋亡的发生。

一般来说，Bcl-2 介导了 Bax 的活化和位移到线粒体外膜上。Bcl-2 和 Bax 蛋白比例决定了细胞的凋亡程度，Bax 的过表达可以促进细胞凋亡的发生，而 Bcl-2 则抑制细胞凋亡，对细胞起保护作用。Bcl-2 表达量的增加引起谷胱苷肽在细胞核内的数量上升，细胞核内过氧化物的氧化性被抑制，改变氧化还原反应的平衡，抑制 caspase 蛋白的活性，减少凋亡的发生，从而对细胞产生保护性作用。模式图见图 1-4。

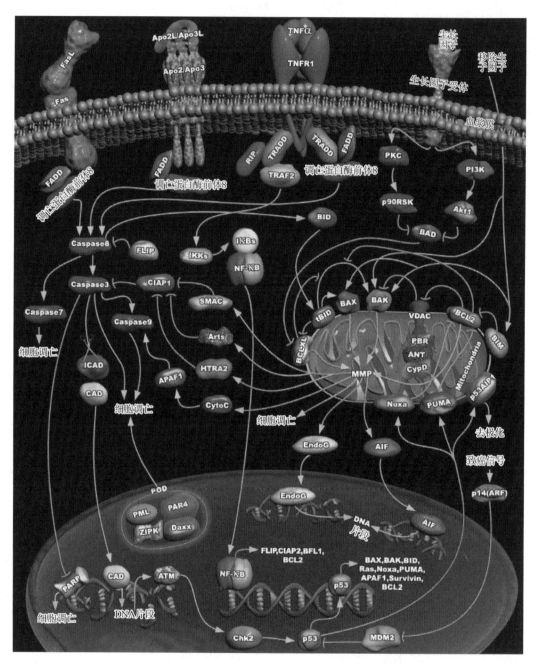

图 1-4 凋亡相关蛋白信号模式图（引自 http://www.biolegend.com/Cellular-Apoptosis-Pathway/pathway/28）

Figure 1-4 The pattern diagram of apoptosis-related protein signaling

综上所述，机体的天然免疫系统是防御病原微生物的重要防线，有效的免疫应答会消灭入侵的金黄色葡萄球菌，抑制感染，自愈损伤，并启动和协同获得性免疫反应。由于金黄色葡萄球菌对于宿主机体的损伤是全面的，相互作用非常复杂，既能够引起炎症使细胞坏死，也能诱导细胞非正常凋亡，还能逃避免疫，避免被免疫细胞杀灭。

因此，对于治疗金黄色葡萄球菌引发的乳腺炎，既要考虑抑制剧烈的炎症，又要防止细胞被金黄色葡萄球菌诱导凋亡，还应该加强天然免疫系统的效能，如增强中性粒细胞对金黄色葡萄球菌本身的杀灭作用，以此综合手段起到保护作用。

（三）表皮葡萄球菌性奶牛乳腺炎的致病机制

有研究表明，表皮葡萄球菌（Staphylococcus epidermidis）是奶牛隐性乳腺炎的重要致病菌（喻华英等，2009），表皮葡萄球菌形成生物膜是其致病性的主要原因，不形成生物膜可以认为不具有致病性（Mack et al.，2007；Otto，2012）。

表皮葡萄球菌是一种凝固酶阴性菌（Rogers et al.，2009），分布在人的皮肤上，在很长一段时间被视为无害，因其不像金黄色葡萄球菌等可以产生大量的毒素。然而随着人们对表皮葡萄球菌的深入研究发现，某些表皮葡萄球菌可以产生一类家族毒素：酚溶性调控蛋白（phenol-soluble modulin，PSM），它可以杀死红细胞和白细胞，还可以溶解人体中的中性粒细胞，增强表皮葡萄球菌的感染能力。虽然只有很少的表皮葡萄球菌能产生 PSM，但某些表皮葡萄球菌还可以产生一种酶：SepA 蛋白酶，具有使抗菌多肽活性降低或退化的功能，也可以促进表皮葡萄球菌的感染（Cheung et al.，2010）。某些表皮葡萄球菌还可以分泌 SesI 蛋白，该蛋白编码前噬菌体，作为表皮葡萄球菌的毒力因子，同样可以提高表皮葡萄球菌的感染能力（Li et al.，2012）。除此之外，表皮葡萄球菌还可以作为基因库将耐药基因同源传递给金黄色葡萄球菌，并使其耐药。瑞典学者研究发现，表皮葡萄球菌是农民在挤奶过程中感染了奶牛（Thorberg，2008），并且已经成为奶牛乳腺炎乳腺内感染（intramammary infection，IMI）的重要致病菌（Thorberg et al.，2009）。生物膜的形成是表皮葡萄球菌致病性的重要方面，也是奶牛隐性乳腺炎难治愈和反复发作的主要原因。生物膜是指黏附于有生命或无生命的物体表面，被自身分泌的胞外大分子包裹的有组织的细菌群体。由此可知，生物膜对细菌来说主要起了保护作用，它可以屏蔽外来物质对细菌的接触，保护细菌免受抗菌药物或先天免疫细胞的杀灭，因此大大增加了治疗难度。生物膜形成主要分为 4 个阶段：初始黏附期、聚集期、成熟期和种子散播期（Melchior et al.，2006）。生物膜的形成是一个复杂过程，涉及许多调控因子，如纤维蛋白原、纤连蛋白和玻连蛋白，在初始黏附阶段发挥了作用（Williams et al.，2002）；聚集阶段，细菌增殖的同时会产生一系列化合物，这些化合物共同作用下形成胞外基质，如产生自溶素溶解细胞，产生大量的胞外 DNA 为膜内细菌提供营养物质，并形成生物膜通道，使生物膜具有多层次的特征结构；种子散播期生物膜通过产生分裂因子（如 PSM）使细胞脱落，释放浮游细菌，浮游细菌可以继续在物体表面定植形成生物膜，诱发感染，周而复始地循环下去（Otto，2014）。

目前已发现，表皮葡萄球菌生物膜主要成分为胞外多糖黏附因子 PIA、胞外 DNA 和胞外蛋白等（Hans-Curt et al.，2007）。PIA 由 ica 基因座调控合成，其中 icaA 基因编码 N-乙酰葡糖胺酰基转移酶，icaD 为该酶活性的必需物，icaC 跨膜蛋白将多糖由胞内释放到胞外，icaB 编码去乙酰基酶（陈伟，2010）。PIA 的合成过程为：icaA 和 icaD 编码合成 N-乙酰葡萄糖胺，在转移酶作用下形成多糖长链。该多糖在 icaC 跨膜蛋白的作用下由胞内释放到胞外，最后在 icaB 去乙酰基作用下脱去乙酰基（Cuong et al.，2004）。

表皮葡萄球菌表面 Zn 依赖的原纤维相关聚集蛋白 Aap 蛋白是非多糖依赖型生物膜形成的主要贡献者。Aap 蛋白与金黄色葡萄球菌中的 SasG 蛋白同源并拥有相同的功能，经研究发现 *SasG* 基因来源于表皮葡萄球菌（Macintosh et al.，2009）。胞外 DNA 在生物膜形成过程中也起着重要作用，是生物膜基质的结构组分，可以加强细胞间的联系。有研究报道，胞外 5′-腺苷三磷酸（extracellular adenosine 5′-triphosphate，eATP）作为信号分子，可以激活机体的炎症和免疫反应（Alain，2009），当宿主细胞察觉到外界的危险或被损伤时，可以通过溶细胞素和非溶细胞素途径分泌 ATP，胞外急速释放的高浓度 ATP 可以作为胞外的、早期的、细胞感觉到压力的"危险信号"，通知免疫系统一个外在或内在的迫在眉睫的危险的存在，并激活免疫应答反应清除病原菌（Virgilio et al.，2009）。然而，病原菌不会坐以待毙，等待宿主细胞将其清除，相反，ATP 不仅对宿主细胞是"危险信号"，对病原菌也起到同样的警示作用，当病原菌感受到危险的存在后，会在 ATP 的调控下，产生大量的胞外 DNA，促进细胞的黏附和生物膜的形成，来保护自身免受宿主先天免疫的迫害（Xi and Wu，2010）。大量的研究证实，DNA 在生物膜形成过程中扮演了重要的角色，但除了依赖溶菌素和 ATP 外，是否存在其他调控机制还需进一步研究（李英，2015）。

（四）链球菌性奶牛乳腺炎的致病机制

通常认为，链球菌"A 群"中的细菌，如化脓链球菌主要对人有致病性，而对动物致病性不强；"C 群"的停乳链球菌主要对各种动物致病，对人致病较少；"B 群"链球菌主要报道的有无乳链球菌，在人医致病比较严重，而在动物上主要导致奶牛乳腺炎的发生；"D 群"链球菌可引起羔羊和小猪的发病，其成员中的猪链球菌也可导致人类的感染（阚威，2014）。动物通过口或伤口感染有致病性的链球菌，从而造成动物组织化脓性、中毒性、变态反应和菌血症等症状，病变组织中的毒素进入血液再到达乳腺组织中引发感染；或致病性链球菌在乳汁和乳导管的表层大量繁殖，直接经乳头感染乳腺，造成乳导管和腺泡上皮细胞的损害，从而使乳腺分泌功能降低或丧失（王会珍和马玉华，2006）。致病性的链球菌具有很强的吸附力、抗吞噬作用及多种致病因子，链球菌的吸附力主要来源于细胞壁上的脂磷壁酸（lipoteichoic acid，LTA）和肽聚糖等成分，提高了链球菌与动物皮肤及黏膜表面亲和力，更容易与宿主口腔等的黏膜上皮部位吸附（孔祥峰，2006）。抗吞噬作用是链球菌的荚膜成分、M 蛋白等，可以增强细菌的致病力（方维焕等，2003）。同时，B 族链球菌（无乳链球菌）具有神经氨酸酶、脂磷壁酸和荚膜多糖抗原等多种毒力因子（车车，2009），其中荚膜多糖在毒力因子中起着最重要的作用，具有抗细胞吞噬作用（刘纪成等，2012）。另外，链球菌的致热外毒素（SPE）、溶血素、M 蛋白、细胞壁上的受体及菌体本身等也都会发挥不同毒力作用（马玉臣，2010）。

而乳房链球菌之所以成为奶牛乳腺炎的另一主要病原菌，是因为其具有荚膜成分，不仅可以使其悬浮于乳汁中，还可以黏附在乳腺上皮细胞上，其分泌的凝血酶原激活因子可刺激乳腺细胞的蛋白酶系统，破坏乳腺的蛋白平衡（阚威，2014）。由此可知，链球菌毒力因子的免疫调节作用与其感染及致病机制息息相关，决定着链球菌感染的方式与致病力的强弱。

（五）肠球菌性奶牛乳腺炎的致病机制

肠球菌主要存在于人和动物机体的肠道内，是人类和动物体内的一种共栖菌，在临床上由于长期、大量地使用广谱抗菌药物，致使其临床耐药菌株不断出现，并导致其耐药机制和耐药特点不断发生新的变化。此类菌对外界环境的适应性、耐受性及抵抗力都强，可在自然界持久存活。而家畜中最常见的是本属菌所致的尿道感染，家禽中鸡则为心内膜炎（李金平等，2011），猪可导致散发性肠炎（文静等，2011），粪肠球菌、屎肠球菌和肠道肠球菌与奶牛乳腺炎相关联。2002年以来我国新疆北部地区的羔羊在春季发生急性死亡的传染病，表现为神经症状及败血症，病原体经分离鉴定发现是具有β溶血的粪肠球菌和屎肠球菌（齐亚银等，2005）。肠球菌的致病机制与以下毒力因子有关，主要有以下几种。

1. 细胞溶血素 肠球菌溶血素（Cyl）又称细胞溶素，能溶解人和动物的红细胞，研究证实，肠球菌溶血素对血小板、精子细胞、中性粒细胞等具有毒性作用。肠球菌溶血素还具有细菌素能力，能增强它在菌群中的竞争优势，这是因为溶血素可以杀死多种革兰氏阳性菌。这些活性在撤去紫外线照射的情况下又可以重新获得，并且还发现肠球菌溶血素基因可以通过质粒相互传递，这说明肠球菌的溶血素是一个强毒力因子，更具有可变性（阚威，2014）。

2. 明胶酶 肠球菌明胶酶（gelatinase，GelE）是一种胞外分泌的锌金属蛋白酶，其大小为72kDa，可以水解明胶、胶原蛋白和酪蛋白，动物实验表明GelE可以提高小鼠腹膜炎、心内膜炎的发生率。

3. 肠球菌表面蛋白 最早是在一株具有高毒力且耐庆大霉素并可以引起菌血症的粪肠球菌中发现肠球菌表面蛋白（Esp）。Esp蛋白与化脓链球菌R28、C透明蛋白、无乳链球菌Rib和金黄色葡萄球菌Bap蛋白结构相似，其结构为球状。在致病性葡萄球菌菌株中Bap蛋白可黏附到非生命物体的表面，具有形成生物膜能力，研究显示，肠球菌的Esp蛋白同样具有生物膜的形成能力（阚威，2014）。*Esp* 基因与粪肠球菌菌株抵抗糖肽类抗生素具有一定的关联性，并在非临床分离菌株中的分布明显低于临床分离菌株。因此，*Esp* 基因是粪肠球菌株的一种重要毒力因子（吴利先和黄文祥，2006）。

4. 肠球菌胶原蛋白黏附素 金黄色葡萄球菌胶原蛋白黏附素（Can）与肠球菌胶原蛋白黏附素（Ace）基因序列具有高度的同源性，其形似矩阵蛋白，主要在肠球菌菌体的表面分泌该种蛋白，其功能是利于肠球菌黏附于宿主细胞表面，并引起宿主细胞感染的发生（马立艳等，2005）。据研究，Ace广泛存在于粪肠球菌和共生菌的致病岛中，是一种很强的毒力因子。当粪肠球菌在46℃培养时，Ace可以黏附层胶原蛋白和Ⅰ型胶原蛋白、Ⅳ型胶原蛋白，而在37℃条件培养时并不出现这一黏附作用（阚威，2014）。

5. 聚集物质 聚集物质（AS）结构主要包括信号肽序列、表面暴露域和跨膜锚定域，分子质量大约为137 kDa。有研究发现，AS在肠球菌致病机制中有多个方面发挥作用，主要体现在：增强肠球菌表面蛋白的疏水性，利于黏附宿主细胞的发生；聚集物质会增大心脏瓣膜赘生物，并导致其他脏器功能的受损；在局部感染中若达到聚集阈值密度时，会使肠球菌造成对宿主细胞的损伤；增强动物机体肠道内多形性核白细胞和巨噬细胞作用下的存活能力；促进动物机体内肠球菌被肠上皮细胞的摄取作用，并在肠腔内跨膜转位。

第六节　牛奶中的体细胞

牛奶中体细胞包括中性粒细胞、巨噬细胞、淋巴细胞和上皮细胞，其主要功能是消除病原体和修复受损组织。体细胞在血液中循环流动，当乳腺感染或损伤时渗入乳腺组织。体细胞是非常特殊的，当发生乳腺炎时只在乳腺升高。因此，影响牛奶体细胞数的因素，都应加以考虑。首先，体细胞数的变化是个体的内在变化，虽然一些环境因素会影响体细胞数的变化，遗传变异也是可能的原因；其次，影响牛奶体细胞数与是否感染乳腺炎相关，感染状态的确定主要是受采样时间、感染数量和病原体种类的影响。

关于牛的年龄对感染率影响：研究表明，体细胞数（SCC）和乳腺炎感染概率随着牛的年龄增加而增加。在泌乳期采样时间的安排也是影响 SCC 的因素。在哺乳期第一天，SCC 明显上升。在泌乳前期 SCC 比泌乳初期高，SCC 升高还发生在泌乳期结束。由于产奶量的降低，在泌乳期结束发生乳腺炎概率更高。产奶量和 SCC 之间相关性略低于产奶量和乳腺炎之间的相关性。一般情况下，奶产量下降，SCC 增加。

一、奶牛体细胞数变化规律的研究

（一）体细胞数

体细胞数是指每毫升牛奶中体细胞数的含量，是衡量奶牛各乳区健康状况和原料奶质量的重要指标之一。SCC 高会降低乳产量、影响原料乳的成分，从而影响原料乳的质量及其制品的风味。西方发达国家在 20 世纪 80 年代就将 SCC 作为检测牛奶质量的指标，从而纳入按质计价的范畴，并制定出相关的等级标准。正常情况下奶牛乳汁中的 SCC 为 0.5 万～20 万/mL。欧盟要求 A 级牛奶中 SCC<40 万/mL，澳大利亚和新西兰则要求 SCC<20 万/mL，加拿大规定 SCC<50 万/mL，美国要求生鲜奶 SCC<75 万/mL（陈建坡等，2013）。

乳汁中体细胞主要来源于腺泡和导管上脱落的上皮细胞、初级巨噬细胞和淋巴细胞。乳腺发生感染时，细菌被乳汁巨噬细胞摄取，释放的酶使上皮内层损伤和松弛，并且释放中性粒细胞趋化因子，它们诱导大量的中性粒细胞从上皮下毛细管和小动脉迁移进入感染部位，反过来中性粒细胞释放的其他炎性因子，进一步促进了中性粒细胞进入感染乳腺抵抗病原微生物。参与炎症早期的细胞因子主要有白介素-2（IL-2）、肿瘤坏死因子-α（TNF-α）等（王小龙，2014）。

国外学者对 SCC 的遗传参数进行了大量研究，认为一个泌乳期 SCC 遗传力为 0.12；后裔测定所得 SCC 的重复力为 0.5～0.6；关于各泌乳期 SCC 之间的关系，奶牛第一泌乳期的 SCC 可能是不同的两个遗传性状之一，因为第一泌乳期的 SCC 与其后泌乳期的 SCC 之间的遗传相关，明显低于第二与第三泌乳期的 SCC 之间的遗传相关。第一泌乳期与其后泌乳期 SCC 之间遗传相关为 0.75，而第二与第三泌乳期的 SCC 之间遗传相关接近 0.9，SCC 与临床性乳腺炎之间的遗传相关在 0.6～0.8（王小龙，2014）。

（二）影响体细胞的因素

1. 胎次对体细胞数的影响　随着胎次的增加，奶牛接触病原微生物的机会也更多，免疫力也有所下降。随着胎次的增加，奶牛乳汁中的体细胞数呈上升趋势，说明隐性乳腺炎发生率随着胎次增加而上升（王相根等，2012）。白瑞景等（2010）报道中国荷斯坦奶牛第三胎 SCC 最高，若不注意奶牛的日常保健管理，随着胎次的增加，乳腺炎发病率也会上升。

2. 泌乳时间对体细胞数的影响 泌乳时间越长，平均 SCC 越高。随着泌乳时间的延长，乳腺细胞合成乳汁的功能减弱，乳产量减少，牛奶中体细胞浓度升高，对病原微生物的易感性也上升。当然 SCC 还与采样时间及乳汁分泌先后有关。

3. 个体及品种对体细胞数的影响 奶牛乳房情况也对体细胞数有一定的影响，如乳房变形或者下垂使之与地面接触的机会增多，乳房感染病原微生物也相应增多，奶牛患乳腺炎概率与 SCC 数量呈正相关，较长的乳头与较高的 SCC 有关。实验研究表明，乳房线性评分与 SCC 之间呈显著相关（$P>0.05$），前乳房附着较强、悬韧带较强较紧密、乳头位置内向及较高后乳房的奶牛有较低的 SCC。石祖星（2010）研究表明，SCC 和前房附着、乳房悬垂、乳房深度及乳头位置之间的表型都呈负相关，具有较高的附着位置和较牢固乳房的母牛 SCC 较低。

4. 产犊季节对体细胞数的影响 夏季产犊奶牛的 SCC 最高，其次是秋季和冬季，春季产犊 SCC 最低。夏季温湿指数最高，母牛受到长期热应激，免疫力下降，对病原微生物易感性增强，夏季牛舍周围环境中的病原菌种类和数量都较其他季节多，所以夏季牛奶中 SCC 最高（王小龙，2014）。王相根（2012）通过对河南郑州某奶牛场 2005 年 9 月至 2010 年 8 月 5 个年度奶牛的奶牛群体改良（dairy herd improvement，DHI）系统数据进行统计分析，在每年的 7~9 月，隐性乳腺炎发病率明显高于其他季节。白瑞景（2010）对河北省某奶牛场 1~6 胎次 DHI 测定，结果表明，产犊季节显著影响着 SCC 的水平，夏季产犊的牛只 SCC 达到了最高水平，与其他季节差异显著。

5. 饲养模式对体细胞数的影响 李晓香等（2013）研究发现，三种饲养模式下 SCC 由高到低为：不完全集约化饲养（IF）＜集约化饲养（CF）＜粗放饲养（EF）。EF 型显著高于 CF 型，CF 型显著高于 IF 型（$P<0.05$）。EF 型奶牛健康状况较差，有较高的罹患乳腺炎的风险，这与 EF 型奶牛饲料结构不合理引起免疫因子水平升高及饲养管理不佳有关。CF 型奶牛 SCC 偏高，可能与其产奶量高生产负荷过大有关。研究发现，饲养管理、环境卫生状况和挤奶规程不当均可能引起 SCC 升高。IF 型奶牛 SCC 较低，可能是营养状况相对较好、产奶量相对偏低、生产负荷相对较低等因素的结果。

6. 营养情况对体细胞数的影响 奶牛营养均衡，饮食科学合理，奶牛自身免疫能力上升，对疾病的易感性减弱，体细胞数会偏低。乳汁中 SCC 与奶牛体内硒水平之间呈负相关，在奶牛日粮中添加益生菌可以降低乳汁中的 SCC（张克春等，2010）。乳酸菌常作为饲料添加剂喂给奶牛，乳酸菌具有增强免疫、调节消化道的微生态环境、提供营养物质等功能，可以降低奶牛乳腺炎的发病率。

二、奶牛体细胞数对乳汁产量和质量的影响

（一）体细胞数对产奶量的影响

有研究表明，体细胞数与日产奶量呈负相关，即随着体细胞数量的增多，牛奶产量下

降（周亚平等，2011）。奶牛产奶量受体细胞数量的影响主要表现在乳腺组织被病原微生物感染后，位于乳腺泡的上皮细胞数目减少，并且其功能亦受到影响，结缔组织代替了部分泌乳上皮细胞，最终导致产奶量降低。有研究表明，以牛奶中体细胞数在 2×10^5/mL 为基础数据，每毫升增加 1×10^5 个体细胞，产奶量会相应减少 2.5%。如果在一个牛群中，体细胞数量高于 4×10^5/mL 的牛达总量的 70% 以上，每头泌乳奶牛每年平均将减少 500 kg 以上的产奶量。一般情况下，产奶量越高的牛越容易感染乳腺炎，并且这一类型的牛正处于产奶高峰期，再加上人工费用、饲养费用等，造成了很大的经济损失，因此体细胞数量高对奶牛生产能力的影响巨大（孙淑霞，2013）。

（二）体细胞数对乳蛋白率的影响

体细胞数过高对乳蛋白率也会产生影响，特别是影响乳蛋白中的酪蛋白数量，但研究者对体细胞数量增多对乳蛋白率的影响报道并不一致。Dosogne 等（2003）研究结果表明，体细胞数量升高时，乳蛋白先升高，当体细胞数在 $5\times10^5\sim10\times10^5$/mL 时，乳蛋白率最高；当体细胞数超过 10×10^5/mL 时，乳蛋白率开始下降，可能的原因是奶牛感染乳腺炎后，乳腺炎病灶破坏了乳腺上皮细胞的通透性，血-乳渗透性增加，使血液中的免疫球蛋白、乳铁蛋白、血清白蛋白等蛋白物质进入乳汁中，导致乳蛋白率升高。而王芳和胡松华（2005）报道，奶中体细胞数的含量达到 10×10^5/mL 时，蛋白质率减少；新西兰国家发表的数值显示，冷藏牛奶中体细胞数为 7.5×10^5/mL 时比为 10×10^5/mL 时酪蛋白减少 20%，用其加工奶粉，奶粉生产量减少 4%，如果达到 12×10^5/mL，奶中酪蛋白减少 25%，用其加工奶粉，奶粉生产量会减少 5%。健康状况下，牛奶中酪蛋白含量占总蛋白含量的 77%～80%，而患隐性乳腺炎的牛奶中酪蛋白含量占 65.8%～73.5%。牛奶中酪蛋白含量降低主要是由于体细胞数偏高的牛奶中蛋白水解酶活性升高，牛奶中乳清蛋白的含量随体细胞数的升高而升高的结果。张慧林等（2010）研究了西安草滩奶牛三场 2004 年处于 2～4 胎即泌乳盛期的荷斯坦奶牛的奶样，测定的指标包括体细胞数、乳脂率、乳蛋白率、日产奶量、胎次、泌乳阶段，共有 341 个样本，结果得出体细胞数量的变化对乳脂率、乳蛋白率的变化不明显，其主要与饲料营养水平、奶牛的消化能力有关。前人研究结果表明，体细胞数量的改变会在一定程度上改变乳蛋白率，但具体的影响还没有一致的定论。

（三）体细胞数对乳脂率的影响

体细胞数量与乳脂率的相关性研究，各研究者报道也不一致。有研究表明，由于奶牛在患乳腺炎期间，机体合成乳腺细胞和分泌能力会下降，含有较高体细胞数的牛奶其乳脂率会明显下降。Shamay 等（2002）报道体细胞数量稍升高对乳脂率并不产生影响，甚至可促进乳脂率轻微升高。虽然至目前为止，牛奶中体细胞数量升高与乳脂率相关性研究还没有一致的结论，但有一点可以肯定的是，由于随着体细胞数的增加，产奶量显著下降，因此乳脂总量会随产奶量的降低而降低。

（四）体细胞数对乳糖的影响

患有乳腺炎的奶牛，其所产牛奶中会损失 10%～20% 的乳糖。这是由于患病牛会感染血管，从而造成血管通透性极度亢进，盐类由血液进入牛奶的数量会增加，其与乳糖分泌的减少相互作用，使乳液渗透压上升（程广龙等，2009）。有研究表明，在冷藏条件下，体细胞数量多的牛奶中游离脂肪酸会增加，是体细胞数含量低的牛奶中的2~3倍。冷藏后期体细胞数量高的牛奶出现较明显的脂肪酸败味，酸度较体细胞数量少的牛奶明显增高，口感变差（孙淑霞，2013）。

（五）体细胞数高低对其他乳成分的影响

健康奶牛所产的牛奶中无机盐含量为0.7%，这些无机盐主要是钙、磷、钠、氯、硫、钾，还包括其他一些微量元素，当牛奶中体细胞数量变化时，其中所含的矿物质也会随之发生变化。当牛奶中体细胞数目偏高时，牛奶中Cl^-、Na^+含量升高，K^+、Ca^{2+}含量降低，牛奶中无机离子随体细胞数量的变化而变化，其原因是乳腺分泌上皮细胞可以选择性吸收血浆中矿物质，牛奶中的Na^+、Cl^-的浓度常低于血液，而Ca^{2+}、K^+浓度常常高于血液。当牛奶中体细胞数量升高后，牛奶中乳糖的合成能力明显下降，而乳糖是泌乳的渗透调节剂，为了维持血液渗透压与牛奶中的渗透压平衡，血液中含量较高的Na^+、Cl^-就渗透到牛奶中，从而导致Na^+和Cl^-含量升高。邢慧敏等（2007）通过对427份牛奶样品中氯、钾、钠、总钙含量、游离钙和SCC测定发现，牛奶中Na^+、Cl^-的含量与SCC的数量呈显著的正相关（$P<0.001$）；游离Ca^{2+}的含量与SCC呈一定负相关（$P<0.05$）；总Ca^{2+}、K^+的含量与SCC的相关性不显著（$P>0.05$）。牛奶中体细胞数和Na^+、Cl^-、K^+、游离Ca^{2+}、总Ca^{2+}含量与月份之间不存在相关性（$P>0.05$）；氯与总钙相关性显著，且呈正相关（$P<0.001$），乳汁中的总钙与游离钙之间，以及氯离子与钠离子之间相关性显著（$P<0.05$）；钠离子与总钙呈显著正相关（$P<0.01$）；其余离子之间相关性不显著（$P>0.05$）。牛奶中盐类比例失调，牛奶热稳定性降低，凝固型酸奶的凝乳松散易碎裂，乳清易析出；过氧化物酶含量升高，乳品风味欠佳；脂肪氧化酶增加，其氧化产物游离脂肪酸含量上升，乳易酸败，大大缩短了乳品保质期；蛋白水解酶尤其是血纤维蛋白溶酶含量和活性增加，造成乳蛋白大量水解，降低干酪产量，牛奶中有害微生物含量增加，抗生素残留的风险增加，降低了乳制品食用安全系数（叶纪梅等，2006）。

现在国内外普遍把奶牛中的SCC作为诊断奶牛乳腺炎的指示性指标。SCC是计量性状，其分布是偏态的，从而使SCC在统计分析中不能应用正态分布的许多统计学特性，因此在实际生产中往往将SCC转换为体细胞评分（somatic cell score, SCS），以克服其在统计分析中的不足。当$SCS>5$，为感染乳腺炎的患病牛，$SCS<5$为正常健康牛。SCC的差异不仅存在于个体间，而且与季节、胎次、泌乳月份等因素相关。毛永江等（2007）分析了南方地区某奶业集团2006年7个牛场2063头中国荷斯坦牛22 377次SCC测定日记录，研究结果表明，牛场、月份、泌乳天数和胎次的不同与乳汁中的体细胞数量的相关性均极显著（$P<0.001$），而产犊季节与乳汁中的SCC的数量相关性不显著（$P>0.05$），随着胎次和泌乳天数增加，SCC数量有升高的趋势。SCC越高，奶牛乳汁中的致病菌和抗生素残留的污染风险越大，对人类健康的危害也越大。

三、引起奶牛乳腺炎的遗传因素

奶牛乳腺炎发病，一部分是由外界环境因素所引起的，如饲养管理方面，另一部分是奶牛的遗传因素引起的。在同一环境和相同管理条件情况下，对乳腺炎的易感性因奶牛个体不同而存在较大差异，Heringstad 等（2001）报道，在同一环境下，育种值上具有最差乳腺炎抗性的 5%的公牛，其下一代母牛的乳腺炎发病率比具有在育种值上最佳乳腺炎抗性的 5%的公牛的下一代母牛高出 10%～15%，这就需要从遗传角度对乳腺炎个体进行选择，但如果从遗传因素考虑选择抗乳腺炎奶牛个体，是比较漫长的一项工作，如果考虑到这项工作对今后奶牛场的影响，它可以长期地改变某一奶牛群体的抗乳腺炎的遗传组成。

最早时期，人们在遗传学方面的育种总是直接选育没有临床乳腺炎或症状较轻的个体奶牛作为育种标准，但经多年的实践表明，将临床乳腺炎作为育种指标并不适合。此后，人们将研究方向转向奶牛抗性性状的相关研究，希望能找到具有与乳腺炎高相关性且遗传潜力高的性状的个体奶牛进行选育。但是抗性性状本身是一个极其复杂的性状，既决定于遗传因素，又受生理状况和环境条件的影响，从遗传角度讲，奶牛乳腺炎平均遗传力为 0.04（Mrode et al.，2003）。另外，不同种间奶牛个体乳腺炎的发病率存在显著的差别，高产奶牛其乳腺炎发病率较高，而地方品种如鲁西黄牛、水牛、牦牛等不易感染乳腺炎。

乳腺炎严重程度升高，体细胞数也随之上升，患乳腺炎奶牛的体细胞数是正常健康奶牛的 3～4 倍，这为以 SCC 作为奶牛乳房健康指标提供了理论依据。丹麦第一胎临床奶牛乳腺炎的遗传力为 0.02～0.06。奶牛临床乳腺炎在第一个泌乳期的遗传力为 0.07，在第二个泌乳期为 0.09。美国奶牛患临床乳腺炎遗传力为：在第一个泌乳期为 0.03～0.25，第二个泌乳期为 0.01～0.19；芬兰奶牛临床乳腺炎的遗传力在第一、第二、第三个泌乳期分别为 0.025、0.046、0.041；欧美 12 个国家奶牛乳腺炎的遗传力为 0.02～0.05。从遗传角度来说，奶牛患乳腺炎的平均遗传力较低，且与乳腺炎抗性相关的遗传基因之间的关联较复杂。因此，进行直接选择奶牛乳腺炎性状的标准难以确定。近年来，随着分子遗传标记和数量性状位点（quantitative trait loci，QTL）技术的应用，为奶牛乳腺炎抗性育种开辟了一个新途径，即奶牛分子遗传育种，奶牛分子遗传育种研究的重点主要集中在主效数量性状基因座育种（肖正中和赖松家，2003）。随着牛基因图谱的完善、遗传多态标记数量和定位技术的提高，基因组数据库系统和试验信息及资源共享体系的完善，可使任何具有多态性的单基因和控制数量性状的 QTL 或主基因在染色体上的定位成为可能（魏伟和苗永旺，2011）。

四、体细胞评分的遗传评定

20 世纪 70 年代末，美国实行体细胞检测计划，即 DHI 计划。一般情况下，高体细胞数，表明乳房健康状况恶化。同样，低体细胞数表明只是轻度感染或未受感染。因此，低 SCC 与低临床乳腺炎的感染概率相关。将体细胞数转化为传统遗传评价的统计方法常用线性的体细胞评分（SCS）。DHI 奶牛记录处理中心把每个个体的体细胞数取以 2 为底的对数，就转化成体细胞评分。总数的 40%左右参与 DHI 计划，每月进行体细胞测试，混合牛奶样品用于体细胞数的检测，按月抽样得分累积成一个泌乳期平均分，并校正年龄和产犊季节。

每月的体细胞计数（SCC）可以作为评价奶牛泌乳期乳房健康状况评定标准。体细胞计数可以用来评定是否存在感染、感染的数量和持续时间、感染的严重程度；体细胞计数增加一倍导致感染程度增加一个单位。相比于体细胞计数，体细胞评分作为评价乳腺炎的标准具有更好的重复性，产生试验方差和误差的方差较小。与体细胞计数相同，体细胞评分与奶牛产奶量间也存在相关性。体细胞评分每增加1个单位，在第一个泌乳期奶产量减少200磅[①]，第二个泌乳期以后减产400磅，乳腺炎造成了巨大的经济损失。体细胞评分每增加一个单位，在50头牛群体中，由乳腺炎造成的经济损失增加2000美元。

用半同胞相关和母女回归分析来自342头父本的682头母牛乳腺炎和SCC之间相关性，发现相关系数分别为0.80和0.98。瑞典黑牛与白牛及瑞典红牛与白牛间4个种群的SCC和乳腺炎之间遗传相关系数为0.46和0.78，平均0.62。体细胞评分与临床乳腺炎平均遗传相关为60%~80%，表明两个性状间有显著遗传相关。相比体细胞计数，细胞评分（SCS）被证明是一个更可靠的指标，因为体细胞评分与产奶量的降低呈线性关系，乳腺炎与SCS的相关系数（0.60）高于SCC和乳腺炎之间相关系数，SCS比SCC有较高的遗传力。

第七节　动物乳腺炎模型的建立

研究乳腺炎的发病机制、筛选敏感的奶牛乳房炎早期诊断指标及寻求更好的防治药物（如中草药）等，都需要一种典型、廉价、易于建立的乳腺炎动物模型。已经有许多乳腺炎动物模型用于研究乳汁电导率、乳产量和乳汁温度等，然而一些动物模型有一定的缺陷，如在整个泌乳期不能及时测定反映乳房健康状况的一系列生理指标。目前的研究主要集中在寻找一种生物学模型，它能系统地用于测定奶牛乳腺炎的早期诊断指标和其他有害因子。国内外已有许多学者通过向奶牛、奶山羊和小鼠等动物乳腺内注入病原菌、内毒素或强制断奶等方法制作乳腺炎模型，也有学者将诱导物不通过乳头管而直接注射到乳房的基部诱导乳腺炎（杨洪森，2011）。

一、小鼠乳腺炎模型

在奶牛乳腺炎的研究中，高昂的成本和管理是一个不可避免的重大问题，出于这个原因，研究者在20世纪70年代建立了乳腺炎的实验小鼠模型。这种实验模型是相对便宜的，因为它只需要标准的动物护理设施和基本的实验动物，目前，小鼠被认为是用于研究乳腺炎的发病机制和对控制牛乳腺内感染的一种合适的实验动物。类似奶牛的乳腺，小鼠也有两对乳腺腺体在腹股沟区域，另有三对在其身体的胸部区域，这是奶牛所不存在的。这两个物种具有的腺体在功能和解剖学上相互独立，此外，每个乳腺只有一个奶头和一个初级导管。小鼠乳腺炎模型还提供牛奶病原体的独特生长环境，并允许与宿主细胞和免疫组分的有机体相互作用。有利于乳腺炎研究的是，在小鼠中观察有关的细菌计数、中性粒细胞

---

① 1磅≈0.454 kg

数和组织学改变与在奶牛中观察相似（肖峰，2015）。

尽管有很多相似性，但两个物种的乳腺之间还是存在一定的差别。例如，奶牛乳腺含有比小鼠乳腺更多的吞噬细胞，乳汁内的蛋白质、脂肪和碳水化合物也有显著的不同。不同于奶牛，小鼠体细胞计数（SCC）分析一般不进行，因为收集小鼠乳汁是一个困难的过程。虽然小鼠乳腺炎模型与奶牛乳腺炎有一定的差异，但从乳腺炎模型得到的结果与牛乳腺炎中得到的有很高的相似性，此外，小鼠感染模型在乳腺炎的研究中提供了更多和非常有吸引力的前景。小鼠乳腺炎模型的成功建立提供了很多便利：动物模型允许筛选大量潜在的抗微生物和免疫调节化合物，这样既降低了成本又节省了时间；有时候当有效药物数量有限并且没有在奶牛身上试验过的药物，可以借助小鼠模型；有些新技术的出现也只能应用于实验室小动物，最重要的是突变、转基因或基因敲除小鼠的产生，对乳腺炎的研究已经产生了深刻影响（Kerr et al.，2001）。

现已有大量研究成果显示，可以通过不同的致病菌诱导不同品系的小鼠建立乳腺炎模型。通过对小鼠进行乳房内注射金黄色葡萄球菌发现，其发病机制与表面蛋白和毒力因子有着密切的联系，如金黄色葡萄球菌表面的α溶血素、凝固酶和蛋白A可以通过化学途径使其发生改变，并将变异后的毒株注射到小鼠乳房当中。通过显微镜观察损伤的组织发现，α溶血素和凝固酶的改变能够使金黄色葡萄球菌毒力下降。同样对β溶血素改变的变异株进行研究，其研究结果与上述相似，β溶血素也与金黄色葡萄球菌的毒力相关。纤连蛋白结合蛋白（FnBP）是金黄色葡萄球菌在乳腺上皮细胞上黏附和定植的重要毒力因子之一，如果葡萄球菌缺少FnBP，虽然也可在乳腺上皮黏附定植，但其黏附能力明显减弱。在小鼠模型中FnBP的变异株可以侵害小鼠的乳腺上皮细胞，所以黏附蛋白的存在并不是金黄色葡萄球菌引起乳房炎的绝对因素。

通过对金黄色葡萄球菌引起乳房炎的研究发现，学者们最终将治疗或者防治的重点都放在了免疫反应上。丁是研究者通过建立BALB/C小鼠无乳链球菌乳腺炎模型在攻菌不同时间段来检测小鼠乳腺局部白细胞群及乳腺中细胞因子的表达情况（Gabriela et al.，2009）。研究发现，细菌在乳腺当中增值并且在攻菌后24 h达到峰值。在感染6 h时可以观察到细菌向全身器官散播，并且伴随有大量细菌增殖、组织坏死及中性粒细胞浸润。在注射后24 h出现严重的炎症反应，在主导管周围出现大量坏死区域和中性粒细胞浸润。HE染色发现大量细菌及被中性粒细胞吞噬的细菌，以及侵入乳腺上皮细胞的细菌。在注射后3天，实质中出现炎性浸润并伴随有多形核淋巴细胞和单核细胞出现。注射后5天与3天时的情况相似。在注射10天时，巨噬细胞与淋巴浆细胞增多，多形核白细胞减少，间质中产生了轻微的纤维素性变性。乳腺当中分泌的细胞因子IL-1β、IL-6、TNF-α从感染后6 h逐渐升高，由此说明早期炎症反应出现。与此同时，在感染后72 h乳腺当中的细胞因子IL-12和IL-10达到峰值，而早期的IL-1β、IL-6、TNF-α表达量逐渐降低。研究表明，乳腺当中一系列细胞因子的变化与细菌增殖的逐渐减少有关。通过流式细胞分析乳腺局部淋巴结中的淋巴细胞可以发现，在感染后5天巨噬细胞、B220$^+$淋巴细胞及CD4$^+$ T细胞、CD8$^+$ T细胞显著升高。感染后10天在乳腺中检测到了无乳链球菌的抗体。通过研究发现，在无乳链球菌引起的乳房炎中天然免疫应答和适应性免疫应答都在感染的不同阶段起着至关重要的作用。

小鼠模型的研究不但有助于弄清楚乳腺的免疫防御机制，更有利于在此基础上研发疫苗。最近的研究发现，IsdH（iron-regulated surface determinant protein H）是一种依赖铁的表面蛋白，它可以结合中性粒细胞，并且与金黄色葡萄球菌感染有着密切的联系（Visai et al.，2009）。给小鼠免疫 IsdH 可以引起持续并且强烈的免疫反应，产生大量的 IgG2。在奶牛体内也产生了同样的效果。这些都应归结于小鼠模型，它不但解决了大动物难于管理和操作的难题，也避免了试剂等方面的浪费。小鼠模型是今后研究乳腺炎和其他疾病疫苗的关键手段之一。

二、家兔乳腺炎模型

家兔和奶牛同属于哺乳动物，应用泌乳家兔建立奶牛乳腺炎模型，能以较低的成本获得更多的试验材料。家兔乳头管灌注金黄色葡萄球菌和大肠杆菌进行乳腺炎的造模成本低、操作简便。

杨洪森（2011）通过家兔乳头管灌注金黄色葡萄球菌和大肠杆菌，观察灌注后家兔的临床病理变化，并在金黄色葡萄球菌和大肠杆菌两种不同的家兔乳房炎中，探讨血清中C反应蛋白（C-reactive protein，CRP）含量，以及组织和血清中肿瘤坏死因子-α（TNF-α）含量的不同点。结果表明，家兔接种金黄色葡萄球菌 6 h 时体温开始显著升高（$P<0.05$），48 h 时缓慢降回接种前水平，接种大肠杆菌 6 h 时显著升高（$P<0.05$），第 5 天时基本恢复到接种前水平，空白对照组 12 h 时显著升高（$P<0.05$），之后缓慢降回接种前水平。母兔接种金黄色葡萄球菌 6 h 时乳房质地开始变硬，到第 5 天时硬度开始下降；接种大肠杆菌 6 h 时开始变硬，第 7 天时硬度开始有所下降。

白细胞总数在接种金黄色葡萄球菌后开始缓慢下降，48 h 时开始显著下降（$P<0.05$），在接种大肠杆菌 6 h 开始显著下降（$P<0.05$），之后缓慢回升，72 h 时恢复到接种前水平，空白对照组变化不明显。金黄色葡萄球菌组与空白对照组在 48 h、5 天和 7 天时差异显著（$P<0.05$），与大肠杆菌在 6 h、12 h、5 天和 7 天时差异显著（$P<0.05$），大肠杆菌组与空白对照组在 6～48 h 差异显著（$P<0.05$）。淋巴细胞在接种金黄色葡萄球菌后显著下降（$P<0.05$），6 h 时下降到最低点之后缓慢回升，72 h 时恢复到接种前水平，之后又显著升高（$P<0.05$）；在接种大肠杆菌后显著下降（$P<0.05$），之后缓慢回升，24 h 时恢复到接种前水平，之后又缓慢下降，5～7 天时显著下降（$P<0.05$），空白对照组水平变化不明显。金黄色葡萄球菌组与大肠杆菌组在 12～48 h、5 天和 7 天时差异显著（$P<0.05$），与对照组在 6～24 h、5 天和 7 天时差异显著（$P<0.05$），大肠杆菌组与对照组在 6 h、24 h、5 天和 7 天时差异显著（$P<0.05$）。中性粒细胞总数在接种金黄色葡萄球菌后开始缓慢升高，24 h 时升高较接种前显著（$P<0.05$），之后缓慢下降，72 h 后显著下降（$P<0.05$），在接种大肠杆菌 6 h 开始显著下降（$P<0.05$），24 h 时下降到最低点之后恢复到接种前水平，空白对照组变化不明显。金黄色葡萄球菌组与空白对照组在 48 h 后差异显著（$P<0.05$），与大肠杆菌在 6～24 h、5 天和 7 天时差异显著（$P<0.05$），大肠杆菌组与空白对照组在 6～48 h 差异显著（$P<0.05$）。

血清中 TNF-α 含量在接种金黄色葡萄球菌后开始下降，和接种前相比，在 12 h 时开始显著下降（$P<0.05$），第 7 天时有所回升，接种大肠杆菌后显著下降（$P<0.05$），48 h

开始回升到接种前水平,第5天时显著升高($P<0.05$),之后显著下降($P<0.05$),对照组虽然也有升高或下降时间点,但总体变化不明显。金黄色葡萄球菌组和大肠杆菌组在6 h、12 h、72 h、5天和7天时差异显著($P<0.05$),和对照组在24 h、48 h和72 h时差异显著($P<0.05$),大肠杆菌组和对照组在6~48 h和7天时差异显著($P<0.05$)。组织中TNF-α在接种金黄色葡萄球菌后显著下降($P<0.05$),48 h时有显著升高($P<0.05$),之后缓慢下降到接种前水平,接种大肠杆菌后24 h时显著升高($P<0.05$),之后缓慢下降,72 h时回到接种前水平,空白对照组变化不明显。金黄色葡萄球菌组与大肠杆菌组在6~24 h时差异显著($P<0.05$),和空白对照组在6~24 h时差异显著($P<0.05$),大肠杆菌组和空白对照组在24 h差异显著($P<0.05$)。

血清中CRP含量在接种金黄色葡萄球菌后开始慢慢上升,24 h和48 h时升高较显著($P<0.05$),之后缓慢下降到接种前水平,在接种大肠杆菌后48 h时显著升高($P<0.05$),之后缓慢下降到接种前水平,空白对照组在接种无菌肉汤后持续升高,72 h时开始升高显著($P<0.05$)。金黄色葡萄球菌组与空白对照组在24~48 h和5~7天时差异显著($P<0.05$),与大肠杆菌组差异不显著($P>0.05$),大肠杆菌组与空白对照组在24~48 h和5~7天时差异显著($P<0.05$)。

健康母兔乳腺上皮排列整齐,部分腺泡可见分泌的乳汁。家兔乳腺内灌注金黄色葡萄球菌后,6 h时乳腺小叶接近正常,腺泡腔分泌物明显增多,间质小静脉充血,少量中性粒细胞浸润;12 h时腺泡腔内和小叶间质内浸润大量中性粒细胞,间质小静脉和毛细血管充血;24 h时腺泡腔内见有较大量浓厚的分泌物,中性粒细胞崩解,数量减少;48 h时腺泡腔显著扩张,腔内充满大量浓厚的分泌物,中性粒细胞数量较前一时间点进一步减少;72 h时腺泡腔显著扩张,腔内充满大量浓厚的分泌物和大量中性粒细胞;第5天时腺泡腔结构紊乱,腔内有数量不等的脱落上皮;第7天时乳腺腺泡腔显著扩张,腔内分泌物的数量仍较多,但浓厚度较72 h时轻。

健康母兔乳腺上皮排列整齐,部分腺泡腔内分泌的乳汁略有增多,兔乳腺灌注大肠杆菌后,6 h时乳腺腺泡腔扩张,腔内分泌物增多,腺泡腔和腺泡壁上均有少量中性粒细胞浸润,小叶间质显著增宽,大量中性粒细胞浸润;12 h时乳腺腺泡腔扩张,腔内分泌物进一步增多,腺泡腔和腺泡壁上均有中性粒细胞浸润,小叶间质显著增宽,出血并见有大量中性粒细胞浸润;24 h腺泡腔显著扩张,充满大量浓厚的分泌物,腺泡腔内见有少量中性粒细胞和脱落的腺泡上皮;48 h时腺泡内见有较多的中性粒细胞,小叶间质显著增宽,充满大量中性粒细胞,间质小静脉扩张充血,并见小灶性出血;72 h时腺泡腔显著扩张,充满多量分泌物,可见蓝染的细胞团块和崩解的中性粒细胞,小叶间质显著增宽,浸润的大量中性粒细胞崩解,形成细胞碎片;第5天时腺泡腔仍充满大量分泌物,中性粒细胞多崩解消失;第7天时腺泡结构正在恢复,部分腺泡腔内有少量分泌物、中性粒细胞和脱落的上皮细胞。

三、奶山羊乳腺炎模型

奶山羊与奶牛在生理特性等诸多方面很相似,以奶山羊为模型动物建立乳腺炎模型,研究取得的结果不仅可直接用于奶山羊乳腺炎防治,同时对奶牛乳腺炎防治也有很好的参考价值。

苗晋锋等（2007）向产后 72 h 的奶山羊乳头管内分别注入磷酸脂多糖（lipopolysaeeharide，LPS）和灭菌生理盐水，与灌注生理盐水相比，LPS 组显示了急性乳房炎的临床症状及乳腺组织的病理变化，乳腺组织中 AG、MPO 活性及 IL-1、IL-6 的含量显著升高（$P<0.05$），TNF-α 含量显著升高（$P<0.05$），证明奶山羊乳头管灌注 LPS 能成功建立奶山羊乳腺炎模型。

牟珊（2010）将 6 只奶山羊分为 2 组（大肠杆菌 O117 型组和金黄色葡萄球菌血清 8 型组），每组 3 只。攻菌前先将乳池中的奶挤完，温水擦洗山羊乳房，以 75%酒精棉擦拭右侧乳头消毒。用灭菌的小号通乳针插入乳头管，将余奶放净后，用无菌注射器吸取菌悬液 1 mL（大肠杆菌量 3×10^3 cfu/mL，金黄色葡萄球菌量 3×10^2 cfu/mL）从通乳针注入乳房中，注射完后立即取出通乳针，轻轻按摩乳房数秒，使菌液分散均匀。左侧乳房消毒后，同上述方法，注入 1 mL 无菌生理盐水。大肠杆菌组奶山羊在 8 h 时乳房开始发热，24 h 时精神沉郁，食欲减退，乳房发红、微烫，乳汁与健康乳样相比略黄；48～72 h 时精神、食欲、乳房外观及乳汁性状逐渐恢复正常。金黄色葡萄球菌组奶山羊在 8 h 时右侧乳房出现不同程度的硬结，发红发热；24 h 精神沉郁，采食量极小，右侧乳房发烫，明显肿大、变硬、有触痛，乳汁发黄、变稠；72 h 精神极度沉郁，卧地不起，拒食，右侧乳房淤青发紫、冰凉，触碰有明显痛感，左侧乳汁黄色、右侧乳汁变红呈脓血状。其中一只羊在 72 h 后死亡。金黄色葡萄球菌组症状严重，为典型的临床型乳房炎；大肠杆菌组开始有轻微症状，2 天后则逐渐消失。攻菌后两组奶山羊泌乳量均降低。大肠杆菌组在 2～3 天降到最低，其后逐渐升高，左侧对照乳区泌乳量基本可以恢复，但右侧攻菌乳区低于攻菌前产奶量。金黄色葡萄球菌组在第一天大幅降低，第二天急剧降低，其后乳质呈脓血状，几乎停乳。

Leena 等（2008）通过在奶牛一个乳区中注入 1500 cfu 的大肠杆菌，14 天后在另一乳区注射同样数量的大肠杆菌来诱发乳腺炎，而后检测血清和乳汁中结合珠蛋白（HaPtoglobin，HP）、淀粉样蛋白 A（serum amyloid A，SAA）和脂多糖结合蛋白的含量，血清和乳汁中 HP、SAA 和脂多糖结合蛋白均升高。Rainard 等（2008）将纯革兰氏阳性菌的脂磷壁酸（lipoteichoic acid，LTA）注入奶牛乳腺的管腔发现，每乳区注射 100 μg 剂量时可以诱发临床乳腺炎，10 μg 时可诱发亚临床型乳房炎，该类型的乳腺炎以乳汁中的中性粒细胞快速升高为特征。由于奶牛直接作乳腺炎模型的成本高、饲养管理不便，因此，山羊、家兔、大鼠和小鼠常作为乳腺炎模型的常用动物。

四、研究展望

在过去的几十年，对于奶牛乳腺炎的研究已经取得了诸多的成果，但是奶牛乳腺炎的发病机制及流行病学特征仍不十分清晰，使得该病的预防和控制仍然十分困难。因此，以后对于奶牛乳腺炎的相关研究应当更加具体、细致、全面，一些关键性的问题应加大研究力度，使其更加清晰明确。

首先，研究发现遗传因素对临床性乳腺炎及隐性乳腺炎发生率影响较低，但世界各大乳品企业仍集中精力，希望通过育种来减少该病的发生继而减少损失。基因差异性出现在不同牛种之间，同时还出现在不同的管理模式。这种基因差异性研究是否有经济价值及基因的改变与环境的关系是怎样的现在还不知道。此外，公牛的选择也对基因育种有重要的作用。遗传因素对于乳腺健康及乳腺炎的发生的相关性还需要深入研究。

其次，研究发现母牛患乳腺炎与其突然从非泌乳状态转换到泌乳状态的应激相关。初产母牛一般容易发生乳腺水肿，易进一步导致乳腺炎的发生。此外，产犊后非酯化脂肪酸（NEFA）、羟丁酸（BHBA）及其他代谢产物的浓度会影响乳腺组织的健康情况。应激因素对乳腺组织代谢产物有何影响，产犊前后机体营养代谢的变化情况及生产应激与乳腺炎发生的密切联系如何，这些问题也应进一步深入研究。

再次，乳腺炎致病菌的种类及感染情况与奶牛乳腺健康的关系是必须要被重视的。乳腺炎患牛产前和产后乳腺致病菌感染率均很高，但产前是如何感染的，以及致病菌的具体来源是皮肤、其他奶牛还是环境，仍不清晰。这对于研究感染性乳腺炎的发生十分重要。应用分子生物学技术去确定细菌的种类、传播方式、感染菌株是否会持续存在、感染来源等仍需要研究。对于乳腺炎致病菌的分离鉴定，以及感染的细菌之间存在的相互作用，如金黄色葡萄球菌感染后会抑制其他细菌的大量繁殖，其他的细菌也有类似的情况，都需要深入研究。此外，感染性乳腺炎病原对于乳汁分泌的影响机制也需要进一步研究。

此外，饲养管理模式与奶牛乳腺炎的发生密切相关，也有待于深入研究。澳大利亚、新西兰及美国的部分地区奶牛为分群养殖，这就导致它们的营养、病原及其他疾病发生之间存在差异，应该分析产生这些差异的原因，消除弊端（郭梦尧，2014）。还有舍饲奶牛的空间及挤奶器的型号、规格等，这些都是潜在诱发奶牛乳腺炎的危险因素，可对此进行研究，以期降低奶牛乳腺炎的发生，保证奶牛业的快速健康发展。

第二章 奶牛乳腺炎的流行病学

奶牛乳腺炎是乳腺组织的一种炎症反应，通常由微生物感染引起，尽管近些年在改善乳房保健方面有了进展，但奶牛乳腺炎仍然是危害当今世界奶牛养殖业的主要疾病之一，严重制约着奶牛业的发展。当现代奶牛场首次在热带地区和亚热带地区建立，研究者就预言乳腺炎是奶牛最重要的疾病之一（Mungube et al.，2004）。近几年世界各国奶牛临床乳腺炎和隐性乳腺炎发病率差异很大，临床乳腺炎奶牛发病率为 1.8%～37.1%，隐性乳腺炎奶牛阳性率为 22.3%～62.9%（Kerro Dego and Tareke，2003；Sori et al.，2005；Getahun et al.，2008）。Lakew 等（2009）报道埃塞俄比亚奶牛临床乳腺炎和隐性乳腺炎发病率分别是 26.5%和 38.1%。而我国奶牛临床乳腺炎奶牛发病率为 17.5%（孔雪旺和陈功义，2005）和 23.1%（邢玫等，2007），隐性乳腺炎奶牛阳性率也在 40%～64.75%（朱贤龙等，2004；邢玫等，2007；李建军等，2008；赛鸿瑞等，2008），朱贤龙等（2004）曾报道广西地区奶牛隐性乳腺炎的发病率是 50.46%。金耀忠等（2011）对浦东地区规模化奶牛养殖场 2～10 岁的泌乳奶牛进行调查，临床乳腺炎平均阳性率为 13%，隐性平均阳性率为 45.2%。王娜和高学军（2011）对哈尔滨市 3 个大型奶牛场、10 户奶牛养殖户 1000 头奶牛的 4000 个乳区进行隐性乳腺炎的调查，其阳性率高达 47.7%，乳区发病率为 27.3%，检测结果远远高于魏成威（2010），说明不同奶牛养殖场的饲养环境及饲养管理的差异，严重影响了奶牛乳腺炎的发病率。李书文等（2012）采用加利福尼亚乳腺炎试验（California mastitis test，CMT）法对贵阳市所属的清镇市、乌当区、花溪区和开阳县 4 个规模化奶牛场 284 头奶牛的 1136 个乳区进行了奶牛隐性乳腺炎的调查，结果显示，乳头发病率为 53.8%，乳区发病率为 26.1%。李会等（2012）报道，在我国华北、西北、东北、西南等地区奶牛乳腺炎的发病率高达 70%～80%，其中奶牛隐性乳腺炎的发病率为 20%～40%，有时高达 50%～80%。金兰梅等（2010）对南京某规模奶牛场 433 头泌乳奶牛隐性乳腺炎发病规律进行调查研究发现，该牧场奶牛隐性乳腺炎发病率为 29.33%，乳区阳性率为 33.60%，病原菌的总检出率为 91.67%，引起隐性乳腺炎发生的致病菌主要为金黄色葡萄球菌和大肠杆菌，分别占总检出率的 37.88%和 16.67%。霍建丽于 2013 年 6～7 月在对青海省乌兰县农户饲养的 158 头泌乳期黑白花杂种奶的随机抽样调查中检出隐性乳腺炎阳性牛 62 头，阳性乳区 114 个，隐性乳腺炎阳性率 39.24%，乳区阳性率 18.63%；其中阳性牛和乳区阳性率均以茶卡镇最高，分别为 40.54%和 23.24%（霍建丽，2013）。王福顺等（2013）采用 LMT 法对天津某奶牛场 500 头泌乳牛进行乳腺炎检测，结果显示，该场奶牛隐性乳腺炎阳性率为 35.60%（178/500），临床乳腺炎阳性率为 14.20%（71/500），乳区阳性率为 49.80%（249/500）。

第一节 奶牛乳腺炎的发病率和主要致病菌调查

奶牛乳腺炎的病因非常复杂，如饲养管理、环境卫生、挤奶技术等因素，均可导致乳腺炎的发生，有报道称乳腺炎病原微生物有 130 余种，包括金色葡萄球菌、真菌、大肠杆菌、链球菌、支原体和藻类等（Bradley，2002）。国内外报道常见的致病菌也有 10 余种，复杂的致病菌给奶牛乳腺炎的防治工作带来了较大难度。为避免乳腺炎的发生，人们研制了一些能够降低该病严重性的一些疫苗，然而这些疫苗仍然不能有效控制乳腺炎的发生发展（Leitner et al.，2003）。另外，已证实无差别的对致病菌使用抗生素治疗、没有任何技术方法和鉴别试验，可促使这些微生物耐药性增加，使得治愈乳腺炎更加困难（Gruet et al.，2001），而且可引起乳汁中药物残留，对乳品加工、公共卫生及食品安全都是一项严峻的挑战。乳腺炎在奶牛的整个泌乳时期均可发生，尤其是泌乳早期发病率最高。

奶牛养殖业正在向规模化、集约化方向迅速发展，为了解规模化奶牛场奶牛临床乳腺炎和隐性乳腺炎的发病率、致病菌及其药物敏感性，并为指导临床合理用药和提高奶牛乳腺炎防治效果提供理论依据，本研究组于 2005～2009 年在某规模化奶牛场调查了临床和隐性乳腺炎的发病情况，并于 2006～2007 年采集临床和隐性乳腺炎乳汁共 276 份，对致病菌的分离鉴定、体外抑菌试验、动物毒性试验和生物被膜的产生情况进行了研究，旨在获取近年来规模化奶牛场奶牛临床乳腺炎和隐性乳腺炎的发病情况、致病菌特征等系统资料，并筛选出有效抗菌药，为指导临床合理用药、提高奶牛乳腺炎防治效果及降低乳腺炎的发生提供科学根据。

一、材料与方法

（一）实验材料

1. 培养基与试剂　　营养肉汤，营养琼脂，革兰氏染色试剂盒，上海隐性乳腺炎诊断试剂和诊断检验盘，各种微量生化发酵管及药敏纸片，草酸铵结晶紫染液、磷酸缓冲盐溶液、沙氏培养基、5%鲜血琼脂培养基、三糖铁培养基、LB 液体培养基、改良爱德华琼脂培养基、麦康凯琼脂培养基、却浦曼琼脂培养基、乳炎消药敏纸片，由实验室按标准配制。

2. 实验设备　　超净工作台，恒温培养箱，高压灭菌器，干燥箱，分析天平，酸度计，生物显微镜，冰箱，可调电炉，96 孔板等。

3. 实验动物和样品　　2005～2009 年规模化奶牛场用于生产的泌乳奶牛，以及 2006～2007 年采集临床乳腺炎奶样 109 份，隐性乳腺炎奶样 167 份。昆明系小白鼠购自广西中医学院实验动物中心，体重 18～25 g。

（二）实验方法

1. 麦氏比浊标准管的配制　　按照表 2-1 配制麦氏比浊标准管。密封并贮存于室温阴暗处。使用前，将比浊管于涡旋振荡器上剧烈振动摇匀，目测其外观浊度应均匀一致。若出现大颗粒，应重新配制。

表 2-1　麦氏比浊标准管的配制及对应细菌浓度

Table 2-1　The preparation of McFarland standards and the corresponding concentration of bacteria

麦氏比浊标准管编号	0.5	1	2	3	4	5	6	7	8	9
0.25% $BaCl_2$/mL	0.2	0.4	0.6	0.8	1	1.2	1.4	1.6	1.8	2
1% H_2SO_4/mL	9.8	9.6	9.4	9.2	9	8.8	8.6	8.4	8.2	8
细菌近似浓度/($\times 10^8$ cfu/mL)	1	3	6	9	12	15	18	21	24	27

2. 奶样的采集　　对患有临床乳腺炎的奶样，临床诊断，采样后直接送回实验室进行细菌检测。临床乳腺炎表现为乳房不同程度的红、肿、热和痛等炎症反应，乳汁有絮片、

凝块或变色（淡黄色、淡红色不等）。

对于奶牛隐性乳腺炎的检测和判定，国际上常用加利福尼亚乳腺炎试验（CMT）法，本实验是使用（上海奶牛隐性乳腺炎诊断试剂，SMT）。SMT与CMT的原理一致，根据牛奶与试剂混合后，与裂解的白细胞反应产生凝块，并依据凝块的多少和程度进行判定得分（表2-2），0=阴性，T=可疑，1=弱阳性，2=阳性，3=强阳性。先用温水清洗乳房，再用碘伏消毒乳头，卫生纸擦干后，弃去前三把奶再将被检乳区的乳汁约2 mL依次挤入诊断盘的4个检验皿中，再用注射器加入等量的SMT，随即作同心圆摇动10～30 s，将诊断液和乳汁充分混合，得分在1以上的乳汁认定是隐性乳腺炎乳汁，其他为阴性，即健康乳汁。当一个以上的乳区检测为SMT阳性时，该头奶牛被认为是隐性乳腺炎奶牛。

表2-2 SMT判定标准

Table 2-2 The judgment standard of Shanghai mastitis test

判断级别	乳汁SCC（单位：10^4/mL）	判定标准
阴性（0）	小于20	倾斜诊断盘时，盘底无沉淀物或絮状物，混合液移动流畅
可疑（T）	15～50	倾斜诊断盘时，混合液移动流畅，盘底出现少量的沉淀物或絮状物，继续旋转摇动时，沉淀物或絮状物消失
弱阳性（1）	40～150	盘底有少量的絮状物或黏性沉淀，但不呈胶状
阳性（2）	80～500	盘底黏性沉淀物量较多，并有胶状物，倾斜诊断盘时，沉淀物有明显黏附于盘底而难以流动的现象
强阳性（3）	大于350	盘底大部分或全部的混合物形成胶冻状，几乎完全黏附于盘底

乳样采集方法：先用温水擦洗乳房皮肤和乳头，再用碘伏消毒乳头，卫生纸擦干后，弃去前三把奶再将奶样用无菌操作方式采集于灭菌的玻璃三角瓶中（15 mL左右），盖紧橡胶塞并注明牛号、采样的乳室及日期，冷藏送回实验室立即进行细菌学检验（Getahun et al., 2008; Yang et al., 2011）。

3. 细菌的分离 参照文献对乳样进行细菌分离（Batavani et al., 2003; Getahun et al., 2008; Yang et al., 2011）。将无菌采集的乳样充分摇匀，挑取2～3铂耳圈划线接种于5%鲜血琼脂培养基，同时接种肉汤一管，37℃恒温培养20～24 h，取出观察并记录5%鲜血琼脂培养基上菌落生长情况及形态特征。根据细菌生长情况可适当延长培养时间。

取可疑单个菌落增菌培养，挑取菌液进行细菌涂片、革兰氏染色、镜检、观察、照相，记录细菌的染色性状及形态，并选择性地接种于5%鲜血琼脂培养基、却浦曼琼脂培养基、改良爱德华琼脂培养基、麦康凯琼脂培养基中培养。根据菌落形态特征、溶血与否、肉汤培养物状态、染色性状及形态，初步判定细菌种类，将其分为链球菌、葡萄球菌、革兰氏阴性杆菌和其他类。接种5%鲜血琼脂斜面培养基，然后进行生化鉴定。在5%鲜血琼脂培养基上未长菌的奶样，用其增菌培养肉汤再次接种，培养48 h后，检查菌落，以免漏检。培养48 h后仍不长菌者，即可认定无菌。

在5%鲜血琼脂上未长菌的奶样，用培养的肉汤再接种5%鲜血琼脂，培养48 h，检查菌落生长情况。若仍未长菌，即可认定无菌。

4. 细菌的鉴定 参照文献对乳样进行细菌鉴定（Batavani et al., 2003; Getahun et al.,

2008；Yang et al.，2011）。当疑似为革兰氏阳性球菌时，可通过观察肉汤培养物的状态和触酶试验进行鉴定。

若触酶试验阳性，肉汤中不形成长链，鲜血琼脂平板上形成较大菌落且其形态与培养特性均符合葡萄球菌特性时，即可鉴定为葡萄球菌属细菌。对于在普通营养琼脂和鲜血琼脂中形成较大、柠檬色、灰白色或金黄色菌落的细菌，可通过凝固酶试验及溶血情况来判定。若凝固酶阳性，且能溶血的即可鉴定为金黄色葡萄球菌；若凝固酶试验阴性，不能溶血的即可鉴定为凝固酶阴性葡萄球菌（CNS，本实验中主要包括表皮葡萄球菌和腐生葡萄球菌）。

若在肉汤中能形成长链，血琼脂上的菌落较小，触酶试验阴性，可鉴定为链球菌属细菌，然后用 CAMP 试验、七叶苷水解试验、马尿酸钠水解试验、甘露醇和山梨醇发酵试验作进一步鉴定。无乳链球菌 CAMP 试验阳性，不水解七叶苷，不发酵甘露醇和山梨醇，水解马尿酸钠呈现深紫色；停乳链球菌 CAMP 试验阴性，不水解七叶苷和马尿酸钠，不发酵甘露醇和山梨醇；乳房链球菌营养肉汤均匀浑浊，CAMP 试验阴性，水解七叶苷和马尿酸钠，发酵甘露醇和山梨醇。

对于疑似为革兰氏阴性杆菌的菌株，可先进行氧化酶试验，若结果阴性，麦康凯琼脂平板上形成粉红色或红色菌落，三糖铁琼脂斜面上划线并穿刺培养，产酸或产气，VP 试验阴性，吲哚试验红色，不分解尿素的即为大肠杆菌。

若在营养琼脂培养基上菌落较小、圆形、光滑、湿润、生长贫瘠、呈半透明或灰白色露滴状，肉汤培养物呈均匀浑浊，管底有少量沉淀物，三糖铁琼脂斜面划线并穿刺培养，斜面为红色，底层为黄色。镜检可见两端钝圆呈明显两极着色的革兰氏阴性中等大小杆菌，可初步怀疑为沙门氏杆菌，可再通过生化试验予以确认。

对于为革兰氏阳性杆菌的菌株，且菌体的一端或两端粗大，呈棒状，排列不规则，在鲜血琼脂上生长不良，为灰白色的小菌落，可初步怀疑为牛棒状杆菌，再通过生化鉴定试验予以确认。

若革兰氏染色阳性，菌体圆形或椭圆形，较大（直径 5～20 μm），周围有肥厚的荚膜，折光性强；墨汁染色可见圆形菌体，周围有一较宽的空白带（荚膜）。菌体细胞常有出芽，但无菌丝或假菌丝。在沙氏培养基上 37℃均可生长，菌落白色或奶白色，黏稠不透明，即可鉴定为新型隐球菌。

若一个奶样鉴定出两种以上细菌时，认为是混合感染。

5. 菌株不同保存方法的比较

（1）琼脂斜面保存法　　将上述实验分离纯化鉴定的奶牛乳腺炎金黄色葡萄球菌、大肠杆菌、无乳链球菌、乳房链球菌和停乳链球菌分别划线接种于 5%鲜血琼脂斜面上，标明菌株名称和日期，37℃培养 16～18 h 后，置 4℃冰箱保存即可。定期检查时接种 5%鲜血琼脂平板，放入 37℃培养箱中培养 24 h 后观察，记录菌落形态和染色镜检结果。

（2）30%甘油生理盐水保存法　　细菌接种鲜血琼脂平板，37℃培养 18 h 后，挑取单个菌落接种营养肉汤或 LB 培养基（适用于大肠杆菌），37℃培养 12～18h，吸取 630μL 菌液加入灭菌过的 1.5mL EP 管中，再加入 370μL 80%灭菌甘油，混匀，即为 30%甘油保存液。每管标明菌株名称和日期,先置4℃预冷2h后转入-20℃冰箱冻存。定期检查时接种5%鲜血琼脂平板,放入37℃培养箱中培养18~24 h后观察,记录菌落形态和染色镜检结果。

6. 药物敏感性试验 本实验采用药敏纸片琼脂扩散法(Kirby-Bauer法或者K-B纸片琼脂扩散法)测定细菌对抗菌药物的敏感性。

(1)测试菌液的制备 挑取1铂耳圈5%鲜血琼脂培养基上纯化培养的细菌接种于含5 mL营养肉汤的试管中,置37℃摇床振荡培养12 h,取出与本实验室自制的5号麦氏标准管浊度(相当于1.5×10^9 cfu/mL)进行比较,若肉汤浊度低于5号麦氏标准管浊度,继续振荡培养至与5号麦氏标准管相同的浊度;若肉汤浊度超过5号麦氏标准管浊度,用生理盐水稀释至与5号麦氏标准管相同的浊度。

(2)药敏试验测定方法 用无菌移液枪头吸取200 μL营养肉汤培养液置于营养琼脂平板上,用灭菌的"L"形玻璃棒轻轻地将菌液涂开,使肉汤培养液均匀分布于营养琼脂平板表面,灭菌棉花吸取多余的肉汤培养液,37℃预热30 min,使营养琼脂表面干燥,然后用灭菌小镊子将药敏纸片分别贴于琼脂表面,一个直径90 mm的培养皿可贴7片药敏纸片,纸片与平皿边缘不少于15 mm,纸片之间的距离不少于24 mm,每个纸片都应按压一下,保证与琼脂紧密接触,将培养皿倒置过来,放置37℃培养箱中培养24 h观察结果。如果效果不明显,可把培养时间延长至36~48 h。用直尺测量抑菌圈直径并记录,计算同一类细菌不同菌株对同一种抗生素的平均抑菌直径。

(3)药敏试验判定标准 药敏试验结果判定标准采用中国药品生物制品检验所国家抗菌药物细菌耐药性检测中心所制定的抗菌药物药敏纸片判定标准(表2-3)。

表2-3 试验用抗生素的种类、药敏纸片及其含药量和抑菌圈直径判断标准
Table 2-3 The kinds of antibiotics employed, antimicrobial sensi-discs and its content of dispersion, the criteria for inhibition result

抗生素类别	抗菌药名及代号	纸片含药量/μg	抑菌圈直径判定标准		
			R	I	S
青霉素类	氨苄西林(ampicillin)AM	10	≤13	14~16	≥17
头孢菌素类	先锋霉素V(cephazolin)CZ	30	≤14	15~17	≥18
氨基糖苷类	卡那霉素(kanamycin)K	30	≤13	14~17	≥18
喹诺酮类	环丙沙星(ciprofloxacin)CIP	5	≤15	16~20	≥21
大环内酯类	红霉素(erythromycin)E	15	≤13	14~22	≥23
	氯霉素(chloramphenicol)C	30	≤12	13~17	≥20
中药制剂	乳炎消	约10	≤10	11~14	≥15

注:R. 耐药,I. 中度敏感,S. 高度敏感,抑菌圈直径单位为mm

7. 乳腺炎致病菌的动物毒性试验 将试验菌普通营养肉汤(含10%血清)37℃培

养 16~18 h，离心沉淀菌体，弃去上面的液体培养基，用灭菌生理盐水悬浮，再用灭菌生理盐水作适当稀释，对比麦氏标准比浊管 3 号管，用 9×10^8 cfu/mL 的细菌悬浮液给随机分组的昆明系小白鼠腹腔注射，接种量为 0.2 mL/只，每株细菌注射 1 只小白鼠，对照组 5 只小白鼠用灭菌生理盐水代替菌液接种，接种后观察 48 h。48 h 后将发病或不发病小白鼠全部剖解或剖杀（刘珍，2005）。

8. 致病菌原的重分离　　小白鼠接种菌液后观察 48 h，死亡小白鼠于死后 4 h 内解剖取肝脏进行致病菌的重分离，未死亡小白鼠于 48 h 后解剖取肝脏进行致病菌的重分离。接种于 5%鲜血琼脂平板后放置于 37℃温箱中培养 18~24 h，若有菌落出现，进行菌落涂片、革兰氏染色和显微镜检查，记录观察结果，核对所分离菌株是否为接种致病菌。若是所接种致病菌，则证明了该致病菌有致病性（刘珍，2005）。

（三）统计分析

用 Microsoft Office Excel 2003 对所获得的数据进行处理分析。

二、实验结果

（一）麦氏比浊标准管的配制

配制的麦氏比浊标准管见图 2-1。

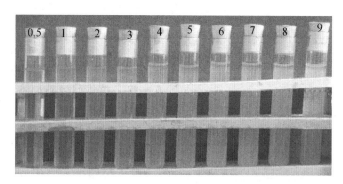

图 2-1　麦氏标准比浊管 0.5~9 号管，分别相当于细菌浓度为 1×10^8 cfu/mL、3×10^8 cfu/mL、6×10^8 cfu/mL、9×10^8 cfu/mL、12×10^8 cfu/mL、15×10^8 cfu/mL、18×10^8 cfu/mL、21×10^8 cfu/mL、24×10^8 cfu/mL、27×10^8 cfu/mL

Figure 2-1　No. 0.5-9 McFarland Standards equal 1×10^8 cfu/mL, 3×10^8 cfu/mL, 6×10^8 cfu/mL, 9×10^8 cfu/mL, 12×10^8 cfu/mL, 15×10^8 cfu/mL, 18×10^8 cfu/mL, 21×10^8 cfu/mL, 24×10^8 cfu/mL, 27×10^8 cfu/mL bacteria, respectively

（二）奶牛乳腺炎发病情况调查结果

1. 2005~2009 年泌乳奶牛的瞎乳头情况　　2005~2009 年泌乳奶牛的瞎乳头情况见表 2-4 和图 2-2。

表 2-4 2005~2009 年瞎乳头奶牛平均发生率

Table 2-4 The average lactating cows had blind quarter（s）of dairy cows from 2005 to 2009

指标	2005	2006	2007	2008	2009	平均数
泌乳奶牛数	973	1032	1105	1151	1184	1089
瞎乳头奶牛数及占总数百分数/%	111（11.4）	113（10.9）	135（12.2）	128（11.1）	139（11.7）	125（11.5）
瞎乳头数及占总乳头数百分数/%	143（3.7）	152（3.7）	170（3.8）	169（3.7）	183（3.9）	163（3.7）
一个瞎乳头的奶牛数及占总数百分数/%	79（8.1）	74（7.2）	100（9）	87（7.6）	95（8）	87（8）
两个瞎乳头的奶牛数及占总数百分数/%	32（3.3）	39（3.8）	35（3.2）	41（3.6）	44（3.7）	38（3.5）

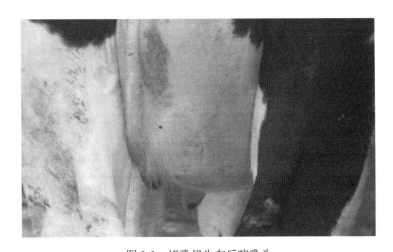

图 2-2 泌乳奶牛左后瞎乳头

Figure 2-2 Left rear blind quarter in a lactional cow

2. 2005~2009 年泌乳奶牛临床乳腺炎发病情况　　2005~2009 年泌乳奶牛临床乳腺炎发病情况见表 2-5、图 2-3 和图 2-4。

表 2-5 2005~2009 年奶牛临床乳腺炎发病情况

Table 2-5 The incidence of clinical mastitis at cow and quarter levels of dairy cows from 2005 to 2009

	2005		2006		2007		2008		2009		平均数	
	总数	阳性数及占总数百分数/%	总数	阳性数及占总数百分数/%	总数	阳性数及占总数百分数/%	总数	阳性数及占总数百分数/%	总数	阳性数及占总数百分数/%	总数	阳性数及占总数百分数/%
奶牛数	973	75（7.7）	1032	84（8.1）	1105	106（9.6）	1151	103（8.9）	1184	107（9）	1089	95（8.7）
乳区数	3749	127（3.4）	3976	141（3.5）	4250	165（3.9）	4435	167（3.8）	4553	173（3.8）	4193	155（3.7）

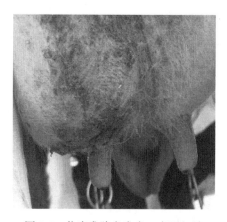

图 2-3　临床乳腺炎乳房，表现红肿
Figure 2-3　Clinical mastitis udder, showing swelling

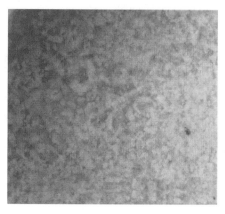

图 2-4　临床乳腺炎乳汁，大量絮状物
Figure 2-4　Clinical mastitis milk, there are a lot of floccus

3. 2005～2009 年泌乳奶牛隐性乳腺炎发病情况　2005～2009 年泌乳奶牛隐性乳腺炎发病情况见表 2-6 和图 2-5。

表 2-6　2005～2009 年奶牛隐性乳腺炎发病情况
Table 2-6　The incidence of subclinical mastitis at cow and quarter levels of dairy cows from 2005 to 2009

	2005		2006		2007		2008		2009		平均数	
	总数	阳性数及占总数百分数/%	总数	阳性数及占总数百分数/%	总数	阳性数及占总数百分数/%	总数	阳性数及占总数百分数/%	总数	阳性数及占总数百分数/%	总数	阳性数及占总数百分数/%
奶牛数	898	405（45.1）	948	439（46.3）	999	547（54.8）	1048	506（48.3）	1077	528（49）	994	485（48.8）
乳头数	3466	621（17.9）	3659	664（18.1）	3859	827（21.4）	4046	739（18.3）	4159	806（19.4）	3838	731（19）

图 2-5　左侧的乳汁是健康乳汁（SMT 检测阴性），右侧的乳汁是隐性乳腺炎乳汁（SMT 检测阳性）
Figure 2-5　The left milk is normal (native for SMT), the right milk is subclinical mastitis (positive for SMT)

4. 不同胎次泌乳奶牛临床和隐性乳腺炎发病情况　不同胎次泌乳奶牛临床和隐性乳腺炎发病情况见表 2-7，各胎次奶牛临床乳腺炎和隐性乳腺炎差异显著性见表 2-8。

表 2-7 不同胎次泌乳奶牛临床和隐性乳腺炎发病率
Table 2-7 The incidence of clinical and subclinical mastitis of dairy cows with different parity

指标	一胎次	二胎次	三胎次	四胎次	五胎次
调查奶牛数	117	91	76	49	43
临床乳腺炎奶牛数及占总数百分数/%	10（8.5）	8（8.8）	7（9.2）	9（18.4）	17（39.5）
隐性乳腺炎奶牛数及占总数百分数/%	27（23.1）	21（23.1）	28（36.8）	26（53.1）	29（67.4）

表 2-8 各胎次奶牛临床乳腺炎和隐性乳腺炎差异显著性
Table 2-8 The difference between each of parities of clinical and subclinical mastitis cows

	临床乳腺炎					隐性乳腺炎				
	一胎次	二胎次	三胎次	四胎次	五胎次	一胎次	二胎次	三胎次	四胎次	五胎次
一胎次	1	0.95	0.87	0.07	3.5×10^{-6}	1	1	0.038	1.6×10^{-4}	1.8×10^{-7}
二胎次			0.92	0.098	2×10^{-5}			0.52	3.4×10^{-4}	7×10^{-7}
三胎次				0.135	7.5×10^{-5}				0.074	1.3×10^{-3}
四胎次					0.024					0.16
五胎次					1					1

5. 2005~2009 年泌乳奶牛乳汁 SMT 得分分布情况 2005~2009 年泌乳奶牛乳汁 SMT 得分分布情况见表 2-9。

表 2-9 2005~2009 年奶牛乳汁 SMT 得分分布
Table 2-9 The distributions of SMT scores of the milk from 2005 to 2009

SMT 得分	各得分乳样数及占乳样总数百分数/%					
	2005	2006	2007	2008	2009	平均数
0+T	2845（82.1）	2995（81.8）	3032（78.6）	3307（81.8）	3353（80.6）	3106（81）
1	320（9.2）	339（9.3）	386（10）	356（8.8）	386（9.3）	357（9.3）
2	167（4.8）	171（4.7）	235（6.1）	205（5）	227（5.5）	201（5.2）
3	134（3.9）	154（4.2）	206（5.3）	178（4.4）	193（4.6）	173（4.5）
乳样总数	3466（100）	3659（100）	3859（100）	4046（100）	4159（100）	3838（100）

注：T 表示可疑（trace，T）

（三）奶牛乳腺炎致病菌分离鉴定结果

1. 奶牛乳腺炎致病菌的检出情况 奶牛乳腺炎致病菌的检出情况见表 2-10。

表 2-10 奶牛临床和隐性乳腺炎乳汁致病菌检出率
Table 2-10 The positive rate of microorganism isolates from clinical and subclinical mastitis milk samples

指标	临床乳腺炎	隐性乳腺炎
乳汁样品总数	109	167
细菌的样品数	105	56
检出率	96%	34%

2. 奶牛乳腺炎致病菌的鉴定结果 奶牛乳腺炎致病菌的鉴定结果见表 2-11、图 2-6~图 2-9。

表 2-11 奶牛临床和隐性乳腺炎病原菌鉴定数及其所占比例
Table 2-11 Number and proportion of microorganism isolates from clinical and SCM milk samples

细菌种类	临床乳腺炎病原菌数及占总分离细菌数百分数/%	隐性乳腺炎病原菌数及占总分离细菌数百分数/%	各病原菌总数占总分离细菌数百分数/%
金黄色葡萄球菌	31（29.5）	18（32.1）	49（30.4）
大肠杆菌	27（25.7）	4（7.1）	31（19.3）
新型隐球菌	17（16.2）	5（8.9）	22（13.7）
无乳链球菌	8（7.6）	10（17.9）	18（11.2）
乳房链球菌	5（4.7）	5（8.9）	10（6.2）
停乳链球菌	3（2.9）	2（3.6）	5（3.1）
凝固酶阴性葡萄球菌	5（4.7）	11（19.7）	16（9.9）
混合感染			
金黄色葡萄球菌+新型隐球菌	3（2.9）	1（1.8）	4（2.5）
大肠杆菌+新型隐球菌	3（2.9）	—	3（1.9）
金黄色葡萄球菌+大肠杆菌	2（1.9）	—	2（1.2）
金黄色葡萄球菌+无乳链球菌	1（1）	—	1（0.6）
总分离细菌数	105（100）	56（100）	161（100）

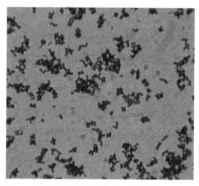

图 2-6 葡萄球菌（革兰氏染色×1000）
Figure 2-6 *Staphylococcus*（Gram stain×1000）

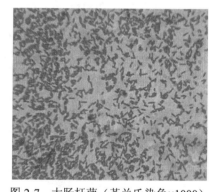

图 2-7 大肠杆菌（革兰氏染色×1000）
Figure 2-7 *Escherichia coli*（Gram stain×1000）

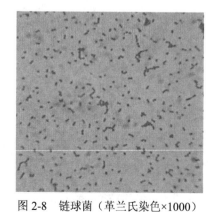

图 2-8 链球菌（革兰氏染色×1000）
Figure 2-8 *Streptococcus*（Gram stain×1000）

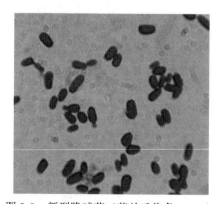

图 2-9 新型隐球菌（革兰氏染色×1000）
Figure 2-9 *Cryptococcus neoformans*（Gram stain×1000）

3. 菌株保存方法的比较结果 经定期在 5%鲜血琼脂平板上分别接种三种保存方法的菌株，观察菌落生长情况和形态以及细菌革兰氏染色镜检特点，核对细菌的数量和保存前的一致性。三种方法的保存时间见表 2-12。

表 2-12 三种保存方法菌株保存时间
Table 2-12 The preservation time of three kinds of preservation method

菌种	保存时间		
	琼脂斜面保存法（4℃）	30%甘油生理盐水保存法（-20℃）	85%甘油生理盐水保存法（-20℃）
金黄色葡萄球菌	4 周	大于 3 年	大于 3 年
大肠杆菌	4 周	大于 3 年	大于 3 年
无乳链球菌	3 周	2 年	大于 3 年
乳房链球菌	2 周	2 年	大于 3 年
停乳链球菌	2 周	18 个月	大于 3 年

（四）奶牛乳腺炎致病菌药物敏感性结果

奶牛乳腺炎致病菌药物敏感性结果见表 2-13。

表 2-13 奶牛乳腺炎致病菌药物敏感性
Table 2-13 *In-vitro* antimicrobial susceptibility test results of bacterial isolates recovered from mastitis milk samples

致病菌种类	数量	氨苄西林（%）		先锋霉素V（%）		卡那霉素（%）		氯霉素（%）		红霉素（%）		环丙沙星（%）		乳炎消（%）	
		高度敏感	敏感	高度敏感	敏感	高度敏感	敏感	高度敏感	敏感	高度敏感	敏感	高度敏感	敏感	高度敏感	敏感
金黄色葡萄球菌	49	8.2	22.4	55.1	38.8	22.5	36.7	24.5	46.9	18.4	49	26.5	49	55.1	38.8
大肠杆菌	29	6.9	13.8	55.2	44.8	34.5	37.9	37.9	41.4	27.6	20.7	48.3	51.7	37.9	55.2
新型隐球菌	22	0	0	0	9.1	0	4.5	0	4.5	0	22.7	0	27.3	36.4	50
无乳链球菌	18	5.6	22.2	38.9	55.5	22.2	50	22.2	55.6	22.2	38.9	27.8	50	44.4	50
乳房链球菌	9	0	44.4	44.4	55.6	11.1	33.3	33.3	44.4	11.1	44.4	22.2	44.5	22.2	77.8
停乳链球菌	4	0	25	75	25	0	50	50	50	25	75	0	50	25	75
凝固酶阴性葡萄球菌	15	0	40	73.3	20	13.3	53.4	26.7	40	6.7	40	40	46.7	40	60

（五）乳腺炎致病菌的动物毒性试验

由表 2-14 可见，56 株金黄色葡萄球菌接种小白鼠后，有 19 只小白鼠于 48 h 内死亡，11 只出现厌食、精神沉郁、被毛粗乱、排脓汁样尿液、腹部皮下接种处呈青紫色等症状；36 株大肠杆菌接种小白鼠后，13 只小白鼠于 48 h 内死亡，9 只出现同上的症状；19 株无乳链球菌接种小白鼠后，1 只小白鼠于 48 h 内死亡，3 只出现同上的症状；10 株乳房链球菌接种小白鼠后,1只出现同上的症状;5株停乳链球菌接种小白鼠后,均未出现临床症状;16株凝固酶阴性葡萄球菌接种小白鼠后,4只出现同上的症状(表2-14)。

表2-14 乳腺炎致病菌对小白鼠的毒性
Table 2-14 The toxicity of mastitis pathogens for mice

菌种	注射只数	注射后状态	致病率/%
金黄色葡萄球菌	56	19只死亡,11只表现异常	54
大肠杆菌	36	13只死亡,9只表现异常	61
无乳链球菌	19	1只死亡,3只表现异常	21
乳房链球菌	10	1只表现异常	10
停乳链球菌	5	未出现临床症状	0
凝固酶阴性葡萄球菌	16	4只表现异常	25

(六)乳腺炎致病菌的重分离结果

接菌2日后对死亡小白鼠和解剖小白鼠肝脏进行致病菌重分离,情况见表2-15。

表2-15 乳腺炎致病菌重分离结果
Table 2-15 The re-isolation results from mice of mastitis pathogens

菌种	接种数	重分离数	分离率/%
金黄色葡萄球菌	56	47	84
大肠杆菌	36	23	64
无乳链球菌	19	6	32
乳房链球菌	10	6	60
停乳链球菌	5	2	40
凝固酶阴性葡萄球菌	16	9	56
对照(生理盐水)	5	0	0

三、分析与讨论

(一)泌乳奶牛瞎乳头情况

规模化奶牛场泌乳奶牛的瞎乳头率平均为3.7%,具有瞎乳头的奶牛所占百分比为11.5%,与戴鼎震等(2002)报道南京地区奶牛的瞎乳头率(3.72%)和毛翔光等(2009)报道昆明地区奶牛的瞎乳头率(3.21%)基本一致,高于2004年朱贤龙等调查广西奶牛瞎乳头率(2.58%)和邢玫等(2007)报道的奶牛水平瞎乳头率(1.54%)。与国外奶牛场相比,本实验与Mungube等(2004)(3.7%)和Almaw等(2008)(3.8%)报道的奶牛的瞎乳头率一致,但高于Kivaria等(2007)(2.1%)和Getahun等(2008)(2.3%)的报道。然而就瞎乳头的奶牛而言,本实验结果高于Kivaria等(2004)报道的5%。导致奶牛瞎乳头的主要原因是临床乳腺炎久治不愈,造成乳池与乳头管狭窄及闭锁,乳腺组织遭受严重的破坏,使泌乳功能严重降低其至丧失。头胎奶牛在怀孕中后期也会散在性发生乳腺炎,常不被技术人员注意，只有在产犊后初次挤奶时被发现，对其终生的产奶量有着极大的影响，严重时即三个乳区产奶。但是国外也有研究表明，奶牛分娩后瞎乳头可重新完全恢复泌乳功能（Vaarst et al.，2006）。

（二）泌乳奶牛临床乳腺炎发病情况

本实验对 2005～2009 年泌乳奶牛进行临床乳腺炎调查，结果表明，临床乳腺炎奶牛的平均发病率为 8.7%，乳区检出率为 3.7%。临床乳腺炎发病率的高低与奶牛场的饲养管理、病原菌致病力强弱、卫生条件、病原菌的致病性强弱、气候季节等多种因素均有直接的关系（寨鸿瑞等，2008）。本实验结果表明，临床乳腺炎发病率与秦璐璐（2009）报道的基本一致，但均低于孔雪旺和陈功义（2005）及邢玫等（2007）分别报道的奶牛临床乳腺炎奶牛发病率 17.5% 和 23.1%。与国外奶牛场相比，本实验临床乳腺炎发病率高于 Mungube 等（2004）（奶牛发病率 6.6%，乳区检出率 2.8%）、Kivaria 等（2004）（奶牛发病率 3.8%，乳区检出率 2.1%）和 Getahun 等（2008）（奶牛发病率 1.8%，乳区检出率 0.51%）的报道。同时也说明了本规模化奶牛场的管理水平和卫生条件在国内可以说良好，但与国外奶牛场相比，仍存在一定程度的高临床乳腺炎发病率和瞎乳头率。因此应进一步加强饲养管理，保持环境卫生的清洁，最大限度地降低环境因素的影响，提高奶牛自身的免疫力，将临床乳腺炎的发病率降到最低。

（三）泌乳奶牛隐性乳腺炎发病情况

本实验对 2005～2009 年泌乳奶牛进行隐性乳腺炎调查，结果表明，隐性乳腺炎奶牛的平均发病率为 48.8%，乳区检出率为 19%。低于 2004 年朱贤龙等（2004）调查广西奶牛隐性乳腺炎的发病率（奶牛发病率 50.46%，乳区检出率 30.34%）；与国外奶牛场相比，本实验隐性乳腺炎发病率（48.8%）与 Mungube 等（2004）（46.6%）、Gianneechini 等（2002）（52.4）报道的隐性乳腺炎发病率一致，但高于 Getahun 等（2008）报道的 22.3%，然而本实验隐性乳腺炎发病率远低于 Kerro Dego 和 Tareke（2003）报道的 62.9% 和 Kivaria 等（2004）报道的 90.3%。本实验隐性乳腺炎阳性乳区检出率（19%）高于 Getahun 等（2008）报道的 10.1%，低于 Mungube 等（2004）报道的 27.8% 和 Gianneechini 等（2002）报道的 27.6%。牛群的大小、管理模式的不同、外来基因的遗传和当地气候等诸多因素致使不同地区、不用时间的调查结果不尽相同。

（四）不同胎次泌乳奶牛乳腺炎发病情况

从表 2-7 可以看出，第一胎次奶牛的临床乳腺炎发病率（8.5%）和第二胎次奶牛的隐性乳腺炎发病率（23.1%）都是最低的，并且随着胎次的增加，发病率都呈现升高的趋势。经卡方（χ^2）检验，各胎次奶牛临床乳腺炎和隐性乳腺炎差异显著性有较大的区别。而到了第三胎次和第四胎次，产奶量最高，乳房的负担加重，导致隐性乳腺炎的发生率较高。特别是五胎次奶牛的乳腺炎发病率显著高于前三胎次的奶牛（表 2-8）。本实验结果与秦璐璐（2009）报道的第二胎次临床乳腺炎和隐性乳腺炎发病率最低有所不同，与王建忠（2007）的报道基本一致。表明奶牛乳腺炎的发病率与胎次之间有着明显的关系，并且随着胎次的增加而升高，主要是由于随着奶牛年龄的增大、分娩次数增多，奶牛机体抗病力减弱，使得环境中的病原菌乘机侵入机体。

（五）泌乳奶牛乳汁 SMT 得分分布

本实验首次使用 SMT 对大批量乳汁进行了得分调查，SMT 得分为 1、2 和 3 的隐性乳腺炎乳汁所占百分比分别是 9.3%、5.2%和 4.5%（表 2-9），本实验阳性得分所占百分比均低于 Contreras 等和 Sung 等报道的山羊的乳汁得分数据，并表明 CMT 检查非感染性乳汁比检查感染乳汁更有效，是由于较多的假阳性的出现。SMT 的判定主要是根据凝集程度的不同判分，具有一定的主观性。整体来看，隐性乳腺炎乳汁的检出率为 19%，2007 年奶牛的隐性乳腺炎检出率最低，5 年之间的隐性乳腺炎乳汁检出率没有显著差异性。

（六）奶牛乳腺炎致病菌分离鉴定

奶牛乳腺炎的发生是病原微生物、机体免疫力、管理等多因素综合作用的结果，其中病原微生物和机体的免疫力是最主要的因素，而病原微生物是导致奶牛乳腺炎难以预防、控制和治疗的一个最主要的原因。目前已经从乳腺内感染的乳汁中分离出超过 135 种不同的病原微生物，但是最主要的致病菌是葡萄球菌、链球菌和革兰氏阴性菌（Bradley，2002）。通常它们被分为接触传染性病原体和环境性病原体两类。前者主要包括无乳链球菌、停乳链球菌、金黄色葡萄球菌等，通过挤奶员的手或挤奶器在乳区和牛群中传播，环境性病原体主要包括大肠杆菌、乳房链球菌、变形杆菌、凝固酶阴性葡萄球菌、真菌或酵母菌等。有报道称引起奶牛乳腺炎的致病微生物中，金黄色葡萄球、链球菌和大肠杆菌占致病微生物分离数的 90%以上（刘营，2008）。

本实验临床乳腺炎乳汁和隐性乳腺炎乳汁中致病菌检出率分别是 96%和 34%，临床乳腺炎乳汁共检出 7 种 114 个菌株，检出最多的致病菌分别是金黄色葡萄球菌、大肠杆菌和新型隐球菌，109 个乳样中混合感染占 8.26%；隐性乳腺炎乳汁共检出 7 种 57 个菌株，检出最多的致病菌分别是金黄色葡萄球菌、凝固酶阴性葡萄球菌和无乳链球菌，167 个乳样中混合感染占感染总数（56）的 1.8%，低于 Almaw 等（2008）报道的隐性乳腺炎致病菌检出混合感染比例（9.9%）。金黄色葡萄球菌和大肠杆菌的检出率在临床乳腺炎中是致病菌检出率的前两位，与杨锐等（2009）和戈胜强等（2008）报道的一致，酵母菌（新型隐球菌）所占比例高达 20.18%，与戈胜强等（2008）报道的基本一致，但高于朱贤龙等（2004）报道的广西奶牛隐性乳腺炎所分离的酵母菌 1.4%。金黄色葡萄球菌的高检出率主要是由于乳房皮肤和乳头广泛分布，可建立慢性和（或）隐性感染，可作为传染源通过挤奶传染给其他健康奶牛（Workineh et al.，2002）。另外，金黄色葡萄球菌由于治愈率极低使其在乳腺中很难清除（Getahun et al.，2008）。真菌的发生与兽医技术人员滥用抗生素有关，在抗生素存在的情况下真菌生长旺盛，越是滥用抗生素对乳腺炎进行治疗，越有利于真菌的感染。金黄色葡萄球菌和新型隐球菌混合感染与牛舍和挤奶不卫生、带菌乳腺内注射、滥用抗生素、不完善的乳腺炎控制程序等因素有关。本实验中金黄色葡萄球菌为优势菌群，与易明梅等（2009）报道的一致，表明饲养管理和环境消毒欠佳等人为因素是引起奶牛临床乳腺炎的主要原因。大肠杆菌和酵母菌都属于条件性致病菌，占检出致病菌的48.25%，说明有效控制由于条件性致病菌引起的乳腺炎是奶牛场管理者的一项重要任务。混合感染的发生与牛场的环境卫生、消毒管理及牛体卫生等也有很大的关系。

对于细菌培养阴性的临床乳腺炎乳汁，国外学者认为临床乳腺炎乳汁细菌的检出阴性率约为11%，本实验结果中有4例（4%）未见致病菌的生长，关于临床乳腺炎无菌生长的原因有以下几个方面：①无菌性乳腺炎，由机械创伤等因素引起的乳腺炎；②机体防卫机能阻止微生物的繁殖，导致菌数过少；③国外报道传染性致病菌常产生持续性感染，有报道称近20%的大肠杆菌性乳腺炎病例的奶样培养阴性（刘珍，2005）；④由于大多数病例均已进行了抗生素治疗，乳汁中含有大量的杀菌药物，使其处于残活状态或滤过性状态，因此在一般培养基上不能生长；⑤使用了不适当的培养基和培养方法，一些临床乳腺炎不是由细菌所引起，在细菌培养基上不能生长，如牛支原体，用常规的细菌培养程序很难得到结果（Wieliczko et al.，2002）。

（七）菌株保存方法比较

链球菌在鲜血琼脂斜面上保存两周就需要转接一次（最好每周转接一次），金黄色葡萄球菌和大肠杆菌可保存1个月（最好每两周转接一次），虽然鲜血琼脂斜面4℃保存细菌菌种方法简单易行，但保存时间较短，需要经常进行转接培养，无形中加大了研究者的工作量。保存时间较短的原因主要是4℃下细菌体内的酶活力减弱，但仍在活动，从而不断消耗菌体内的能量，导致细菌的衰老死亡。

-20℃下细菌的能量消耗和细菌体内酶的活性都很低，而且30%甘油生理盐水能够在-20℃下给细菌提供一定的能量，但是30%甘油生理盐水在-20℃冻结，85%甘油生理盐水的凝固点为-30℃，因此在-20℃不会冻结，这样-20℃情况下85%甘油生理盐水所提供的保存细菌环境就使细菌处于潮湿低温的状况，使细胞基本上处于休眠状态，生长繁殖受到抑制，但又不至于死亡，减少细菌的能量消耗和细菌体内酶的活动，从而使细菌长期存活（方光远等，2004）。本实验结果也表明，85%甘油生理盐水可以将金黄色葡萄球菌、大肠杆菌和链球菌保存3年以上。因此，85%甘油生理盐水保存法操作简便，效果理想，适合一般实验室中长期保存菌株。

冷冻真空干燥法是迄今最佳的菌种保存方法，但是需专用的仪器，一般实验室难以配备（王丹敏等，2006）。常用的保护剂有甘油、脱脂牛奶、脱纤维羊血等，链球菌营养要求比较高，需要在液体培养基中加入15%~20%的血清以延长其保存时间（王英等，2006），金黄色葡萄球菌和大肠杆菌对营养的要求没有链球菌那么高，故相同保存条件下金黄色葡萄球菌和大肠杆菌的保存时间比链球菌长。

（八）奶牛乳腺炎致病菌药物敏感性

分离的乳腺炎致病菌对本实验选用的不同抗生素的敏感性呈现比较大的差异（表2-13），可能与使用抗生素治疗临床乳腺炎有关，抗生素的大量使用是细菌产生耐药性的直接原因，因此，进行细菌分离与药敏试验是提高临床治疗效果的有效途径。

本研究发现，所有致病菌（除了新型隐球菌）对所有的抗生素（除了氨苄西林）都比较敏感，这些抗生素中，先锋霉素Ⅴ、卡那霉素、氯霉素和环丙沙星和乳炎消（主要成分：紫花地丁、金银花、当归、白芷和白蜡）等广谱抗生素的抗菌效果较好，与易明梅等（2009）的报道基本一致，可作为有效治疗奶牛临床乳腺炎的候选药物。

抗生素在奶牛乳腺炎的治疗过程中起了重要作用，但长期使用抗生素导致耐药性菌株的出现，对奶牛乳腺炎的控制是一项严峻挑战。在治疗奶牛临床乳腺炎时应尽量采用交叉用药，防止耐药性菌株的产生，尽量少用抗生素，并降低抗生素产生的危害，如使用市场上已经出现的多种中草药制剂（油剂、膏剂等）。目前由于抗生素的广泛应用，耐药菌株数量越来越多，并且某些病原菌能侵入乳房上皮细胞，并在细胞内持续生存，抗生素很难在用药后进入细胞内并达到杀菌浓度，形成慢性持续性感染，并且作为整个奶牛场的传染源。

抗生素治疗奶牛临床乳腺炎失败的常见原因为：①长期使用一种或两种抗生素治疗临床疾病，造成耐药菌株的产生。例如，已经证实葡萄球菌属可以产生β-内酰胺酶，该酶能够解开β-内酰胺类抗生素的β-内酰胺环。②在发生炎症和感染乳区，机体免疫功能受到抑制，对致病菌的抵抗能力大大减弱，很易发生重新感染。③在抗生素治疗后，如果没有完全清除细菌，停药后又大量繁殖。④抗生素对机体天然防御功能的影响，研究发现，用抗生素治疗可影响吞噬细胞的形态和存活率，并干扰吞噬细胞的功能及其对细菌的处理能力（刘朝，2007）。

随着抗菌药物的广泛应用，耐药菌株逐渐增多，不仅给奶牛乳腺炎治疗的选择用药带来很大困难，而且存在于质粒或染色体上的耐药基因以接合的方式在菌株间传播，当这些耐药菌株感染人发病时，很多抗菌药物将无效。随着人们对绿色消费需求的提高，相关部门应加强奶制品检测，尽量少用或不用抗菌药物，开辟其他的用药新途径，如使用疫苗、纯中药制剂、利用溶菌酶等防治奶牛乳腺炎（孙怀昌等，2004；刘朝，2007）。

（九）乳腺炎致病菌的动物毒性试验

毒力强的菌株在一定菌液浓度的条件下可导致小白鼠接种后48 h内死亡，而毒力相对较弱的菌株在相同菌液浓度的条件下在接菌后48 h后仍未见有发病的临床症状出现。本实验中接种金黄色葡萄球菌的56只小白鼠死亡19只，11只具有临床反应，并从所有死亡和临床反应的小白鼠肝脏中分离到金黄色葡萄球菌，未死亡小白鼠在剖检后可见腹水增多、浑浊，肝充血肿大，表面有大小不一的脓肿结节；脾肿大、坏死，表面和切面有大小不等的坏死灶；肾肿大、色淡；肠系膜和浆膜的小血管扩张充血，肠黏膜有出血点。说明多数致病菌的致病力不强，但它们在乳房内生产繁殖，并产生大量的毒素、代谢产物等，可造成乳房出现炎症。大肠杆菌株接种后，部分发病急、死亡快，发病率为61%，并全部分离到大肠杆菌，其在乳房内繁殖，细菌死亡后崩解，释放出内毒素（脂多糖），导致奶牛发生急性乳腺炎；链球菌属和凝固酶阴性葡萄球菌发病慢，对小白鼠的致病性较差，死亡率几乎为零，致病率低于30%。对照组的5只小白鼠均未见任何临床症状和肉眼病变。本实验金黄色葡萄球菌的再分离率为84%，高于乌兰巴特尔（2007）报道的47.62%，乌兰巴特尔攻毒小白鼠用10^6 cfu/mL和10^7 cfu/mL浓度的菌液0.2 mL/只，本实验使用9×10^8 cfu/mL的菌液0.2 mL/只，故分离率较高。

四、小结

（1）2005～2009年瞎乳头奶牛发生率平均是11.5%，瞎乳头率平均是3.7%，临床乳腺炎奶牛发生率平均是8.7%，临床乳腺炎乳区发生率平均是3.7%，隐性乳腺炎奶牛发生率平均是48.8%，隐性乳腺炎乳区发生率平均是19%。

（2）奶牛临床乳腺炎和隐性乳腺炎乳汁致病菌检出率分别是96%和34%，临床乳腺炎主要致病菌是金黄色葡萄球菌、大肠杆菌和新型隐球菌，隐性乳腺炎主要致病菌是金黄色葡萄球菌、凝固酶阴性葡萄球菌和无乳链球菌。

（3）各种致病菌对不同抗生素的敏感性呈现很大差异，所有致病菌（除了新型隐球菌）对所用的抗生素（除了氨苄西林）都比较敏感，61%的大肠杆菌对小白鼠具有致死作用，54%的金黄色葡萄球菌对小白鼠具有致死作用。

第二节 热应激对不同泌乳阶段奶牛日产奶量的影响

荷斯坦奶牛以产奶量高而著称，原产于夏季凉爽的荷兰（7月份平均气温为17℃），具有耐寒怕热的生物学特征。荷斯坦奶牛产奶的适宜温度一般认为在5～25℃，在此温度范围内，奶牛产热和散热可维持动态平衡，生理指标正常，生产能力最高，饲料利用最经济（莫放，2003）。近年来，随着奶牛业集约化养殖的迅速发展和全球气候变暖，夏季热应激造成的奶牛生产性能下降、繁殖率降低等问题越来越严重。湖北省夏季高温高湿天气对奶牛尤其是高产泌乳奶牛的生产性能产生的影响更为明显，该规模场高温高湿天气主要集中在7月和8月，白天的平均气温在35℃以上，平均湿度在60%左右，这样的气候特点极易造成奶牛热应激反应，降低机体免疫力、产奶量及奶品质。Bohmanova等（2007）研究表明，在低湿度的条件下湿度对奶牛热应激的影响较小，而在高湿度条件时湿度对热应激的影响较大，因此在高温高湿地区用温湿指数（THI）来评价奶牛热应激更为确切。

当THI超过72时，就会导致奶牛产生热应激反应（郭时金等，2013）。奶牛热应激时表现为体温升高、呼吸及心率加快、食欲下降、采食量降低，导致奶牛机体免疫力及生产性能下降；严重者机体代谢会受到影响，出现脱水、休克，甚至死亡（周传社等，2006）。Ingraham等利用THI判断热应激程度时认为，当THI小于72时，奶牛无热应激反应；当THI为72～78时，奶牛会产生轻微热应激；当THI为79～89时，奶牛会产生中度热应激；当THI在90以上时，奶牛会产生严重热应激反应。田萍等（2002）认为，当THI大于70时奶牛处于热应激，产奶量开始下降，且二者呈负相关（$r=-0.8478$）。

当环境温度为25.0～28.6℃时，每头牛的日均产奶量将会减少0.5～2.5 kg，乳脂率会下降0.3%～0.5%。按持续高温90天计算，一个产奶量中等的万头奶牛群，仅因产奶量降低所带来的经济损失就达30多万元（熊琪等，2011），这还不包括热应激引起的奶品质降低、疾病发病率升高等造成的经济损失。

热应激是导致我国南方夏季高温高湿地区奶牛生产性能、繁殖性能和免疫力低下的主要因素之一，因此降低奶牛热应激是规模化奶牛场夏季饲养管理必须抓好的工作。本研究以某规模化奶牛场为例分析了2013年4～10月牛舍THI和奶牛产奶量的相关性，为规模化奶牛场夏季生产管理作指导。

一、材料与方法

(一) 实验动物

实验动物为某规模化奶牛场的1247头临床健康处于泌乳期的荷斯坦奶牛,采取舍饲半拴系饲养,每天饲喂两次(6:00,15:00),采用全混合日粮(TMR)饲喂,每次喂料结束后散放至运动场自由活动,自由采食苜蓿干草,饮水槽供应足够的清洁饮水。每日两次(6:30,15:30)在挤奶厅采用机器挤奶。

(二) 实验方法

1. 实验分组 本实验的数据记录时间为2013年4月19日至10月26日,根据这段时间的天气,将实验数据分为以下三个时间段:奶牛非热应激的一个月(30天)(Ⅰ)、奶牛中度热应激的一个月(30天)(Ⅱ),以及奶牛热应激恢复的一个月(30天)(Ⅲ)。再根据同一个时间段内奶牛的泌乳期分别分为三个小组:泌乳前期组(产后15~100天)、泌乳中期组(产后112~175天)和泌乳后期组(产后200~230天)(温雅俐,2011;何钦,2012)。

2. 牛舍温度和湿度测定 在牛舍中间部分牛床上方2 m处悬挂干湿球温度计,保证温度计的有效通风,避开阳光直射和雨淋。实验期间每天于10:00、14:00、17:00分别记录干球温度和湿球温度,求出当天三次测量结果的平均值。依据公式:温湿指数(THI)=0.72×(T_d+T_w)+40.6,其中T_d为当天干球平均温度,T_w为当天湿球平均温度,计算得出当天的THI。

3. 奶牛产奶量测定 实验期间对1247头泌乳奶牛采用挤奶器挤奶,每天两次(6:30,15:30),并记录单次的产奶量,以同一天采集的两次产奶量之和为当天产奶量,视为一个实验样本。

(三) 统计方法

实验数据的统计采用软件SPSS 17.0的t-检验进行组内差异显著性检验和相关性分析,干球温度、THI、日产奶量用平均值±标准差来表示,其中$P<0.05$表示处理组间差异显著,$P<0.01$表示处理组间差异极显著。

二、实验结果

(一) 实验阶段划分及牛舍的温度与THI

根据2013年4月19日至10月26日的干湿球温度和THI,将实验数据分为以下三个时间段:2013.5.1~2013.5.30为实验Ⅰ期(非热应激期,THI小于72),2013.7.23~2013.8.21为实验Ⅱ期(中度热应激期,THI为79~89),2013.9.26~2013.10.25为实验Ⅲ期(热应激恢复期,THI小于72)。

实验期间三个时间段牛舍的干球温度与THI的平均值及相关系数见表2-16,变化曲线见图2-10,实验Ⅰ期牛舍内日均温度和日均THI分别为25.2℃和71.2,实验Ⅱ期牛舍内日均温度和日均THI分别为34.9℃和84.7，实验Ⅲ期牛舍内日均温度和日均THI分别为23.2℃和68.7。

表2-16 实验期间牛舍内的干球温度与温湿指数
Table 2-16 The dry bulb temperature in the cow barn and temperature and humidity index（THI）during the experiment

实验时间	干球温度/℃	THI	相关系数
2013.5.1～2013.5.30（Ⅰ）	25.2±3.27^{Aa}	71.2±3.79^{Aa}	0.98
2013.7.23～2013.8.21（Ⅱ）	34.9±0.95^{B}	84.7±1.51^{B}	0.96
2013.9.26～2013.10.25（Ⅲ）	23.2±3.33^{Ab}	68.7±4.25^{Ab}	0.99

注：同列肩标大写字母不同表示差异极显著（$P<0.01$），同列肩标小写字母不同表示差异显著（$P<0.05$）

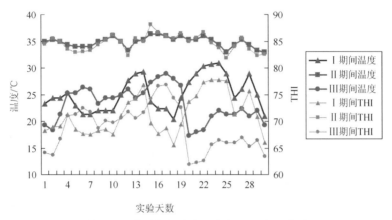

图2-10 实验期间牛舍温度与THI变化曲线

（二）THI对奶牛产奶量的影响

实验期间三个时间段的奶牛日产奶量及不同泌乳阶段的日产奶量见表2-17，数据表明，以全部奶牛样本为研究对象，实验Ⅰ期奶牛的日产奶量与Ⅱ期和Ⅲ期奶牛的日产奶量无统计学差异显著性（$P>0.05$），Ⅱ期奶牛的日产奶量显著低于Ⅲ期奶牛的日产奶量（$P<0.05$）；以泌乳早期奶牛样本为研究对象，实验Ⅰ期和Ⅲ期奶牛的日产奶量极显著高于实验Ⅱ期奶牛的日产奶量（$P<0.01$）；以泌乳中期奶牛样本为研究对象，实验Ⅰ期奶牛的日产奶量显著高于Ⅱ期奶牛的日产奶量（$P<0.05$）；以泌乳后期奶牛样本为研究对象，三个时间段奶牛的日产奶量之间无统计学差异显著性（$P>0.05$）。

表2-17 实验期间三个时间段的奶牛日产奶量及不同泌乳阶段的日产奶量
Table 2-17 The cows' daily milk yield and milk production in the three different stages of lactation period during the experiment

实验阶段	全部		泌乳早期		泌乳中期		泌乳后期	
	样本量	日产奶量/kg	样本量	日产奶量/kg	样本量	日产奶量/kg	样本量	日产奶量/kg
Ⅰ	16 282	17.6±3.36^{ab}	3 871	22.34±2.65^{A}	1 987	18.17±1.92^{a}	10 424	15.73±1.59
Ⅱ	6 764	14.58±2.17^{a}	3 593	16.08±1.51^{B}	2 546	13.19±1.33^{b}	625	11.66±1.06
Ⅲ	11 819	20.55±4.54^{b}	7 993	23.3±2.2^{A}	2 747	15.76±1.7^{Bab}	1 079	12.38±1.4

注：同列肩标大写字母不同表示差异极显著（$P<0.01$），同列肩标小写字母不同表示差异显著（$P<0.05$）

三、分析与讨论

（一）三个实验阶段牛舍的温度与THI

本研究规模场的夏季白天的平均气温在35℃以上，湿度在60%左右，这样的季节极易造成奶牛热应激反应，降低机体免疫力、产奶量及奶品质。本实验选取2013.5.1～2013.5.30为实验Ⅰ期，牛舍平均温度为25.2℃，略高于Igono和Johnson所推荐的产奶牛的适宜温度（4～24℃），平均THI为71.2，可见在此期间奶牛几乎没有受到热应激的影响；2013.7.23～2013.8.21为实验Ⅱ期，为2013年该规模场夏季持续高热的一段时间，牛舍平均温度和THI分别为34.9℃和84.7，THI为79～89，奶牛处于中度热应激状态；2013.9.26～2013.10.25为实验Ⅲ期，牛舍平均温度和THI分别为23.2℃和68.7，奶牛热应激状态基本结束。

（二）THI对奶牛产奶量的影响

李建国等认为，热应激降低牛泌乳力的机制是高温引起牛代谢机能和酶活性改变，T3和皮质醇分泌减少，代谢机能减弱，采食量下降，营养摄入不足，体内维生素合成减少，呼吸频率增加，体温升高，造成体液中电解质及维生素的损失，产奶量下降。此外，夏季牛体内免疫球蛋白合成下降，血清中抗坏血酸水平降低，机体抵抗力减弱，乳腺炎发病率上升，影响牛泌乳性能的发挥。

杨毅等（2009）发现，热应激程度越高泌乳奶牛的产奶量受热应激的影响越严重，泌乳奶牛在轻微和中度热应激期时产奶量分别下降13.8%和26.9%（$P<0.01$），产奶量与牛舍THI有极显著的负相关关系，THI每升高1个单位，产奶量则降低1.9%。在本次实验中，实验Ⅱ期奶牛处于持续中度热应激状态，结果显示泌乳早期、中期和后期的奶牛日产奶量分别下降了28%、27.4%和25.7%，均高于李建国等的研究结果，其报道显示奶牛处于慢性热应激，泌乳前、中、后期奶牛的日产奶量分别显著降低19.3%、15.88%、13.83%。

West等（2003）报道，当环境THI超过72时，奶牛采食量和产奶量显著下降，并且THI每增加1个单位，产奶量就降低0.2 kg。魏学良等（2005）研究发现，在热应激期奶牛泌乳性能下降25.6%～33.5%。韩玉婷等（2011）认为，当环境THI每增加1个单位，产奶量下降0.6 kg，泌乳性能下降26.7%。杜红芳等（2007）报道，在夏季炎热的时候，每头奶牛的日均产奶量减少0.5～2.5 kg。在本次实验中，实验Ⅰ期比Ⅱ期THI高13.5个单位，THI每增加1个单位，泌乳早期、泌乳中期和泌乳后期的奶牛日产奶量分别下降0.46 kg、0.37 kg和0.3 kg；在奶牛热应激结束后，也就是Ⅲ期，THI比Ⅱ期THI降低了16个单位，此期间THI每降低1个单位，泌乳早期、泌乳中期和泌乳后期的奶牛日产奶量分别升高0.45 kg、0.16 kg和0.04 kg。由此可见，泌乳早期的奶牛受到热应激的影响程度大于泌乳中期的奶牛，泌乳后期的奶牛受到应激的影响程度最小。然而，当奶牛从热应激状态恢复至适宜环境后，日产奶量虽然有所恢复，但是并未恢复到热应激之前的日产奶量。因此，从热应激方面考虑，需要合理安排奶牛的配种与产犊时间，建议我国南方高热高湿地区的奶牛场通过合理安排奶牛的配种与产犊时间，在上一年的10月份、11月份和

12月份尽量不给奶牛配种，避开母牛在7月份、8月份和9月份产犊，从而避开热应激期间奶牛处于泌乳早期和中期。

（三）奶牛场主要降温措施评价

本研究中奶牛场采取主要降温措施有：采用隔热性能好的彩钢复合板建造成倾斜式的顶棚；牛舍和奶牛活动区周围种植了树木；在夏季来临前对奶牛饲料结构进行调整和改善，提高日粮蛋白质和脂肪含量，添加了矿物质和B族维生素，达到增进奶牛营养物质进食量，增强奶牛抗热应激能力的目的；在牛舍内和挤奶厅等牛群集中活动的区域安装了数量足够、风量较大的电风扇，加速牛舍内气流速度和牛体散热等。然而实验结果表明，该牛场在实验Ⅱ期（2013.7.23~2013.8.21）的THI为79~89，表明牛场奶牛持续处于中度热应激状态下，缓解奶牛热应激的措施亟待更进一步的改善。

侯引绪等（2012）对吹风、喷淋两种物理性抗应激措施的应用效果进行了研究。结果表明，单一的持续喷淋措施，可使牛舍内温度比舍外（露天）降低2℃，热应激指数比舍外下降2.2；采用持续吹风同时持续喷淋措施，可使牛舍内温度比舍外下降7.5℃，热应激指数比舍外下降6.0。刘海林等（2013）采用风扇通风和喷淋系统降温，提高日产奶量2.66 kg（$P<0.05$），比未采用该措施提高了19.79%。本次实验中发现实验牛场存在奶牛饲养密度过大，而且又未安装喷雾装置的问题，建议通过合理调整奶牛牛群密度，在高温高湿天气对牛舍进行喷淋降温，由此降低牛舍温度，提高奶牛的舒适度，从而减弱夏季热应激使奶牛日产量降低的不良影响。

四、小结

我国南方夏季的高温高湿天气造成奶牛不同程度的热应激，严重影响了奶牛的日产奶量，泌乳早期的奶牛受到热应激的影响程度大于泌乳中期的奶牛，泌乳后期的奶牛受到热应激的影响程度最小。然而，当奶牛从热应激状态恢复至适宜环境后，日产奶量虽然有所恢复，但是未能恢复到热应激之前的日产奶量。因此从热应激方面考虑，除改进夏季牛场降温措施外，最好能合理安排奶牛的配种与产犊时间，在上一年的10月份、11月份和12月份尽量不给奶牛配种，避开母牛在7月份、8月份和9月份产犊，从而避开热应激期间奶牛处于泌乳早期和中期。

第三章 奶牛乳腺炎致病性金黄色葡萄球菌病原学特征

第一节 金黄色葡萄球菌研究进展

金黄色葡萄球菌（Staphylococcus aureus）不仅是一种常见的人类共生细菌，而且也是一种临床上最常见的感染性病原菌。金黄色葡萄球菌可以引起的疾病非常广泛，从损伤较弱的皮肤和软组织感染到危及生命的菌血症、败血症、心内膜炎和肺炎等。金黄色葡萄球菌也是一种重要的动物病原菌。它是引起奶牛乳腺炎的主要致病菌之一，对奶牛产业产生了严重的危害。

一、金黄色葡萄球菌生物学特性及其致病性

金黄色葡萄球菌是奶牛乳腺炎首要的病原菌，可引起超急性、急性和亚临床性乳房炎，慢性和亚临床性是该菌所致乳腺炎的主要形式。金黄色葡萄球菌产生的透明质酸酶使其对组织具有极大的侵害能力，其表面蛋白 A 可抵抗吞噬细胞的吞噬，糖醛酸磷壁酸可抵御免疫系统对它的免疫性。金黄色葡萄球菌为兼性的细胞内病原体，可在吞噬细胞内生存，在乳腺组织中可形成脓肿，最终导致纤维化。某些金黄色葡萄球菌菌株由于基因的突变表现出对抗生素的抗性（宋战辉，2014）。

（一）金黄色葡萄球菌的生物学特性

金黄色葡萄球菌隶属葡萄球菌属，是一种革兰氏阳性菌，直径 0.5～1.5 μm，菌落排列方式有单个、成对或呈葡萄串状，在液休培养基或浓汁中呈双球状或短链状排列、无芽孢、无鞭毛，有的形成荚膜，能分解甘露醇，血浆凝固酶呈阳性。需氧或兼性厌氧，普通培养基生长良好，最适温度 35～40℃，最适 pH 7.0～7.5。平板培养时菌落厚、圆形有凸起、有光泽，直径 1～2 mm。血平板培养时菌落周围形成透明溶血环。金黄色葡萄球菌具高度耐盐性，在 10%～15% NaCl 肉汤中可生长。可产生卵磷脂酶及甘油酯和溶性磷酸胆碱，在 Baird-Parker 平板上培养时可生成黑色菌落，菌落周围有浑浊带，外层有透明圈。产生的凝固酶可使血浆蛋白酶原变成有活性的血浆蛋白酶，使血浆发生凝固。本菌对干燥有较强抵抗力，对龙胆紫等碱性染料敏感，消毒净、新洁尔灭及洗必泰等可杀死本菌。本菌对青霉素、红霉素、土霉素及磺胺类药物敏感，易产生耐药性。

（二）金黄色葡萄球菌的致病性

金黄色葡萄球菌可引起局部化脓、肺炎、伪膜性肠炎及心包炎等，甚至可引起败血症、脓毒症等全身感染，其致病性主要是由于它产生多种毒素和酶，这些毒力因子影响金黄色葡萄球菌吸附、入侵组织和逃避宿主防御的能力。感染过程中，首先黏附素与细胞表面因子结合并黏附到乳腺上皮细胞表面，继之定居并大量繁殖，与此同时，释放大

量毒素破坏细胞，并依靠荚膜多糖逃避宿主吞噬作用，在各种酶的作用下导致炎症加重扩大（宋战辉，2014）。

肠毒素（staphylococcal enterotoxin，SE）是由血浆凝固酶或耐热核酸酶阳性菌株产生的一组结构相似抗原性不同的可溶性蛋白质，分子质量为26～34 kDa，抗酸耐热，100℃加热30 min或胃蛋白酶水解而不被破坏。根据抗原的差异性可分为A、B、C1、C2、D、E、G、H、I、J、K、L、M、N共14个型，细菌能产生一个以上的类型，其中A、D型最为常见，B、C型次之，E型出现率最低，A型毒力最强，B型耐热性最好。肠毒素可引起急性肠胃炎，极少量即可致动物出现呕吐、腹泻、发烧、血小板减少、肺水肿及心跳加快等症状。

溶血素（haemolysin）按抗原不同分为α、β、γ、δ、ε五种类型，能破坏溶酶体，损伤血小板，引发机体局部缺血和坏死。α溶血素（α-haemolysin）又名α毒素（Alpha toxin），分子质量约为34 kDa，是一种蛋白质，结构基因为 *hla*，不耐热，65℃加热30 min可将其破坏，为金黄色葡萄球菌分泌的一种外毒素，对人类有致病作用，对多种哺乳动物红细胞有溶血作用，与牛羊的坏疽性乳腺炎有关。β溶血素常见于动物分离株，具有神经磷脂酶活性，裂解细胞膜上的鞘磷脂使细胞膜渗漏而导致细胞溶解。γ溶血素是双组分蛋白，可破坏中性粒细胞、巨噬细胞、哺乳动物红细胞及哺乳动物细胞膜，一般作为细胞膜表面活性剂使用，通过溶解目标细胞的膜以达到裂解细胞的目的。

杀白细胞素（staphylococcal leukocidin，SL）由基因 *luks-pv* 和 *lukf-pv* 共同编码，属synergohymenotropic毒素家族，是一种31 kDa左右不耐热蛋白质，由17种氨基酸组成，分为F组分和S组分，能杀死人和兔子的巨噬细胞及多形核粒细胞。S或F单一组分没有杀白细胞的作用。SL具有抗原性，其产生的抗体能阻止金黄色葡萄球菌复发。

血浆凝固酶（coagulase，cal）是金黄色葡萄球菌产生的能使含枸橼酸钠或肝素抗凝剂的人和动物的血浆发生凝固的酶类物质，常作为葡萄球菌有无致病性的重要标志。其分为两类，一种分泌至体外，称为游离性凝固酶（Free coagulase），作用类似凝固酶原；另一种结合于菌体表面，在菌株表面起纤维蛋白原特异受体作用，称为结合凝固酶（bond coagulase）或聚集因子（clumping factor）。凝固酶具备抗原性，有助于致病菌抵御宿主体内吞噬细胞和杀菌物质，但易被蛋白酶分解。

耐热核酸酶（heat stable nuclease）是一种细胞外酶，由致病菌产生，为金黄色葡萄球菌特有，且不同菌株间有较高保守性，常作为金黄色葡萄球菌检测的重要指标之一，其编码基因 *nuc* 常作为PCR检测引物。在100℃条件下作用15 min不失活性。其在金黄色葡萄球菌感染部位的组织细胞和白细胞崩解时被释放出，能加速核酸分解，有利于病原菌的扩散。

表皮溶解素（epidermolytic toxin）又名表皮剥脱毒素（exfoliatin），分A、B两个型，是由噬菌体Ⅱ型金黄色葡萄球菌产生的一种蛋白质，分子质量为24 kDa，有抗原性，被甲醛脱毒后变成类毒素。其可引起烫伤样皮肤综合征，常见于新生畜和婴幼儿，病死率20%左右。

毒性休克综合征毒素-1（toxicshocksyndrometoxin1，tsst-1）由噬菌体Ⅰ群金黄色葡萄球菌产生，是一类蛋白质，可增加毛细血管的通透性，引发机体器官功能紊乱或毒性休克综合征（TSS）。

Staphopain A/B 是一种外分泌型半膀胱氨酸蛋白水解酶，有 A、B 两个类型，由 *scpA* 和 *scpB* 基因分别编码。在金黄色葡萄球菌中高度保守，与致病性有关。

黏附素（adhesion factors or protein，Adhesin）于微生物表面识别黏附基质成分分子（microbial surface components recognizing adhesive matrix，MSCRAMM），是金黄色葡萄球菌表面表达的一种可特异性识别上皮细胞表面的蛋白，是金黄色葡萄球菌感染早期最重要的致病因子。纤连蛋白结合蛋白（FnBP）、凝集因子（ClfA）和胶原蛋白结合蛋白（Can）是目前研究较多的三种黏附素。成功黏附到上皮细胞被认为是金黄色葡萄球菌感染过程中关键性的步骤。

荚膜多糖（capsular polysaccharide，CP）为金黄色葡萄球菌的 O 抗原，分子质量小，免疫原性弱，需与大分子蛋白结合以增强其免疫原性，为疫苗研制指明方向。共有 11 个血清型，5 型和 8 型是临床分离到的绝对优势型，占 80%左右。5 型和 8 型荚膜多糖具有抗中性粒细胞吞噬的作用，同时其抗体还被认为具有调理素的作用，从而增强细菌毒力，提高其对动物机体的侵袭力。

二、金黄色葡萄球菌感染性疾病研究进展

（一）金黄色葡萄球菌肺炎研究进展

目前，金黄色葡萄球菌肺炎已成为一种罕见的社区获得性肺炎（CAP），其发病率占 CAP 的 1%~5%。金黄色葡萄球菌肺炎主要发生在流感患者中。此外，金黄色葡萄球菌也是一种重要的医院获得性肺炎致病菌。近几十年来，由于 MRSA 广泛流行，金黄色葡萄球菌肺炎引起人们的广泛关注。目前发现，MRSA 引起的肺炎占医院获得性肺炎和呼吸机相关性肺炎的 20%~40%。近年来，社区获得性 MRSA（CA-MRSA）成为一种新的引起肺部感染的病原菌，它携带Ⅳ型 SSC mec（Rubinstein et al.，2008）。在法国发现 16 例肺炎是由 CA-MRSA 引起的，这些 CA-MRSA 含有Ⅳ型 SSC mec 及杀白细胞素（PVL），PVL 常常存在于 CA-MRSA 中。目前认为，PVL 是一种重要的肺炎相关的毒力因子。同时，溶血素和蛋白 A 等金黄色葡萄球菌毒力因子也增加了 CA-MRSA 和医院获得性 MRSA（HA-MRSA）菌株的毒力，增加了金黄色葡萄球菌肺炎的发病率和死亡率（Bubeck et al.，2008）。

（二）金黄色葡萄球菌菌血症研究进展

金黄色葡萄球菌是一种非常重要的人类致病菌。它也是一种重要的共生菌，它可以持续性定植在 20%人群中，甚至间歇性定植在 60%人群中，终身携带细菌。金黄色葡萄球菌主要引起机体的皮肤和软组织感染。金黄色葡萄球菌可以从鼻腔转移到创伤和静脉注射等引起的皮肤伤口，从而引起机体皮肤和软组织感染。目前，治疗金黄色葡萄球菌引起的皮肤和软组织感染（SSTI），是一件非常容易的事情。但是，在某些情况下，金黄色葡萄球菌可以经过血液循环进入到外周组织从而引起心内膜炎、器官脓肿、败血性关节炎、脊椎性骨髓炎和脑膜炎等继发性感染。近年来发现，FnBPA、ClfA 和 ClfB 等表面蛋白在金黄色葡萄球菌菌血症过程中具有黏附、侵入、血小板聚集和逃避机体免疫等生物学功能。

(三) 金黄色葡萄球菌乳腺炎研究进展

乳腺炎是指乳腺的炎症性反应，它可以由革兰氏阳性菌、革兰氏阴性菌和支原体等多种病原引起。其中，金黄色葡萄球菌是引起乳腺炎的主要致病菌之一。研究表明，多种毒力因子参与到金黄色葡萄球菌引起的乳腺炎过程中。纤连蛋白结合蛋白（FnBP）在金黄色葡萄球菌黏附和侵入牛的乳腺上皮细胞过程中起着重要的作用，从而加重了牛的乳腺炎。金黄色葡萄球菌生物被膜基质蛋白（Bap），不仅可以促进金黄色葡萄球菌黏附牛源乳腺上皮，而且可以通过结合宿主受体GP96抑制了牛源乳腺上皮内吞金黄色葡萄球菌，使金黄色葡萄球菌在乳腺组织中持续感染（Jaione et al.，2012）。溶血素和凝固酶是两个重要的金黄色葡萄球菌毒力因子。研究表明，溶血素和凝固酶在金黄色葡萄球菌引发的小鼠乳腺炎感染过程中起着重要的作用。由于抗菌药物的过度滥用，导致病原菌对抗菌药物发生耐药。目前，细菌对抗菌药物的耐药性引起人们的广泛关注，金黄色葡萄球菌对甲氧西林类药物尤为严重，甚至对多种药物发生耐药。氨基糖苷类药物万古霉素曾经是治疗耐甲氧西林金黄色葡萄球菌引起感染的首选药物。近年来发现，金黄色葡萄球菌对万古霉素中度甚至完全耐药。由于抗菌药物易引起金黄色葡萄球菌发生耐药，我们需要寻找新的方法及途径来预防和治疗金黄色葡萄球菌的感染。

三、金黄色葡萄球菌耐药性研究进展

(一) 乳腺炎源性葡萄球菌对抗生素耐药性

细菌对抗生素极易产生耐药性。目前，细菌耐药性已成为全球关注的热点。耐药细菌所致的感染性疾病对新世纪感染治疗是极大的挑战。因此及时准确地掌握细菌的耐药机制，对指导临床合理用药和新的抗菌药物的研制具有重要意义。在兽医领域，虽然很多中药正在被开发和应用，但是抗生素还是治疗疾病的基础用药，抗生素的耐药性是治疗感染性疾病的一大挑战。在奶牛乳腺炎治疗中，青霉素、红霉素、四环素等抗生素应用非常普遍，因此对葡萄球菌，尤其是金黄色葡萄球菌的耐药监测是非常必要的（于恩琪，2014）。

青霉素、红霉素、四环素是奶牛场兽医经常使用的抗生素。正是由于这些药在牛群中的广泛应用和滥用，一些病原菌对这些药均表现出高耐药性。目前，很多来自不同国家的研究者已经报道乳腺炎源性金黄色葡萄球菌对青霉素具有较高的耐药性，但在不同地区存在着差异（Sahebekhtiari et al.，2011）。Osteras等已经证实，在奶牛干奶期用青霉素进行治疗会对金黄色葡萄球菌的青霉素耐药性产生影响。关于金黄色葡萄球菌对四环素的耐药性报道在我国也经常见到，王登峰等（2011）报道，新疆、浙江、山东、内蒙古和上海等地的金黄色葡萄球菌对青霉素的耐药率高达90%以上、红霉素80%以上、四环素96%以上。邓海平等（2010）报道，甘肃地区的金黄色葡萄球菌对青霉素耐药率达70.4%，甘肃和贵州地区的对金黄色葡萄球菌对四环素耐药率为96.3%和81.6%。王凤等（2013）报道，北京、河南、天津和广西等地区的金黄色葡萄球菌对青霉素的耐药率为87.50%、红霉素为31.25%、四环素为6.25%。燕霞等（2012）报道，浙江省的金黄色葡萄球菌对青霉素的耐药率为90%、红霉素为50%、四环素为60%。王桂琴等（2011）报道，宁夏地区的金黄色葡萄球菌对青霉素的耐药率为86%、红霉素为60%、四环素16%。易明梅等（2009）

报道，上海地区的金黄色葡萄球菌对青霉素的耐药率为 91.6%、红霉素为 18.3%、四环素为 26.8%。苏洋等（2012）报道，内蒙古地区的金黄色葡萄球菌对青霉素的耐药率为 94.7%、红霉素为 5.3%、四环素为 18.4%。张宇等（2013）报道，金黄色葡萄球菌对青霉素的耐药率为 73.8%、红霉素为 61.9%、四环素为 7.1%。

1. 金黄色葡萄球菌对 β-内酰胺类抗生素的耐药机制 金黄色葡萄球菌对 β-内酰胺类抗生素的耐药机制主要有 2 种：①金黄色葡萄球菌产生了 β-内酰胺酶，使此类抗生素失去作用。金黄色葡萄球菌 β-内酰胺酶可分为 4 种类型，为 A～D 型。β-内酰胺酶最多的是 A 型，其次为 C 型，B、D 型较少见。A 型 β-内酰胺酶由 blaZ 基因编码，blaZ 基因位于较大质粒的转座子中。β-内酰胺酶共涉及 3 个调节基因：blaI、blaR1 和 blaR2。其中，blaI 编码一种拟制因子，对 β-内酰胺酶的编码基因的转录进行负调节；blaR1 编码的蛋白包括：一个细胞外的区域，用来感受 β-内酰胺酶类药物对其的作用，一个细胞内区域，负责产生一种信号，导致 blaZ 基因的去抑制；blaR2 基因可使 β-内酰胺酶的表达下调，但是，目前机制尚不明确。②青霉素结合蛋白（PBP）的作用位点发生了改变或产生了新的对 β-内酰胺类抗生素不敏感的 PBP 蛋白。PBP 是金黄色葡萄球菌细胞壁合成所必需的酶。正常的金黄色葡萄球菌都会产生 4 种内在性 PBP：PBP1、PBP2、PBP3 和 PBP4。β-内酰胺类抗生素能与细胞膜上的 PBP 结合。产生耐药性的金黄色葡萄球菌会合成一种新的 PBP2a，这种 PBP2a 几乎很少或完全不能与 β-内酰胺类抗生素相结合，PBP2a 因为不能被结合，可以一直保持酶活性，合成细菌壁，从而使金黄色葡萄球菌产生耐药性（于恩琪，2014）。

2. 金黄色葡萄球菌对四环素的耐药机制 金黄色葡萄球菌对四环素的耐药机制主要有 4 种：①所有 tet 外排泵基因都编码与膜相关的外排泵蛋白，外排泵蛋白可将四环素主动泵出胞外，从而产生耐药性。这些外排泵基因编码的外排泵蛋白根据氨基酸的同源性分为 6 个群：第 1 群蛋白包括 tetA、tetB、tetC、tetD、tetE、tetG、tetH、tetZ。除了 tetZ 蛋白基因在革兰阳性细菌存在，该群其他的蛋白基因只在革兰阴性细菌存在。第 2 群蛋白包括 tetK 和 tetL，这些蛋白主要在革兰阳性菌中存在。tetK 和 tetL 蛋白基因主要存在于小的传递性质粒中，有很小的概率可以整合入染色体或葡萄球菌的大质粒中。tetK 小质粒是一个大小为 4.4～4.7 kb 的家族。pT181 质粒是这一族的代表，已测出全序。第 3 群蛋白包括 otrB 和 tcr3。第 4 群蛋白包括 tetA（P）。第 5 群蛋白包括 tetV。第 6 群蛋白包括从纹带棒状杆菌分离到的未命名的蛋白。②四环素结合细菌核糖体 30S 亚基后，蛋白质合成中的肽链延伸就无法进行。核糖体保护蛋白结合至核糖体 30S 亚基，依赖 GTP 的水解释放的能量将四环素从 30S 亚基上释放出来，核糖体保护蛋白随即也从核糖体脱离，使核糖体重新进入正常的肽链延伸循环中。已知的核糖体保护蛋白有 9 种，研究最多的是 tetM 和 tetO。革兰阳性菌中常见的是 tetM 和 tetO68。③灭活或钝化四环素的酶，tetX 基因是唯一通过产生灭活四环素的酶而耐药的。例如，tetX 产生的胞浆蛋白在氧和 NADPH 存在时可化学修饰四环素，序列分析表明这个酶与其他 NADPH 需要的氧化还原酶有同源性。④其他（未知）耐药机制，tetU 蛋白有低水平的耐四环素能力（于恩琪，2014）。

3. 金黄色葡萄球菌对红霉素耐药机制 金黄色葡萄球菌对红霉素产生耐药的机制主要有 3 种，外排机制、核糖体突变机制和甲基化转移酶机制。这些耐药机制单独或联合作用导致金黄色葡萄球菌对红霉素产生耐药。①在金黄色葡萄球菌中，有两种外排系统，第 1 种是由 msrA 基因编码的外排系统；第 2 种是 mef 基因，靠质子动力外排红霉素；但在金黄色葡萄球菌比较少见。②甲基化转移酶主要由基因编码产生，该基因可以在染色体上，也可以与质粒和转座子相连。甲基转移酶家族催化核糖体的特定位置，使其发生单甲基化或双甲基化，从而引起核糖体构象发生改变，降低红霉素与核糖体的亲和力，进而导致细菌产生耐药性。这种甲基化转移酶修饰药物作用位点的耐药机制，被认为是引起金黄色葡萄球菌耐红霉素药物的最主要机制。③核糖体发生点突变或部分碱基发生突变或缺失，导致细菌耐药。这种机制在革兰阳性细菌中较少见（于恩琪，2014）。

（二）耐甲氧西林金黄色葡萄球菌研究进展

在 20 世纪 60 年代初期，甲氧西林在临床上被广泛用于治疗金黄色葡萄球菌感染。金黄色葡萄球菌不久就被发现对甲氧西林药物产生了耐药性，这类耐药的金黄色葡萄球菌被称为耐甲氧西林金黄色葡萄球菌（methicillin-resistant *S. aureus*，MRSA）。到 80 年代后期，MRSA 在世界各地已成为最常见的医源性感染的致病菌之一，这对医疗系统产生了强烈的冲击。20 世纪 80 年代，第 1 株社区获得性 MRSA（community-acquired MRSA，CA-MRSA）被报道。虽然这 1 株 CA-MRSA 从社区患者中分离后获得，但是，后来被证实是医院获得性 MRSA（hospital-acquiredMRSA，HA-MRSA）引起的感染。随后，CA-MRSA 在世界各地广泛流行。金黄色葡萄球菌细胞壁肽聚糖是由 *N*-乙酰葡萄糖胺和 *N*-乙酰胞壁酸交联形成的肽聚糖链；肽聚糖链之间的连接需要青霉素结合蛋白（PBP）催化的转肽作用。PBP 催化肽聚糖中甘氨酸五肽与邻链上的 D-丙氨酰-D-丙氨酸之间的连接，使胞壁酸短肽链之间相互连接，从而形成金黄色葡萄球菌细胞壁的肽聚糖多层网状结构。甲氧西林等 β-内酰胺类抗生素的 β-内酰胺环与 D-丙氨酰-D-丙氨酸在结构上非常相似，能与 D-丙氨酰-D-丙氨酸竞争性结合 PBP 活性位点，β-内酰胺环的亲和能力要比 D-丙氨酰-D-丙氨酸强的多，当 β-内酰胺类抗生素浓度大于或等于 MIC（最低抑菌浓度）时，可使 PBP 失去催化活性，抑制细胞壁的合成，造成细菌的死亡。MRSA 之所以对甲氧西林产生耐药性主要是因为金黄色葡萄球菌可以产生一种新的 PBP，即 PBP2a。PBP2a 的转肽酶结构域（TPase domain）亲和 β-内酰胺类抗生素的能力非常弱，在 β-内酰胺类抗生素环境下，金黄色葡萄球菌的 PBP 被抗生素结合而失去活性，PBP2a 能够代偿 PBP 的功能，继续合成肽聚糖，使金黄色葡萄球菌在 β-内酰胺类抗生素环境下存活（袁文常，2012）。PBP2a 是由 *mecA* 基因编码的，*mecA* 基因、调节基因 *mecR1* 和抑制基因 *mecI* 一起构成 *mec* 基因复合体。在正常情况下，抑制基因 *mecI* 表达 MECI 蛋白。MECI 蛋白与 *mecA* 基因的启动子区相结合，抑制 *mecA* 基因的转录。在甲氧西林等 β-内酰胺类抗生素环境下，*mecR1* 基因可以表达调节蛋白 MECR1，解除 MECI 对 *mecA* 基因的阻遏作用，使 *mecA* 基因表达 PBP2a。这些基因位于一个可移动的基因岛上，该基因岛被称为金黄色葡萄球菌的盒式染色体 mec 盒（SCCmec）（邱家章，2012）。某些 SCC 类型也包括其他基因组件，例如，*Tn554* 基因编码抗大环内酯类抗生素、克林霉素和链霉素的蛋白；*Pt181* 基因编码抗四环素类药物的蛋白，使金黄色葡萄球菌对多种抗生素产生耐药。根据 SCCmec 的大小和组成，将其分成 5 种类型，即 SCCmecⅠ、SCCmecⅡ、SCCmecⅢ、SCCmecⅣ和 SCCmecⅤ。HA-MRSA

主要是 SCCmecⅠ、SCCmecⅡ和 SCCmecⅢ，但是，CA-MRSA 主要是 SCCmecⅣ和 SCCmecⅤ；与前 4 种类型相比，SCCmecⅤ比较小，仅含有 *mecA* 基因。与 HA-MRSA 相比，CA-MRSA 生长速度较快，这就可以解释为什么前者在社区中更容易存活。杀白细胞毒素基因（PVL）也常常存在于 CA-MRSA 中。PVL 在 1932 年首次被发现，是一种 synergohymenotropic 毒素，可以破坏白细胞和介导组织坏死。PVL 参与金黄色葡萄球菌引起的儿童和青壮年的皮肤和软组织感染以及坏死性肺炎（陈福广，2014）。

对于治疗 MRSA 等耐甲氧西林病原菌引起的感染来说，糖肽类抗生素万古霉素是目前临床上唯一治疗有效的药物。但是，由于在临床上对这类抗生素过度依赖，导致了耐万古霉素金黄色葡萄球菌的出现。1997 年，在一家日本实验室里，第 1 株异源性万古霉素耐药金黄色葡萄球菌（heterogeneous vancomycin-resistant *S. aureus*，hVRSA）被发现，hVRSA 对万古霉素易感性减弱。hVRSA 是从由 MRSA 引起的肺炎的老人的唾液中分离出来的，该 hVRSA 被命名为 MRSA Mu3。在无抗生素培养基上，MRSA Mu3 生长后形成细菌亚群，对万古霉素的耐药性发生变化，这证实了该菌株对万古霉素的易感性发生自然抑制性或可变性。在万古霉素存在的条件下，Mu3 生长后形成金黄色葡萄球菌亚群；该亚群对万古霉素的耐药性与使用的万古霉素浓度成正比。

随后，中度耐万古霉素金黄色葡萄球菌（vancomycin-intermediate *S. aureus*，VISA）菌株在日本被发现。第 1 株 VISA 菌株是从一个 4 个月大的婴儿手术创口处分离出来的；这 1 株 VISA 菌株被命名 MRSA Mu50。通过肉汤稀释法，证实 MRSA Mu50 对万古霉素的 MIC 值为 8 mg/mL。虽然 VISA 菌株很少被发现，但是，根据一份土耳其调查报告（1998～2002 年），256 例 MRSA 分离菌株（MRSA 主要来源于患者的血液和浓汁）中，46 例（占 18%）为 VISA，这意味着 VISA 病例呈现上升趋势（Banu et al.，2005）。近年来，耐万古霉素金黄色葡萄球菌（vancomycin-resistant *S. aureus*，VRSA）被广泛发现；VRSA 对万古霉素完全发生耐药，这引起了人们的广泛关注。迄今为止，4 株 VRSA 临床分离株在美国发现。在 2002 年 7 月，第 1 株临床分离株是从 40 岁女子做血液透析通路时的出口中分离出来的；该患者患有高血压、糖尿病、外周血管疾病和慢性肾衰竭。在这之前，患者因慢性足部溃疡接受过治疗，并且接受过多个疗程的万古霉素抗菌治疗。在 2002 年 4 月，患者因坏疽性脚趾而进行截肢手术，随后感染 MRSA，最后接受万古霉素和利福平治疗。金黄色葡萄球菌菌株对万古霉素（MIC＞128 mg/L）和苯唑西林（MIC＞16 mg/L）产生耐药性。在 2002 年 9 月，第 2 株 VRSA 临床分离株在美国宾夕法尼亚州被发现，通过肉汤稀释法测定这株临床分离株的 MIC 是 32 mg/L。这株临床分离株是从 70 岁患有病态肥胖高血压男性患者的慢性足溃疡中分离出来的，患者曾经接受过多个抗生素（万古霉素除外）疗程，表现连续数周嗜睡，间歇热，发冷，萎靡，夜间盗汗和呼吸困难病史。第 3 株 VRSA 临床分离株是从一名长期接受抗生素治疗的老人的尿液样本分离的，通过肉汤稀释法测定，这株 VRSA 临床分离株的 MIC 是 32 mg/L。第 4 株 VRSA 在 2005 年 3 月被报道（陈福广，2014）。

万古霉素和替考拉宁等糖肽类抗生素能够与金黄色葡萄球菌肽聚糖前体 D-Ala-D-Ala 结合，阻碍肽聚糖合成过程中的转糖基和转肽作用，抑制了细菌细胞壁的生物合成从而发挥杀菌效应。目前认为，细胞壁增厚和遗传物质转移是金黄色葡萄球菌对万古霉素产生耐药的主要原因。研究表明，细菌细胞壁改变是金黄色葡萄球菌对万古霉素发生耐药的重要机制。VISA 菌株可以通过增加肽聚糖前体 D-Ala-D-Ala 的数量来合成额外的肽聚糖；这些肽聚糖前体 D-Ala-D-Ala 可以结合万古霉素从而有效的阻断万古霉素到达其作用靶点，使金黄色葡萄球菌对万古霉素发生耐药。金黄色葡萄球菌 Mu50 对万古霉素产生耐药性，这与肽聚糖的快速合成、细胞壁增厚、非酰胺化肽聚糖的数量增加和交联的肽聚糖数量减少密切相关。细胞壁的增厚，使万古霉素结合到细胞壁外层从而有效的阻断万古霉素与细胞壁内层结合。临床分离株 Mu3 对万古霉素产生耐药性，这与细胞壁增厚和肽聚糖合成活性增强有关；但是，临床分离菌株 Mu3 的细胞壁厚度和肽聚糖合成速度明显小于临床分离菌株 Mu50。此外，细胞壁磷壁酸的结构和代谢发生变化从而减少细胞壁降解速度，也使金黄色葡萄球菌减少了对万古霉素的敏感性（Sieradzki and Tomasz, 2003）。大量证据证实，VRSA 主要是由于遗传物质转移至金黄色葡萄球菌中引起的。通过遗传分析的方法，发现美国密歇根州 VRSA 分离菌株的万古霉素抗性基因从耐万古霉素粪肠球菌转移到 MRSA 后产生的（Weigel et al., 2003）。从这株 VRSA 临床分离株成功获得 *vanA* 基因，这是通过共分离的耐万古霉素粪肠球菌种间转移 Tn1546 后产生的。Tn1546 是一种 *vanA* 转座子，Tn1546 存在于多种耐药质粒中。该 VRSA 临床分离株不能合成正常的肽聚糖前体 D-Ala-D-Ala，而是合成异常的肽聚糖前体 D-Ala-D-Lac。在低浓度万古霉素压力下，该 VRSA 临床分离株可以合成新的肽聚糖前体，因为这株新的肽聚糖前体 D-Ala-D-Lac 对万古霉素的亲和力较弱，所以金黄色葡萄球菌可以继续合成细胞壁。

四、乳腺炎致病性金黄色葡萄球菌毒力基因研究进展

金黄色葡萄球菌是兼性厌氧菌，革兰氏阳性球菌，显微镜观察呈现葡萄串状排列，血液琼脂平板上菌落为圆形、金黄色，并有溶血现象，金黄色的外观是该细菌的名字的词源，*aureus* 在拉丁语的意思是"金黄色"。金黄色葡萄球菌也是临床最难对付的病原菌之一，金黄色葡萄球菌被称为"a well-armed pathogen"（防卫精良的致病菌）或者"Superbug, super genome"（超级细菌，超级基因）（Lindsay and Holden, 2004）。金黄色葡萄球菌作为一个"a well-armed pathogen"的病原菌主要得益于其复杂的结构和毒力因子，这些毒力因子（毒素和酶）能协助金黄色葡萄球菌侵入机体组织细胞、逃避机体免疫系统的清除、产生耐药性等。随着金黄色葡萄球菌许多毒力和抗性基因这些可动的遗传因子在细菌之间进行的横向转移，导致相关亲近菌株的快速分化（Lindsay and Holden, 2004），即超级细菌。

金黄色葡萄球菌毒力因子的种类（virulence determinant，毒力决定子）超过 30 种，毒力的强弱主要取决于三方面：产生的毒素、侵袭性酶和细胞结构成分（细胞壁/膜），这些毒力因子在金黄色葡萄球菌侵袭组织的各个阶段发挥重要作用（Cheung et al., 2002）。关于金黄色葡萄球菌各种毒力基因检测一直以来都是国内外研究的热点，临床意义较大的毒力因子有以下几种。

聚集因子 A（clumping factor A，clfA）或者**纤维蛋白原结合蛋白**（fibrinogen-binding protein），能够与血浆中的纤维蛋白原交联使菌体快速凝聚。

肠毒素（enterotoxin）是金黄色葡萄球菌分泌的一类具有超抗原活性的细菌毒素，性质稳定，能抵抗许多蛋白质酶的消化，如胃蛋白酶和胰蛋白酶，并且十分耐热，用

巴氏消毒不能破坏食品中的肠毒素，在131℃下加热30 min后才能被破坏。若食品被肠毒素污染，用一般蒸煮的方法不能将其破坏（李慧，2007）。因其引起人和哺乳动物胃肠道毒性反应而得名。由于肠毒素是一类超抗原蛋白，无需抗原递呈细胞的加工，就直接与抗原递呈细胞的MHCH及T细胞的TCR结合并激活T细胞，极少的量即可激活大量的T细胞，使淋巴细胞产生强烈的细胞毒性作用和高水平的细胞因子（徐君英，2009）。

中毒休克毒素-1（toxic shock syndrome toxin 1，TSST-1），由噬菌体 I 群金黄色葡萄球菌产生的一类蛋白质，其可增加毛细血管通透性，引起心血管系统等多个器官系统的功能紊乱，从而导致休克或发生毒性休克综合征。

耐热核酸酶（heat stable nuclease），是致病性金黄色葡萄球菌产生的一种细胞外酶，该酶在100℃下15 min或者97℃下60 min其活性仍然稳定，并且在5～37℃的各种食品中该酶的活性至少可以维持10个月以上。当机体发生感染时，组织细胞和白细胞崩解释放出核酸，使渗出液黏性增加，阻止细菌和炎症的扩散，而致病性金黄色葡萄球菌产生的耐热核酸酶能迅速分解核酸，使渗出液黏性降低，以利于病原菌的扩散，因此，将检测耐热核酸酶的产生情况作为鉴定致病性金黄色葡萄球菌的重要指标之一。

纤连蛋白结合蛋白（fibronectin-binding protein，FnBP）能特异性地黏附于纤连蛋白（Fn）上，从而使细菌黏附于宿主组织中，介导细菌对宿主细胞的黏附（Dinges et al.，2000；Renzoni et al.，2004）。细菌的黏附虽然对组织不造成直接的受损，但对致病菌感染组织的选择和感染的严重程度却有重要的影响。纤连蛋白结合蛋白与机体纤连蛋白的结合被认为是细菌感染的先决条件，细菌就不能在宿主组织中定居和繁殖。FnBP还可将机体的可溶性纤连蛋白大量结合在细菌的表面，使细菌菌体完全被宿主纤连蛋白和其他胶原蛋白包被，从而不被宿主的免疫系统所识别（贺宽军，2006）。目前分离到的大多数金黄色葡萄球菌都能够与细胞外基质上的纤连蛋白特异性地结合。

溶血素（hemolysin）是金黄色葡萄球菌重要的毒力因子之一，协助细菌侵入机体并逃避机体的免疫反应（Silva et al.，2005）。其中α溶血素是研究最多的金黄色葡萄球菌毒素之一，导致机体发生溶血、皮肤坏死和神经毒性，其作用机制是溶血素插入到红细胞细胞膜疏水区，形成微孔，造成细胞内渗透压失调，并破坏细胞膜的完整性，使细胞凋亡或发生融解。α溶血素具有良好的抗原性，经甲醛处理后可制成类毒素，接种绵羊和山羊后，对金黄色葡萄球菌所致的乳腺炎有部分的保护作用。较大剂量的α溶血素可引起大脑生物电的迅速停止而导致死亡，还会引起乳腺平滑肌收缩、麻痹，引发牛的坏疽性乳腺炎。β溶血素是一种神经磷脂酶，对牛和绵羊红细胞的活性极高。δ溶血素也诱导细胞膜微孔的形成，溶解马红血细胞的活性很高（Silva et al.，2005）。

荚膜多糖（capsular polysaccharide，CP）是金黄色葡萄球菌的菌体抗原，由于分子质量比较小，免疫原性差，属于半抗原，需偶合大分子蛋白才可以增强其免疫原性。CP5和CP8具有抗中性粒细胞吞噬的作用，并且有研究者认为抗荚膜多糖特异性抗体具有调理素的作用（Tollersrud et al.，2001）。对于金黄色葡萄球菌来说，其荚膜越厚，抗细胞吞噬作用越强,从而增强其致病性。因此 CP5 和 CP8 被认为与金黄色葡萄球菌的毒力和抗噬菌作用有关(Han et al.,2000)。从患乳腺炎的奶牛中分离出 94%~100%金黄色葡萄球菌是具有荚膜的。

随着青霉素的广泛应用,某些金黄色葡萄球菌产生了青霉素酶,该酶能够水解 β-内酰胺环,表现出对青霉素的耐药。此后又研究出一种能耐青霉素酶的新的半合成青霉素——甲氧西林(methicillin)。但是仅仅两年后,就发现了对甲氧西林产生抗药性的耐甲氧西林金黄色葡萄球菌(methicillin resistant *Staphylococcus aureus*,MRSA),MRSA 除了对甲氧西林具有耐药性以外,对其他所有与甲氧西林有着相同结构的 β-内酰胺类抗生素和头孢类抗生素均有耐药性。MRSA 的主要耐药机制与其 *mecA* 基因大量表达与青霉素结合的蛋白 PBP2a 有关,PBP2a 与 β-内酰胺类抗生素结合,可替代高亲和力的 PBP 行使功能。因此利用 PCR 技术检测 *mecA* 基因的存在就可以鉴定 MRSA,*mecA* 基因的 PCR 检测还是目前鉴定 MRSA 的"金标准"。MRSA 还能通过产生修饰酶,改变抗生素的作用靶位,降低膜的通透性从而产生大量对氨基苯甲酸等机制,对氨基糖苷类、四环素类、大环内酯类、磺胺类、氟喹诺酮类、利福平产生不同程度的耐药性,仅对万古霉素仍表现敏感。MRSA 的耐药机制的来源可分为两种:固有性耐药和获得性耐药,固有性耐药是由染色体基因介导的耐药性,该耐药性的产生与细菌产生的青霉素结合蛋白有关,获得性耐药是由质粒介导的耐药性(房强,2009)。

第二节 奶牛乳腺炎致病性金黄色葡萄球菌分子特征性分析

金黄色葡萄球菌引起的奶牛乳腺炎不易控制,且感染菌株分布广泛,本实验通过采用聚合酶链式反应(PCR)方法对规模化奶牛场奶牛乳腺炎金黄色葡萄球菌的多种致病因子和荚膜类型,以及 MRSA 进行检测,为进一步研究和分析金黄色葡萄球菌(*S. aureus*)致病性和有效防治提供参考。

一、材料与方法

(一)实验材料

1. 菌株来源 本实验室从奶牛乳腺炎乳样分离到的致病菌,经形态学、染色镜检及血浆凝固酶实验等生化检验已经鉴定出 39 株金黄色葡萄球菌。

2. 试剂和药品 细菌基因组 DNA 提取试剂盒,DNA Marker DL 2000,DNA Marker Ⅰ、50 bp DNA Ladder 和 100 bp DNA Ladder,2×*Taq* PCR MasterMix [0.1 U *Taq* Polymerase/μL,500 μmol/L dNTP,20 mmol/L Tris-HCl(pH 8.3),100 mmol/L KCl,3 mmol/L $MgCl_2$,其他稳定剂和增强剂],琼脂糖,ddH_2O,无水乙醇,溴化乙锭(EB),Tris,冰醋酸,EDTA,胰蛋白胨,酵母浸出物,NaCl,NaOH。

50×TAE:242 g Tris,57.1 mL 冰醋酸、100 mL 0.5 mol/L EDTA(pH 8.0),加三蒸水定容至 1000 mL,工作液为 1×。

LB 液体培养基:胰蛋白胨 10.0 g,酵母浸出物 5.0 g,NaCl 10.0 g,溶于 1000 mL 蒸馏水中,完全溶解后用 5 mol/L NaOH 调 pH 至 7.0~7.2,121℃高压灭菌 20 min,室温保存。

3. 实验设备 　　台式高速离心机（Centrifuge 5810R），PCR 扩增仪，电泳仪，电泳槽，紫外凝胶成像系统，微波炉，电子分析天平（XS205 Dual Range），超净工作台，恒温摇床，全自动高压灭菌器，冰箱，微量移液器。

（二）实验方法

1. 金黄色葡萄球菌模板提取及含量测定 　　将 39 株金黄色葡萄球菌株接种在 LB 液体培养基中，37℃摇菌过夜，按照细菌基因组 DNA 提取试剂盒说明书对 39 株金黄色葡萄球菌进行基因组 DNA 提取。用紫外分光光度计测量 A_{260}、A_{280} 吸光值，计算 A_{260}/A_{280} 值。

2. 引物设计与合成 　　引物序列参照 Salasia 等（2004）的报道，设计 23S rRNA、*clfA*、*sea*、*fnbA*、*fnbB*、*hla*、*hlb*、*cap5*、*cap8* 基因的引物；参照 Johnson 等的报道，设计 *tsst-1* 基因的引物；参照 Brakstad 等的报道，设计 *nuc* 基因的引物；参照 Murakami 等的报道，设计 *mecA* 基因的引物（表 3-1）。

表 3-1　奶牛乳腺炎致病性金黄色葡萄球菌扩增的毒力基因序列

Table 3-1　The primers of virulence determinants genes for amplification of the *Staphylococcal aureus* which caused cows mastitis

基因	上游引物序列（5'→3'）	下游引物序列（5'→3'）	产物大小/bp	退火温度/℃
23S rRNA	ACG GAG TTA CAA AGG ACG AC	AGC TCA GCC TTA ACG AGT AC	1250	57
clfA	GGC AAC GAA TCA AGC TAA TAC AC	TTG TAC TAC CTA TGC CAG TTG TC	719	57
sea	AAA GTC CCG ATC AAT TTA TGG CTA	GTA ATT AAC CGA AGG TTC TGT AGA	216	55
tsst-1	ATG GCA GCA TCA GCT TGA TA	TTT CCA ATA ACC ACC CGT TT	350	54
nuc	GCG ATT GAT GGT GAT ACG GTT	ACG CAA GCC TTG ACG AAC TAA AGC	279	56
fnbA	GCG GAG ATC AAA GAC AA	CCA TCT ATA GCT GTG TGG	1279	51
fnbB	GGA GAA GGA ATT AAG GCG	GCC GTC GCC TTG AGC GT	812	54
hla	GGT TTA GCC TGG CCT TC	CAT CAC GAA CTC GTT CG	534	55
hlb	GCC AAA GCC GAA TCT AAG	GCG ATA TAC ATC CCA TGG C	833	53
cap5	ATG ACG ATG AGG ATA GCG	CTC GGA TAA CAC CTG TTG C	880	53
cap8	ATG ACG ATG AGG ATA GCG	CAC CTA ACA TAA GGC AAG	1147	53
mecA	AAAATCGATGGTAAAGGTTGGC	AGTTCTGCAGTACCGGATTTGC	533	55

3. PCR 反应条件 　　20 μL PCR 反应体系：2×*Taq* PCR MasterMix 10 μL，上游和下游引物各 1 μL，模板 1 μL，ddH$_2$O 8 μL。

扩增条件：94℃预变性 3 min 后，按以下参数进行 30 个循环：94℃变性 40 s，退火（复性）30 s，72℃延伸 1 min，最后一个循环于 72℃延伸 5 min。退火温度依据每种基因的退火温度而定。

4. DNA 琼脂糖凝胶电泳 　　取 5 μL PCR 扩增产物在 2%琼脂糖凝胶中，DL-2000 Marker 为 DNA 标准（2000 bp，1000 bp，750 bp，500 bp，250 bp，100 bp），以 1×TAE 缓冲液，100 V（5 V/cm）电压电泳 40 min，溴化乙锭染色 10 min，紫外凝胶成像系统拍照保存，并记录结果。

二、实验结果

（一）病原菌的 DNA 模板提取结果

39 株金黄色葡萄球菌基因组提取的含量见表 3-2。

表 3-2 金黄色葡萄球菌基因组提取含量测定结果
Table 3-2 The levels of *S. aureus* genome extracted

菌株号	A_{260}/A_{280}	含量/(ng/μL)	菌株号	A_{260}/A_{280}	含量/(ng/μL)
1	2.16	136.1	21	1.87	283.5
2	2.02	149.5	22	2.19	81.4
3	1.9	183	23	2.23	108.5
4	2	167.9	24	1.88	145.4
5	2.04	87.1	25	2.06	238
6	2.26	129	26	2.22	121.6
7	2.2	134.5	27	1.9	163.7
8	1.91	266.2	28	2.14	195.6
9	2.16	159.1	29	2.19	99.9
10	1.87	278.8	30	1.95	193.9
11	2.07	216.5	31	2.11	85.5
12	1.99	273.5	32	2.24	88.3
13	2.13	200.6	33	1.86	245.1
14	2.12	182	34	2.31	82.9
15	2.08	231.1	35	2	137.3
16	1.98	269.7	36	1.92	266.7
17	2.12	162.1	37	1.83	148.8
18	2.13	152.5	38	2.29	91
19	2.1	162	39	2.28	108
20	2.05	162.6			

（二）金黄色葡萄球菌基因检测结果

本实验通过对 39 株生化鉴定为金黄色葡萄球菌 23S rRNA 序列的检测，全部在目的带 1250 bp 处出现亮带（图 3-1），检出率 100%，从分子水平也证实了这 39 株细菌确定为金黄色葡萄球菌。通过对这 39 株金黄色葡萄球菌毒力基因的 PCR 检测，全部检测出毒力基因，总检出率达 100%（39/39），见表 3-3。检出了 *clfA*、*tsst*-1、*nuc*、*fnbA*、*hla*、*hlb*、*cap5*、*cap8* 和 *mecA* 共 9 种基因（扩增结果见图 3-2、图 3-3、图 3-4、图 3-5、图 3-6、图 3-7、图 3-8、图 3-9 和图 3-10，扩增片段大小分别是 719 bp、350 bp、279 bp、1279 bp、534 bp、833 bp、880 bp、1147 bp 和 533 bp），但是所有菌株均未检测出 *sea*、*fnbB* 基因。由表 3-3 可见，实验菌株以 *fnbA*、*nuc*、*hla*、*hlb*、*clfA* 和 *cap5* 毒力基因检出率最高。

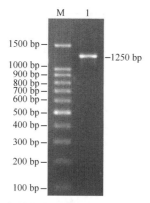

图 3-1　金黄色葡萄球菌 23S rRNA 扩增产物（1250 bp）

Figure 3-1　PCR typical amplicon of *S. aureus* 23S rRNA（1250 bp）

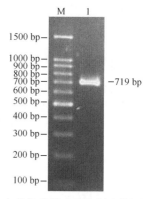

图 3-2　金黄色葡萄球菌编码聚集因子基因的 PCR 扩增产物（719 bp）

Figure 3-2　PCR typical amplicons of the genes encoding clumping factor A（*clfA*）of *S. aureus* with size of 719 bp

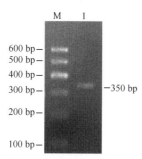

图 3-3　金黄色葡萄球菌编码中毒休克毒素-1 基因的 PCR 扩增产物（350 bp）

Figure 3-3　PCR typical amplicons of the genes encoding toxic shock syndrome toxin 1（*tsst-1*）of *S. aureus* with size of 350 bp

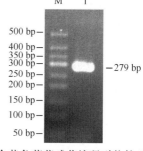

图 3-4　金黄色葡萄球菌编码耐热核酸酶基因的 PCR 扩增产物（279 bp）

Figure 3-4　PCR typical amplicons of the genes encoding heat stable nuclease（*nuc*）of *S. aureus* with size of 279 bp

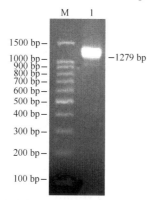

图 3-5　金黄色葡萄球菌编码纤连蛋白结合蛋白 A 基因的 PCR 扩增产物（1279 bp）

Figure 3-5　PCR typical amplicons of the genes encoding fibronectin-binding protein（*fnbA*）of *S. aureus* with size of 1279 bp

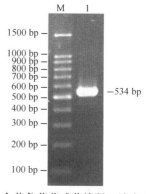

图 3-6　金黄色葡萄球菌编码 α 溶血素基因的 PCR 扩增产物（534 bp）

Figure 3-6　PCR typical amplicons of the genes encoding α-hemolysin（*hla*）of *S. aureus* with size of 534 bp

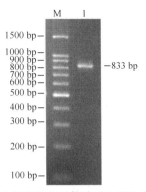

图 3-7 金黄色葡萄球菌编码 β 溶血素基因的 PCR 扩增产物（833 bp）

Figure 3-7 PCR typical amplicons of the genes encoding β-hemolysin（*hlb*）of *S. aureus* with size of 833 bp

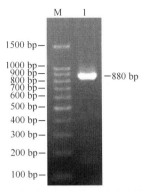

图 3-8 金黄色葡萄球菌编码荚膜多糖 5 基因的 PCR 扩增产物（880 bp）

Figure 3-8 PCR typical amplicons of the genes encoding capsular polysaccharide 5（*cap5*）of *S. aureus* with size of 880 bp

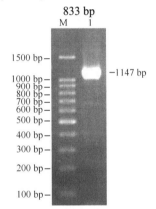

图 3-9 金黄色葡萄球菌编码荚膜多糖 8 基因的 PCR 扩增产物（1147 bp）

Figure 3-9 PCR typical amplicons of the genes encoding capsular polysaccharide 8（*cap8*）of *S. aureus* with size of 1147 bp

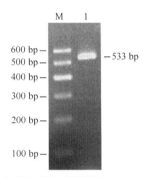

图 3-10 金黄色葡萄球菌编码 *mecA* 基因的 PCR 扩增产物（533 bp）

Figure 3-10 PCR typical amplicons of the genes encoding *mecA* of *S. aureus* with size of 533 bp

表 3-3 金黄色葡萄球菌基因检测结果

Table 3-3 The results of *S. aureus* genetic tested

菌株号	23S rRNA	clfA	sea	tsst-1	nuc	fnbA	fnbB	hla	hlb	cap5	cap8	mecA
1	+					+		+	+	+	+	
2	+					+		+	+	+	+	
3	+					+						
4	+	+			+	+		+	+	+		
5	+	+			+	+		+	+			
6	+				+	+		+				
7	+			+	+	+		+			+	
8	+	+			+	+		+	+			
9	+	+			+	+		+			+	
10	+					+	+					

续表

菌株号	23S rRNA	clfA	sea	tsst-1	nuc	fnbA	fnbB	hla	hlb	cap5	cap8	mecA
11	+				+	+		+	+	+		
12	+	+			+	+		+	+	+		
13	+				+	+						
14	+	+			+	+		+	+	+		
15	+					+						
16	+	+			+	+		+	+	+		
17	+	+			+	+		+	+		+	
18	+				+	+		+	+			
19	+	+			+	+		+	+			
20	+	+			+	+		+	+		+	
21	+				+	+		+	+	+		
22	+	+			+	+		+	+			
23	+	+			+	+		+	+	+		
24	+	+			+	+		+	+	+		
25	+				+	+		+	+			
26	+	+			+	+		+		+		
27	+				+							
28	+				+							
29	+				+	+						+
30	+	+			+	+		+	+		+	
31	+	+			+	+		+	+		+	
32	+	+			+	+		+	+		+	
33	+	+			+	+		+	+	+		
34	+	+			+	+		+	+	+		
35	+	+			+	+		+	+			
36	+	+			+	+		+	+			
37	+	+			+	+		+	+			
38	+	+			+	+		+	+			
39	+	+						+	+	+		
阳性数及占总数百分数	39 (100%)	24 (62%)	0 (0%)	1 (3%)	35 (90%)	38 (97%)	0 (0%)	33 (85%)	32 (82%)	23 (59%)	9 (23%)	1 (3%)

三、分析与讨论

（一）金黄色葡萄球菌 DNA 模板的提取

DNA 浓度测定原理：在一定范围内，DNA 的光密度 OD_{260} 与其含量成正比。当使用 1 cm 比色杯，$OD_{260}=1$ 时，寡核苷酸浓度约为 30 μg/mL。纯度测定原理：当 DNA 样品中含有蛋白质、酚或其他小分子污染物（如 RNA 等）时，会影响 DNA 吸光度的准确测定。一般情况下同时检测同一样品的 OD_{260} 和 OD_{280}（280 nm 是蛋白质的吸收峰），计算其比值来衡量样品的纯度。纯 DNA：$OD_{260}/OD_{280} \approx 1.8$（大于 1.9 表明有 RNA 污染或 DNA 降解；小于 1.6 表明有蛋白质或酚类污染）。

由表 3-2 可见，所测样品的 A_{260}/A_{280} 为 1.83～2.31，说明所提取的大部分 DNA 的纯度不够纯，受到 RNA 的污染，但是做普通的基因检测基本上不受纯度的影响。

经 Microsoft Office EXCEL 2003 统计分析，所测得的 A_{260}/A_{280} 与 DNA 含量之间的相关系数是 $r=-0.662$，查相关系数临界值表可知，A_{260}/A_{280} 与 DNA 含量极显著相关 [$r>0.371_{(n-2=37, α=0.01)}$]。说明提取的 DNA 直接受到 RNA 的影响，RNA 越多，A_{260}/A_{280} 值越大，DNA 含量越少。

（二）金黄色葡萄球菌 23S rRNA 保守序列的检测

在细菌的生物进化过程中，rRNA 基因的进化相对保守，被广泛认为是衡量生命进化历史最理想的标尺。RNA 操纵子是由 5S rDNA、16S rDNA 和 23S rDNA 三部分组成，这个操纵子作为 1 个单位进行转录。转录后处理成为成熟的 5S、16S 和 23S rRNA。尤其是核糖体 16S rRNA 的基因最为保守，在种属间的变异性常用于细菌的分类，通常用来鉴定所有细菌的亲缘关系。23S rRNA 作为细菌另一核糖体核酸基因，是细菌核糖体大亚基的组成部分，其片段较大（3 kb），其变异性高于 16S rRNA（洪帮兴等，2004）。研究表明，其在细菌鉴别诊断中的优越性高于 16S rRNA，更适合临床常见致病微生物的鉴别诊断（洪帮兴等，2004）。

本实验对 39 株细菌的 23S rRNA 序列进行检测，全部确定为金黄色葡萄球菌，与之前生化鉴定结果完全一致。国内外已有人利用 23S rRNA 对种的保守性，通过 PCR 技术鉴定金黄色葡萄球菌（Stephan et al.，2000；Gillespie and Oliver，2005；杨波等，2009）。

（三）金黄色葡萄球菌毒力基因检测

金黄色葡萄球菌是引起人和动物发生感染性疾病主要的病原菌之一，被称为"超级"细菌（Lindsay and Holden，2004），金黄色葡萄球菌这种"超强"的感染能力主要归因于其复杂而特别的结构和超过 30 种的毒力因子（Cheung et al.，2002），这些毒力因子能够协助金黄色葡萄球菌侵袭组织细胞，并且逃避机体免疫系统的清除，或者对抗生素产生耐药性等，因此在国外金黄色葡萄球菌被称为"a well-armed pathogen"。

金黄色葡萄球菌超抗原基因的检测常采用 PCR 方法，一些研究者为提高工作效率，就使用多重 PCR 方法同时检测多个毒力基因，但方法的可靠性有一定的争议，为了得到准确的结果，我们采用了单重 PCR 的方法。

金黄色葡萄球菌携带的大量的毒力因子，Dewanand 等报道 clfA 和 nuc 是奶牛隐性乳腺炎金黄色葡萄球菌主要的毒力基因，没有检测到 entA 基因（Kalorey et al.，2007）。本研究结果显示，金黄色葡萄球菌毒性基因检出率 100%，并且一株细菌含有多个毒力基因，fnbA、nuc、hla、hlb、clfA 和 cap5 毒力基因检出率最高，说明了这些毒力因子是引起规模化奶牛场奶牛乳腺炎的主要致病力因子。检出毒力为 100%，与杨波等（2009）报道的基本一致，但是高于巴西 Cardoso 等和德国 Zschöck 等（2000）分别报道的金黄色葡萄球菌毒力基因的检出率 65% 和 36.2%。CP5 和 CP8 基因的检出率分别是 59% 和 23%，总检出率是 77%，低于杨宏军（2007）报道的 94%～100%。不同国家和地区奶牛乳腺炎金黄色葡萄球菌检出毒素基因的报道不尽相同，说明奶牛乳腺炎金黄色葡萄球菌毒力基因的流行具有一定的地域性。

本次调查发现，奶牛乳腺炎金黄色葡萄球菌全部检测出毒力基因，总检出率达100%，携带一种以上毒力基因的菌株比较普遍，提示菌株间毒力基因连锁流行的情况也比较普遍。当前奶牛隐性乳腺炎发生率在50%左右，患有隐性乳腺炎奶牛的乳汁均未废弃，仍进入加工和流通环节，其中的毒素对消费者的健康有较大的损害，因此开展乳源金黄色葡萄球菌毒素基因的流行病学研究，确保食品安全十分重要。

四、小结

（1）金黄色葡萄球菌的生化鉴定和PCR鉴定完全一致。

（2）39株金黄色葡萄球菌毒力基因的检出率达100%。

（3）金黄色葡萄球菌 $fnbA$、nuc、hla、hlb、$clfA$ 和 $cap5$ 毒力基因检出率最高，未检测出 sea、$fnbB$ 基因。

第四章　奶牛隐性乳腺炎诊断技术

前面讲到隐性乳腺炎奶牛的乳房和乳汁外观均健康正常,不能用触诊检查或肉眼直接观察,但是乳汁的理化性质已发生变化,因此需要通过对乳汁进行实验室检测才能作出诊断。乳腺组织感染后,由于炎症的发生,乳腺血管扩张,渗透性增大,血液中的白细胞、缓冲物质(如碳酸氢钠)、蛋白质等成分渗入腺泡,使乳汁成分发生改变,如乳汁中的体细胞数、乳汁的 pH、导电率和氯化物的含量均升高,所以需要通过乳汁的生理生化指标的变化,成为间接快速诊断隐性乳腺炎的标志。必要时可进行乳汁的细菌学检查,为药物治疗隐性乳腺炎提供依据。

第一节　奶牛隐性乳腺炎常用的诊断技术

一、体细胞计数法

体细胞数(somatic cell counts,SCC)是指每毫升乳汁中含有的体细胞数量。体细胞包括多种类型的细胞,如中性粒细胞、巨噬细胞、淋巴细胞、嗜酸性粒细胞及各种乳腺上皮细胞。关于每种细胞的所占比例有不同的报道。在健康奶牛乳腺中,大部分活性体细胞是巨噬细胞和淋巴细胞。乳汁中一定数量的白细胞有抵御外界微生物感染的作用。当乳腺发生感染后,血液中大量的白细胞(如巨噬细胞、中性多核白细胞等)游走到乳腺组织,产生溶菌酶、乳铁蛋白和补体成分,吞噬和消化病原微生物,同时,引发抗体的形成。乳腺腺泡上皮组织也会出现病变,上皮细胞脱落增多,从而使乳汁中的体细胞数增多。然而乳汁中的体细胞数受感染状态、年龄和泌乳阶段、应激或季节、挤奶刺激等因素的影响。SCC 是目前来说鉴别奶牛乳区、牛只和牛群乳腺炎的最有效方法,可以确定乳汁中体细胞数的范围,精确地判断隐性乳腺炎的炎症程度(Pyörlä,2003;杨德英等,2008),因此乳汁体细胞计数对奶牛隐性乳腺炎的诊断具有重要意义。普遍认为,正常乳汁的体细胞计数低于 100 000 个/mL,但是国内一般认为体细胞数低于 250 000 个/mL 的乳区都是健康的(赵兴绪,2006)。常用的有乳汁中体细胞直接计数法和间接计数法。

乳汁体细胞直接计数是监测奶牛隐性乳腺炎的一种传统、可靠的方法,可通过确定乳汁中体细胞数的大致范围,来判断隐性乳腺炎的炎症程度。直接显微镜计数法检测成本低,但因操作人员经验水平和所处环境条件的不同,导致数据结果误差较大,且耗时长,不易掌握,不适合大批量快速检测。现在已经可以采用专门的荧光显微自动测定仪来检测乳汁中的体细胞数,如丹麦福斯公司生产的 Fossomatie 5000 系列奶牛体细胞测定仪,全自动计算机操作,准确、快速,每小时可检测样品 200 个以上,很快在世界范围内推广应用(伊岚等,2006)。目前较先进的乳汁体细胞计数法还有采用荧光流式细胞技术对牛奶中的体细胞进行荧光染色处理的自动计数法。优点是快速、准确、误差小、重复性好,电脑自动计数并打印结果,现已有商业的仪器出售,该方法的缺点是需要专门的仪器设备,对于一些规模较小的养殖场(户)会增加成本(董全等,2010)。

现在世界各国，包括我国在内，普遍采用加利福尼亚州乳腺炎试验（California mastitis test，CMT）方法间接测定乳汁中体细胞数，该方法操作简易，效果准确，可在牛旁进行。CMT 法是 1957 年 Scalm 和 Noorlander 首创的化学检验法，也称加利福尼亚隐性乳腺炎检测法。其原理是，CMT 中的表面活性物质与乳汁中的细胞作用，使细胞破裂，释放出核酸，一定量的核酸与碱性物质反应可以产生一定量的沉淀或者凝块，根据核酸与碱性物质反应后产生的沉淀和凝块的多少，间接判定核酸的量，进而判断乳汁中体细胞的数量，乳汁中的体细胞数越多，产生的沉淀和凝块也越多，因而采用这种方法可以诊断奶牛隐性乳腺炎。判定标准是按 CMT 试剂与乳汁的反应产生凝胶的程度，分为 N、T、1、2、3 级共 5 个级别，分别所对应的体细胞数为 N 级为小于 20 万/mL，T 级为 20 万～40 万/mL，1 级为 40 万～120 万/mL，2 级为 120 万～500 万/mL，3 级为 500 万/mL 以上（刘峰和迟玉杰，2005）。CMT 可用于隐性乳腺炎的检查，对于急性临床乳腺炎没有诊断意义。另外，初乳和泌乳末期的乳汁 CMT 值偏高，不适合用于产后初期和泌乳末期隐性乳腺炎的诊断。

间接测定体细胞数的方法除了 CMT 外，还有根据其原理研制的兰州乳腺炎试验（LMT）、杭州乳腺炎试验（HMT）和北京乳腺炎试验（BMT）等。目前广西区各大型奶牛场使用上海光明乳业生产的上海乳腺炎试验（SMT）。应用 CMT 挑选隐性乳腺炎奶牛进行干奶期治疗，其检测准确率高达 75%～80%。CMT 的优点是对奶牛隐性乳腺炎检测效果好、经济实惠和实时结果。

二、细菌学检查

乳腺炎主要是由病原微生物引起，常见的病原菌包括金黄色葡萄球菌、大肠杆菌、无乳链球菌、乳房链球菌、停乳链球菌和酵母菌等。如果在乳汁中能检出病原微生物，就可大致诊断出隐性乳腺炎，但是对于非特异性隐性乳腺炎，不仅乳房和乳汁肉眼检查正常，而且乳汁中也检不出病原菌。另外一种情况是有些病原菌虽经常存在于乳房内，但不一定引起发病。因此有报道称根据乳汁中有无病原菌判断奶牛是否患有隐性乳腺炎并不完全可靠，但大多数学者认为可以根据检出的病原菌的种类和数量大致作出判断（Babaei et al.，2007；杨德英等，2008）。因此对于奶牛隐性乳腺炎病例，细菌学检查乳汁的同时还应结合乳汁体细胞数计数，才能最终作出确诊。

三、其他检查方法

（一）乳汁酶类活性检测

乳汁中酶类活性检测是目前奶牛隐性乳腺炎诊断方法研究的热点之一，动物机体细胞结构损伤后会引起血液和乳汁中酶类活性的改变，乳腺炎的发生会增强乳汁中一些酶的生物化学活性发生改变，如乳酸脱氢酶（LDH）、碱性磷酸酶（ALP）、黄嘌呤氧化酶、过氧化氢酶、N-乙酰基-β-D-氨基葡萄糖苷酶、β-葡糖苷酸酶和乳酸过氧化物酶（LP）等，其活性随着 SCC 的升高而升高。其中乳汁中一些酶活性的变化特点已用于诊断奶牛隐性乳腺炎及评价奶牛乳房的健康程度，如过氧化氢酶、NAGase 和抗胰蛋白酶的活性变化（杨德英等，2008），以及 LDH 和 ALP 的活性变化（Babaei et al.，2007）。

乳汁中的髓过氧化物酶（MPO）主要参与依赖氧的杀菌活动（Zhang et al.，2002），其在哺乳动物体内抵抗微生物的作用已得到证实（郭昌明等，2006）。奶牛的 MPO 是中性粒细胞、单核细胞、巨噬细胞释放的嗜天青颗粒的主要成分，并在炎症过程中扮演吞噬和溶菌的重要角色（郭昌明等，2006）。实验已经证明了乳汁中的 SCC 可以作为奶牛隐性乳腺炎的诊断依据，同样乳汁中的 MPO 的高低也可作为奶牛隐性乳腺炎的诊断依据，且具有很规律的正相关性，乳汁中的 SCC 和 MPO 的相关系数为 0.91。乳汁中 MPO 的 ELISA 结果显示具有很高的敏感性和特异性，敏感性的变异总和为 10%，由此判断奶中的 MPO 稳定性很好；85%的奶样中随着 MPO 的增加，分离的微生物的量也提高；在没有体细胞或很少体细胞的乳样中用 ELISA 没有检测出 MPO，有 90%样品没有分离出微生物，结果清楚地说明，分析乳汁中 MPO 的量，可以确定奶牛是否患隐性乳腺炎（Raulo et al.，2002；郭昌明等，2006）。

有报道称，奶牛隐性乳腺炎乳汁中 LDH 和 ALP 的活性发生改变，并用于奶牛乳腺感染的诊断（Babaei et al.，2007）。LDH 活性升高的原因是患乳腺炎奶牛的乳汁中出现了白细胞，另一个原因是炎症过程中乳腺上皮细胞和间皮细胞遭到破坏，将细胞中的 LDH 释放出来。黄权利等也认为 LDH 在健康乳汁和隐性乳腺炎乳汁中活性不同，可作为诊断奶牛乳腺损伤和性乳腺炎发生的一个重要指标。乳腺发生感染时增加了微血管的通透性，血-乳屏障遭到感染的破坏，ALP 可由血液进入乳汁中，使得乳汁中 ALP 活性显著升高。乳汁中 LDH 的活性检测已经被认为是由疾病引起的乳腺功能发生改变的敏感指标，ALP 活性的检测被认为是早期诊断奶牛隐性乳腺炎的可靠指标（Babaei et al.，2007）。

虽然先前的研究评价了乳汁中一些酶类活性的变化，并可能作为奶牛隐性乳腺炎的诊断指标，如 ALP、LDH 和天冬氨酸氨基转移酶（AST）的活性变化诊断奶牛乳房的感染情况（Babaei et al.，2007），但是至今几乎没有人研究乳汁中丙二醛（MDA）浓度和谷胱甘肽过氧化物酶（GPx）、超氧化物歧化酶（SOD）活性的改变与奶牛隐性乳腺炎的关系。乳腺内感染后感染细胞通过分泌许多种化学物质增加微循环的通透性，如组织胺、前列腺素、细胞分裂素和氧自由基等。MDA 是脂质过氧化反应的最终产物，也是氧化应激最可靠和最广泛使用的指标之一。体细胞数较高的乳汁中有更多的中性粒细胞浸润，会引起氧化反应增加（Su et al.，2002），测定牛奶中 MDA 的含量通常用来评价乳汁在不同情况下的过氧化状态（Cesa，2004），以及乳汁 SCC 与 MDA 浓度呈正相关（Suriyasathaporn et al.，2006）。奶牛乳汁 SCC 的升高将会引起 MDA 升高，使乳汁极大程度上降低了乳汁的质量（Suriyasathaporn et al.，2006）。

奶牛乳汁中 GPx 的功能还没有完全清楚，但是 GPx 已经被国内外学者确定是含硒酶，可以结合机体硒总量的 30%，硒是人类饮食中的重要微量元素，并催化许多种过氧化物的还原反应，保护机体的细胞免受氧化损伤，充足的 GPx 能够保护乳汁的脂类不被氧化。乳汁中的 GPx 水平随物种（人＞羊＞牛）和饮食的不同而变化（Fox and Kelly，2006）。乳腺炎奶牛与健康奶牛相比，血清脂质过氧化水平升高，血液中谷胱甘肽过氧化酶的水平下降。已经有研究测定了乳汁中 GPx 的活性，为 12~32 IU/mL，并且与硒的浓度呈显著正相关（Przybylska et al.，2007）。与健康奶牛相比，发生乳腺炎的奶牛乳清的脂质过氧化反应增加，血液中 GPx 活性降低。Atroshi 等研究发现，乳腺炎奶牛 GPx 的降低与脂质过氧化反应和前列腺素的生成有关。人乳汁中 GPx 的活性随着泌乳时间的增加而降低。

首次报道乳汁中 SOD 的是 Holbrook 和 Hicks，并认为 SOD 在乳汁的氧化稳定中起着重要的作用，并有报道称乳汁中 SOD 的活性在奶牛个体间和品种间变化较大（Lindmark-Mansson and Akesson，2000）。牛奶中 SOD 的活性不受奶牛的泌乳阶段和（或）年龄的影响（Przybylska et al.，2007）。然而至今仍没有人研究关于乳汁中 MDA 浓度与牛奶隐性乳腺炎的关系。

（二）乳汁导电率

采用导电法是国内外学者诊断奶牛隐性乳腺炎较好的方法之一，它是根据隐性乳腺炎乳汁成分的变化，特别是无机盐离子的变化，使乳汁的导电性能发生改变的原理来检查的。乳腺发生感染后，血-乳屏障的渗透性改变，致使奶牛分泌的乳汁与正常乳有很大差异，如钠离子、氯离子、HCO^- 离子的浓度增加，引起电导率增加，而钾离子和乳糖的浓度下降，影响乳汁中无机盐的平衡，导致一些离子活性的改变及离子强度的改变，直接影响酪蛋白胶体的稳定性，并且可作为牛奶分泌过程中的特殊影响作用而一直存在（郝建国，2002；郭庆等，2009）。因此，在临床上可以通过检测乳汁的电导率来诊断奶牛的隐性乳腺炎。

从 20 世纪 40 年代开始，国内外学者研究利用隐性乳腺炎乳汁的导电性来检测隐性乳腺炎。新西兰研究者发明的 AHI 是世界上最早的导电诊断仪，但是它仅能显示隐性乳腺炎阴性、阳性和可疑，不能显示炎症轻重程度。国内一些单位在研制乳汁电导率值测试仪器方面做了很多工作，如西南民族学院、西北农业大学、山西农业大学等。

（三）PCR 检测法

通过传统的病原学对病原菌的分离鉴定来检测奶牛隐性乳腺炎费时费力，难以广泛应用。随着分子生物学技术的发展，许多细菌全基因组 DNA 测序工作已经完成，PCR 技术在检测病原菌方面得到广泛应用，并且展现出了良好的应用前景（肖颖等，2010）。基于细菌 rRNA 区域序列的 PCR 技术是一种新的在分子水平上进行细菌分类与鉴定的方法，该方法已在医学和兽医学临床诊断中得到了广泛应用（曹随忠等，2005）。PCR 技术的优点是细菌不经过培养，直接应用样品就可以进行细菌学鉴定，是一种经济有效、快速、准确、特异性强和具有很重要应用价值的检测手段（马保臣等，2006）。多重聚合酶链式反应检测方法可以直接从乳汁中对常见致病菌进行同步鉴定，如金黄色葡萄球菌、无乳链球菌、停乳链球菌和乳房链球菌等（Gillespie and Oliver，2005；曹随忠等，2005）。张善瑞等（2008）通过对金黄色葡萄球菌、链球菌和大肠杆菌 3 种乳腺炎致病菌设计了 3 对特异性引物，建立了多重 PCR 检测方法。研究结果表明，此种方法的特异性为 100%，检测的最低浓度为 1.25×10^3 cfu/mL，表明了该方法具有快速、准确和特异等优点。澳大利亚的 Phuektes 等（2001）研制的多重 PCR 技术，可以同时检测乳腺炎奶样中的 4 种病原菌，包括金黄色葡萄球菌、乳房链球菌、无乳链球菌和停乳链球菌。加拿大的 Riffon 等（2001）研制出的 PCR 技术，可以同时检测引起乳腺炎发生的 6 种病原菌，包括金黄色葡萄球菌、大肠杆菌、无乳链球菌、停乳链球菌、乳房链球菌和副乳房链球菌。

(四) ELISA 检测法

ELISA 检测法是常用的检测与鉴定特定病原和疾病的方法,该方法在检测抗原方面特异性好,灵敏度很高,应用广泛。奶牛乳腺炎发生时,乳汁中可出现大量的炎症物质、病原生物或抗体,而这是正常乳汁中不存在的,ELISA 检测法可以方便地检测这些物质,为诊断提供可靠依据(伞治豪,2014)。目前国内外已广泛地应用 ELISA 检测乳腺炎和其致病菌。在检测金黄色葡萄球菌、链球菌、支原体、炎症反应抗原等方面,ELISA 检测结果精确可靠。

(五) 电子计数仪检测法

在发达国家,一些牛场(如美国的威斯康星州牛场)利用电子计数仪对奶牛乳汁中细胞数按月进行调查,然后利用计算机技术进行体细胞分析。这种检测法能够迅速、准确地测出牛奶中所含体细胞数量,且结果比常用的隐性乳腺炎检测法更为准确、客观。缺点是用这种方法需要较高的费用购买检测设备,且需要经常对检测结果进行校正,还有测定的乳样比较集中,相对来说在生产中不如 CMT 法实用(鲍士兵,2014)。

(六) 4%氢氧化钠凝乳试验法

4%氢氧化钠凝乳试验法同样是一种间接判断奶牛乳腺炎的方法,其原理为在氢氧化钠的作用下,牛奶中的白细胞会发生变性凝结反应。具体检测时先滴 5 滴被检乳汁在深色背景的载玻片上,再加入 2 滴 4%氢氧化钠溶液,均匀混合后作结果判断(鲍士兵,2014)。标准见表 4-1。

表 4-1 4%氢氧化钠凝乳试验法检测奶牛隐性乳腺炎判定的标准
Table 4-1 Detection of subclinical mastitis of 4% caustic soda curd experimental method to detemine the standard

检测结果	被检乳样的特征	总细胞数量/(个/mL)
阴性(-)	混合液呈浅灰色,有不透明沉淀析出	小于 50 万
可疑(±)	有细小的沉淀物出现	50 万~100 万
弱阳性(+)	混合后的反应物稍显透明,且出现小的凝块	100 万~200 万
阳性(++)	混合后的反应物为透明水样状,且出现较大凝块	200 万~500 万
强阳性(+++)	混合后的反应物为全透明状,凝结成一块	大于 500 万

(七) 被毛微量元素检测法

田风林等在研究微量元素含量与乳腺炎的相关性时发现,患有乳腺炎的奶牛乳汁中 Cu、Fe、Co 和 Zn 含量显著性低于正常乳汁($P<0.05$ 或 $P<0.01$),乳腺炎与微量元素含量具有相关性。还有人研究发现,隐性乳腺炎奶牛被毛中的锌、钴、锰的含量与正常牛只有一定相关性,隐性乳腺炎病牛被毛中锌、钙显著低于正常牛被毛中含量。同时,由于季

节、营养、个体差异等的影响，该诊断方法不太准确，有待进一步研究（丁丹丹，2014）。

（八）蛋白质组学检测法

蛋白质组学（proteomics）是基因组学的重要组成部分，在功能基因组学研究中占据重要地位。近年来对白细胞蛋白质组学、病原微生物蛋白质组学、差异蛋白质组学等的研究有了突破性进展，使得蛋白质组学成为诊断乳腺炎的一个新的研究热点（丁丹丹，2014）。

第二节 奶牛隐性乳腺炎乳汁中丙二醛浓度和酶类活性分析

乳腺炎是乳腺组织的一种炎症反应，会造成产奶量降低、治疗后抗奶引起的乳汁废弃、增加治疗费用、淘汰和死亡，致使奶牛场遭受巨大的经济损失，由于隐性乳腺炎没有任何临床症状，因此在疾病的早期能够有效地检测隐性乳腺炎的发生，显得极其重要。隐性乳腺炎阳性奶牛要达到两个条件：细菌检测阳性和炎症反应。通常情况下体细胞数（SCC）用来表示乳腺的炎症反应程度（Babaei et al.，2007）。定量评价乳汁中体细胞数的方法是直接显微镜计数或者使用上海乳腺炎检测（SMT）进行间接评价（Babaei et al.，2007）。在乳汁检测实验室，计算 SCC 最常用的方法是荧光电子显微镜计数法（Sierra et al.，2006），但是荧光电子显微镜计数法费用昂贵、所需仪器先进，并不是在每个实验室都可以使用。SMT 已经用于现场间接诊断泌乳奶牛隐性乳腺炎，敏感性和特异性已经有报道（Pyörlä，2003）。

丙二醛（MDA）是脂质过氧化反应的最终产物，也是氧化应激最可靠和最广泛使用的指标之一。体细胞数较高的乳汁中有更多的中性粒细胞浸润，会引起氧化反应增加（Su et al.，2002），测定牛奶中 MDA 的含量通常用来评价乳汁在不同情况下的过氧化状态（Cesa，2004），以及乳汁 SCC 与 MDA 浓度呈正相关（Suriyasathaporn et al.，2006）。牛奶中的谷胱甘肽过氧化物酶（GPx）的功能还没有完全清楚，但是可以结合30%硒的总量，硒是人类饮食中的重要微量元素。乳汁中的 GPx 水平随物种（人＞羊＞牛）和饮食的不同而变化（Fox and Kelly，2006）。乳腺炎奶牛与健康奶牛相比，血清脂质过氧化水平升高，血液中谷胱甘肽过氧化酶的水平下降。多位研究人员在牛奶中发现了超氧化物歧化酶（SOD），认为 SOD 在乳汁的氧化稳定中起着重要的作用。然而至今仍没有人研究乳汁中 MDA 浓度、GPx 和 SOD 活性与牛奶隐性乳腺炎的关系。乳汁中乳酸脱氢酶（LDH）和碱性磷酸酶活性（ALP）已经被用来作为奶牛隐性乳腺炎的指标（Babaei et al.，2007）。

本项实验的研究目的：①调查亚热带奶牛场奶牛隐性乳腺炎的发病率和致病菌情况；②研究牛奶中的 MDA 浓度，以及 GPx、SOD、LDH、ALP 和天冬氨酸氨基转移酶（AST）的活性是否可以作为奶牛隐性乳腺炎的诊断指标。

一、材料与方法

（一）实验材料

1. 实验动物 在亚热带规模化奶牛场从 124 头经产泌乳荷斯坦奶牛采集了 124 份

乳汁样品，随机从4个乳区采集乳汁样品，所有的乳汁样品在奶牛产后第2~10周采集，所有实验奶牛的年龄4~5岁，临床表现健康，没有临床乳腺炎或其他疾病的症状。

2. 药品和试剂 营养肉汤，营养琼脂，5%鲜血琼脂培养基、三糖铁培养基、改良爱德华琼脂培养基、却浦曼琼脂培养基、麦康凯琼脂培养基，革兰氏染色试剂盒，上海隐性乳腺炎诊断试剂和诊断检验盘，各种微量生化发酵管，MDA、GPx、总SOD、LDH、ALP和AST检测试剂盒。

3. 实验仪器 超净工作台，恒温培养箱，灭菌器，干燥箱，分析天平，高速冷冻离心机，冰箱，紫外可见分光光度计。

（二）实验方法

1. 乳汁的采集 早晨挤奶前将乳区用温水清洗后擦干，乳头用碘伏擦拭消毒，弃去前三把乳汁，采集10 mL装在无菌试管中，然后放置在冰盒中送回实验室进行处理。

2. 上海奶牛隐性乳腺炎诊断（SMT） 同第二章第一节中的相关方法。

3. 乳汁的致病菌的分离 同第二章第一节中的相关方法。

4. 乳汁的致病菌的鉴定 同第二章第一节中的相关方法。

5. 隐性乳腺炎的定义 临床上奶牛乳房和乳汁都正常，细菌学阴性和SMT检测阴性的乳汁是健康乳汁，而细菌学阳性和SMT检测阳性的乳汁是隐性乳腺炎乳汁（Batavani et al.，2003）。

6. 乳汁的预处理 健康和隐性乳腺炎乳汁用高速冷冻离心机在4℃下12 000 r/min离心20 min进行脱脂，吸取上清（乳清），用于MDA浓度和酶活性的检测。

7. 乳汁中MDA浓度和酶的活性的测定 乳清中MDA浓度以及GPx、总SOD、LDH、ALP和AST的活性的测定按照试剂盒的说明书进行操作，采用分光光度计测定分析技术（Yang et al.，2011）。

（三）统计分析

SMT检测结果的敏感性和特异性用标准的2×2列联表计算，采用95%置信区间。乳清MDA浓度和酶的活性数据用平均数±标准误（SEM）表示，经Office Excel 2003和SAS 9.1数据统计软件对试验结果进行统计分析，t-检验评价隐性乳腺炎乳清和健康乳清的差异显著性，以$P<0.01$为差异显著。

二、实验结果

（一）奶牛隐性乳腺炎的发病情况

在泌乳奶牛产后第2~10周共采集了124个乳汁样品，SMT检测阳性65个（52.4%），细菌检测阳性45个（36.3%），根据上述隐性乳腺炎的定义，此次调查中奶牛隐性乳腺炎的发病率为26.6%（$n=33$）。SMT检测奶牛隐性乳腺炎的特异性、敏感性、阳性预测值（PPV）和阴性预测值（NPV）分别是59.49%、73.33%、50.8%和79.7%（表4-2）。Kappa值（$\kappa=0.3$）表明SMT和乳汁细菌培养之间较差的相同性。

表 4-2 2×2 列联表表示 124 个乳汁样品 SMT 和乳汁细菌培养之间相同性

Table 4-2 Two-way contingency table to investigate agreement between bacteriological and the SMT results for 124 milk samples

	培养+	培养−	总数
SMT+	33	32	65
SMT−	12	47	59
总数	45	79	124

注:"+"表示阳性,"−"表示阴性

(二)隐性乳腺炎乳汁致病菌分离情况

从隐性乳腺炎乳汁分离的致病菌是:金黄色葡萄球菌(47%),凝固酶阴性葡萄球菌(CNS)(27%),大肠杆菌(9%),无乳链球菌(9%),乳房链球菌(4%)和新型隐球菌(4%)(表4-3)。

表 4-3 隐性乳腺炎乳汁分离的致病菌情况

Table 4-3 Frequency distribution of microorganisms isolates from subclinical mastitis samples

致病菌	分离数及占总数百分数/%
金黄色葡萄球菌	21(47)
凝固酶阴性葡萄球菌	12(27)
大肠杆菌	4(9)
无乳链球菌	4(9)
乳房链球菌	2(4)
新型隐球菌	2(4)
总 数	45(100)

(三)乳汁中 MDA 浓度和酶类活性测定结果

健康乳汁和隐性乳腺炎乳汁中 MDA 的平均浓度、GPx、SOD、LDH、ALP 和 AST 的平均活性见表 4-4。隐性乳腺炎乳汁中 MDA 的平均浓度、LDH 和 ALP 的平均活性均极显著高于健康乳汁($P<0.01$);而隐性乳腺炎乳汁中 GPx 的平均活性极显著低于健康乳汁($P<0.01$);隐性乳腺炎乳汁与健康乳汁中 SOD 和 AST 的平均活性没有显著差异($P>0.05$)。

表 4-4 健康乳汁和隐性乳腺炎乳汁中 MDA 平均浓度、GPx、SOD、LDH、ALP 和 AST 的平均活性

Table 4-4 Mean level of MDA and activities of GPx, SOD, LDH, ALP and AST in normal and subclinical mastitis milk samples of lactation cows

指标	健康乳汁(47)	隐性乳腺炎乳汁(33)
MDA/(nmol/mL)	24.37 ± 0.9^A	28.45 ± 0.96^B
GPx/(IU/mL)	32.81 ± 1.41^A	26.41 ± 2.03^B
SOD/(IU/mL)	1.82 ± 0.11	1.6 ± 0.12
LDH/(IU/L)	177.94 ± 12.55^A	724.49 ± 34.91^B
ALP/(IU/L)	71.85 ± 3.71^A	116.58 ± 6.18^B
AST/(IU/L)	151.99 ± 11.56	140.98 ± 11.76

注:同行内不同上标大写字母表示差异极显著($P<0.01$)

三、分析与讨论

(一) 奶牛隐性乳腺炎的发病率

根据本实验的调查结果,奶牛隐性乳腺炎发病率(26.6%)与 Getahun 等(2008)报道的(22.3%)一致,但是高于 Mungube 等(2004)和 Gianneechini 等(2002)报道的奶牛隐性乳腺炎发病率(分别是 46.6%和 52.4%),低于第二章中奶牛隐性乳腺炎的发病率(48.8%)。这些报道中奶牛隐性乳腺炎发病率的不同受到许多因素的影响,特别是调查的国家地区、年份季节、奶牛的饲养管理、隐性乳腺炎的定义和诊断标准。此外,第二章检测奶牛隐性乳腺炎的方法是单一的 SMT 检测,未进行细菌学检查,导致这两次实验奶牛隐性乳腺炎发病率的差异。

SMT 在奶牛业已经标准化用于隐性乳腺炎的诊断,该方法仅与核酸 DNA 发生反应形成凝聚物(Batavani et al.,2003),乳汁的细菌培养是确定奶牛乳腺内感染的诊断方法,SMT 检测阳性率(52.4%)高于乳汁的细菌培养阳性率(36.3%)。已经有学者报道 SMT 的敏感性和特异性,Sargeant 等(2001)报道乳腺内感染的敏感性和特异性分别是 57%和 56%。另外,Vijaya Reddy 等报道敏感性和特异性分别是 71%和 75%,本次实验中,SMT 在检测奶牛隐性乳腺炎的敏感性和特异性分别是 73.33%和 59.49%,Kappa 值($\kappa=0.3$)表明,SMT 和乳汁细菌培养之间的相同性较差,因此在亚热带奶牛场诊断奶牛隐性乳腺炎,SMT 并不是一个非常好的诊断方法,还要依靠实验室细菌培养。特异性(59.49%)较低和隐性乳腺炎发病率(26.6%)较低使得阳性预测值(PPV)较低;当乳汁中体细胞数较多但细菌培养阴性时,使 SMT 检测的假阳性的出现和特异性(59.49%)较低;当乳汁细菌培养阳性但体细胞数较少时,使 SMT 检测的假阴性的出现和敏感性(73.33%)较低。

(二) 隐性乳腺炎乳汁致病菌分离

本次实验从奶牛隐性乳腺炎乳汁分离的最多的致病菌是金黄色葡萄球菌(47%)和凝固酶阴性葡萄球菌(27%),这些致病菌是引起非临床乳腺内感染的主要因素(Getahun et al.,2008)。由于乳腺内和乳房乳头的皮肤广泛存在金黄色葡萄球菌,导致其检出率较高(Workineh et al.,2002)。传统上认为凝固酶阴性葡萄球菌是次要的乳腺炎致病菌,它单独引起奶牛隐性乳腺炎和(或)临床乳腺炎的病例不是很多,也有报道称凝固酶阴性葡萄球菌普遍存在于牛舍的空气和灰尘中,很容易在奶牛个体之间发生传染,动物机体在环境中与凝固酶隐性葡萄球菌接触时能够定居在皮肤,并通过舔舐或挤奶过程感染乳腺,若是致病菌的话将会引起隐性乳腺炎的发生(Batavani et al.,2003)。与细菌培养阴性的乳汁和主要致病菌引起的感染相比,凝固酶阴性葡萄球菌引起的奶牛乳腺内感染对乳汁体细胞的影响适中(Schukken et al.,2009)。因此 SMT 检测由凝固酶阴性葡萄球菌引起的乳腺内感染率较高的奶牛时其敏感性较低。其他致病菌依次是大肠杆菌、无乳链球菌、乳房链球菌和新型隐球菌,都是一些常见的致病菌。

(三) 乳汁中 MDA 浓度和酶类活性测定结果

之前已经有一些研究评价了乳汁中 ALP、LDH 和 AST 的活性变化诊断奶牛乳房的感染情况（Babaei et al.，2007），但是几乎没有研究关于乳汁中 MDA 浓度和 GPx、SOD 活性的改变与奶牛隐性乳腺炎的关系。乳腺内感染后感染细胞通过分泌许多种化学物质增加微循环的通透性，如组织胺、前列腺素、细胞分裂素和氧自由基等。

MDA 是脂质过氧化反应的最终产物，并且常用于脂质过氧化反应的指标。本实验中隐性乳腺炎乳汁中的 MDA 平均浓度 [（28.45±0.96）nmol/mL] 极显著高于健康乳汁 [（24.37±0.9）nmol/mL]（$P<0.01$），因此 MDA 有可能作为诊断奶牛隐性乳腺炎的敏感指标。隐性乳腺炎乳汁中 MDA 的浓度高于健康乳汁说明了隐性乳腺炎乳汁的自动氧化活性高于健康乳汁。关于乳汁中 MDA 的类似结果也有报道，乳腺炎的乳汁质量差，不适合人们饮用消费，另外，由于 SCC 引起 MDA 升高的乳汁极大降低了乳汁的质量（Suriyasathaporn et al.，2006）。实际上 MDA 也是一种可疑的致癌物质，能够使机体的 DNA 发生变异。

GPx 已经被国内外学者确定是含硒酶，并催化许多种过氧化物的还原反应，保护机体的细胞免受氧化损伤，充足的 GPx 能够保护乳汁的脂类不被氧化（Fox and Kelly，2006）。已经有研究测定了乳汁中 GPx 的活性，为 12～32 IU/mL，并且与硒的浓度呈显著正相关（Przybylska et al.，2007）。与健康奶牛相比，发生乳腺炎的奶牛乳清的脂质过氧化反应增加，血液中 GPx 活性降低。本实验中健康奶牛的乳汁 GPx 的平均活性 [（32.8±1.41）IU/mL] 极显著高于乳腺炎的乳汁 [（26.41±2.03）IU/mL]（$P<0.01$），由于 GPx 通过分解过氧化氢作为细胞的防御功能，因此相比低活性 GPx 的隐性乳腺炎乳汁，高活性 GPx 的健康乳汁可能是乳制品制作的较好的来源。Atroshi 等研究发现，乳腺炎奶牛 GPx 的降低与脂质过氧化反应和前列腺素的生成有关。人乳汁中 GPx 的活性随着泌乳时间的增加而降低，但是至今没有关于牛奶中 GPx 活性变化的参考文献。

首次报道乳汁中 SOD 的是 Holbrook 和 Hicks（1978），他们认为 SOD 在乳汁的抗氧化能力方面起着重要的作用。本实验健康奶牛乳汁中 SOD 的平均活性是（1.82±0.11）IU/mL，而隐性乳腺炎乳汁中 SOD 的平均活性是（1.6±0.12）IU/mL，有报道称乳汁中 SOD 的活性在奶牛个体间和品种间变化较大。牛奶中 SOD 的活性不受奶牛的泌乳阶段和（或）年龄的影响（Lindmark-Mansson and Akesson，2000）。本实验也证明了健康乳汁的 SOD 活性与隐性乳腺炎乳汁没有显著差异（$P>0.05$）。

动物机体细胞结构损伤后会引起血液和乳汁中酶类活性的改变，本实验结果显示隐性乳腺炎乳汁中 LDH 和 ALP 的平均活性极显著高于健康乳汁（$P<0.01$），但是隐性乳腺炎乳汁中 AST 的平均活性与健康乳汁之间没有显著性差异（$P>0.05$）。与先前的研究报道（Batavani et al.，2003；Babaei et al.，2007）基本一致。LDH 活性升高的原因是患乳腺炎奶牛的乳汁中出现了白细胞，另一个原因是炎症过程中乳腺上皮细胞和间皮细胞遭到破坏，将细胞中的 LDH 释放出来。黄权利等也认为 LDH 在隐性乳腺炎病理中起着重要作用，可作为诊断隐性乳腺炎和乳腺损害的一个重要指标。乳腺发生感染时增加了微血管的通透性，血-乳屏障遭到感染的破坏，ALP 可由血液进入乳汁中，使得乳汁中 ALP 活性显著升高。乳汁中 LDH 的活性检测已经被认为是由疾病引起的乳腺功能发生改变的敏感指标，ALP 活性的检测被认为是早期诊断奶牛隐性乳腺炎的可靠指标（Symons and Babaei et al.，2007）。

四、小结

（1）SMT 在诊断亚热带奶牛隐性乳腺炎方面具有一定的价值。

（2）金黄色葡萄球菌和凝固酶隐性葡萄球菌是引起本次调查奶牛隐性乳腺炎的主要致病菌。

（3）测定乳汁中的 MDA 浓度、GPx、LDH 和 ALP 活性是诊断奶牛隐性乳腺炎发生的一种方法，并且在一定程度上为探明奶牛乳腺炎的发病机制提供依据。

第五章　奶牛乳腺炎的治疗措施研究

对奶牛乳腺炎的治疗主要有物理疗法和药物疗法两种。物理疗法主要包括按摩和热敷。按摩可以促进乳腺的血液循环和淋巴循环，以促进乳汁排出。每次挤奶过程中按摩2～3次，每次大约15 min，能缓解乳腺炎症状。这种方法并不适用于化脓性乳腺炎或出血性乳腺炎，因为按摩可能损伤乳腺组织，破坏乳腺细胞，而加剧疼痛。热敷法对于奶牛乳腺炎和其他同类炎症都具有明显效果，热敷的方法很多，如将毛巾浸于40～50℃热水中，在浸泡过程中加入适量的乙醇、食醋和硫酸镁等可以增加效果，待3～5 min，取出拧干覆盖于患处，尽量覆盖全面，用塑料布包裹，固定，每次40 min，每天2～3次（方磊和郭玉江，2012）。奶牛乳腺炎治疗过程中应该避免使用利尿剂，尽管利尿剂有助于排放体内的液体，缓解乳房肿胀，但这种效果的代价也会造成钾的流失、电解质平衡的破坏，以及葡萄糖形成障碍，影响奶牛健康。

除了辅助的物理治疗以外，药物疗法是目前最为常用和有效的方法。20世纪40年代，青霉素的发明使得抗生素在对抗细菌感染中占据重要作用，而且抗生素在抗菌消炎方面发挥了重要作用，也是控制和治疗奶牛乳腺炎的首选方案。在以抗生素为基础的研究中，人们逐渐发现，通过将抗生素配合使用，可以产生快速、高效、经济的防治效果。正是基于这一目的，研发出多种用于治疗乳腺炎的药物，如青霉素类、大环内酯类、四环素类、氯霉素类、氨基糖苷类以及磺胺类药物等。抗生素的应用已经具有数十年的历史，且已证实在治疗多种病原菌引起的急性感染、慢性感染和败血症方面，具有极其重要的作用。但抗生素的广泛应用是一把双刃剑，滥用抗生素会导致耐药菌株增多。据报道，在一些地区，青霉素、氨苄西林、恩诺沙星、盘尼西林等抗生素所引发的细菌耐药性问题已经十分严重，奶牛乳腺炎的一些主要致病菌产生了相当的耐药性，使治疗难度大大增加。另外，残留的大量抗生素还能进入乳汁中，也给人类的健康带来了危害（伞治豪，2014）。

由于抗生素治疗的缺陷，人们逐渐注意到中草药是纯天然物质，很多中草药不仅能营养保健、调节机体免疫功能，还有杀菌、消炎、安全低毒的特征，作用于细菌并不易使其产生耐药性。因而，中草药逐渐进入人们的视线，人们尝试应用中草药替代或者辅助抗生素来防治奶牛乳腺炎，这一崭新的课题，越来越被国内外所认同和关注。近年来，学者们对于各种中草药疗效的评估、评定进行了广泛研究。郭庆等（2012）选用金银花、连翘等8味中草药对多种病原菌进行体外抑菌试验，结果表明，金黄色葡萄球菌对金银花和连翘较为敏感，大肠杆菌和无乳链球菌则对芙蓉花较为敏感。黄远全等（2012）在治疗奶牛临床型和隐性乳腺炎中，采用自制中草药制剂对其进行治疗并取得了理想的疗效，其对隐性乳腺炎的治愈率可达到80%。Mukherjee等（2010）研究表明，青牛胆茜草提取物可提高动物对病原菌的防御能力，而此种防御能力与其提高机体白细胞介素-8和髓过氧化物酶的含量密切相关。除了单纯使用中草药，也可以将其与酶复合物共同使用，疗效显著。吕平等（2012）研究了酶复合物与中草药提取物的共同作用，对感染乳腺炎的常见病原菌大肠杆菌和金黄色葡萄球菌进行了体外抑菌试验。结果表明，中草药对其有较强的抑制作用。可见，中草药预防治疗乳腺炎有广阔的应用前景，但目前，中草药在乳腺炎治疗领域还存在一些弊端，如验方及临床观察，对于以上这些问题，尚缺乏深层次的理论依据。对于如何提高中草药的疗效，如何将中草药与西药抗生素进行配伍，以及中草药方剂的研究，以期开发出高效、稳定、高效安全的中草药制剂，是潜力巨大的课题。

众多研究证明，中草药在治疗奶牛乳腺炎中的疗效可与抗生素相当，甚至优于抗生素。朱永平等（2011）选用金银花、蒲公英、紫花地丁、连翘、黄芪等组方，煎煮后灌服，每日1剂，治疗患牛116头，治愈99头，好转12头，总有效率达95.7%，而且产奶量提高了9.1%。黄远全等（2012）将患有临床型乳腺炎的30头奶牛随机分成3组，分别进行了抗生素乳腺灌注、抗生素静脉滴注和乳腺涂抹自制中草药消炎药膏（由黄芩、金银花、桃仁、蒲公英等组成）处理，3种处理方法的治愈率分别为20%、30%和60%，试验结果表明，自制中草药制剂涂抹疗效最好。同时，该课题组在隐性乳腺炎治疗中也进行了抗生素和涂抹自制中草药消炎药膏的疗效比较，结果均显示中草药制剂涂抹治疗的治疗效果更好，治愈率达80%，总有效率为90%，明显高于抗生素治疗组的疗效。龚平阳等（2009）应用中草药复方透皮吸收搽剂治疗临床型奶牛乳腺炎18例，治愈16例，好转2例，治愈率达88.9%，有效率为100%。徐京平等（2010）用酒知母、酒黄柏、生地、连翘、桔梗、甘草等中草药组方，治疗奶牛乳腺炎56例，治愈53例，治愈率达95%。

第一节　奶牛乳腺炎治疗措施研究进展

乳腺炎作为奶牛饲养中反复发生的疾病，给奶牛养殖业带来了严重的经济损失，所以我们要进一步在抗菌疗法的新技术方面进行研究。消灭病原微生物，控制感染的发生，减轻和治愈炎症，提高奶牛抵抗力，防止败血症发生是治疗的基本原则。控制乳腺炎表面措施通常分为两种：直接措施，感染后相应的诊断（细菌学，临床观察）；间接措施，在炎症症状的基础上基于对乳房细菌学状态的预测（体细胞数量，乳汁导电性）。

一、抗生素治疗

在奶牛养殖中，临床实践治疗奶牛乳腺炎的首选药物就是抗生素，特别是抗生素对急性临床乳腺炎的治疗能够起到非常明显的治疗效果，对于急性临床型乳腺炎转为慢性甚至机体败血症都能起到有效的治疗作用。抗生素是治疗和控制奶牛乳腺炎的基本药物，目前被广泛应用，尤其对治疗急性、多发性和亚急性乳腺炎有重要的作用。通过实验研究发现，肯尼亚最常见的细菌为金黄色葡萄球菌（58.8%），庆大霉素是医治的首选药物（Ondiek et al., 2013）。常用的抗生素主要有青霉素、链霉素、卡那霉素、先锋霉素、庆大霉素、四环素、磺胺类、喹诺酮类等。如果亚临床乳腺内感染的细菌固化，可以增加时间，进行持续治疗（高海慧，2014）。

抗生素治疗仍然是控制奶牛乳腺炎的重要手段，在控制乳腺炎上发挥着重要作用，尤其是防止败血病以及防止其他细菌感染方面。虽然抗生素能够有效治疗奶牛乳腺炎，但其往往对动物机体造成一些副作用，如机体免疫的功能下降、动物食欲的减退、动物机体内菌群的失调等。长期大量使用抗生素也存在难以解决的问题，主要表现在以下几方面。

(1) 细菌耐药性　　应用抗生素治疗，对所用药物敏感的细菌得以消除，耐药的菌株却生存繁殖下来。耐药性可以平行地通过质粒介导从一种耐药菌株转移到另一种细菌，耐药质粒在微生物间可以通过转化、转导、易位等方式进行转移，耐药因子增多造成了多重耐药性，降低治疗效果。Güler 等（2005）检测了 265 株分离牛奶的金黄色葡萄球菌，其中 1.8%对复方新诺明产生耐药性，27.9%对四环素不敏感，63.3%对青霉素和阿莫西林存在耐药性，对诺氟沙星、卡那霉素耐药的菌株没有分离到。

(2) 乳腺的二次感染　　抗生素使得奶牛免疫功能降低，容易引起二次感染。治疗时，如果宿主没有清除被抗菌药物抑制的病原菌，停药后细菌可以大量繁殖。

(3) 抗菌药物不能保证在患处达到有效的药物浓度　　由于给药剂量较小、给药时间间隔太长、疗程短等原因造成在一定时间内药物不能维持抗菌浓度水平；吸收、消除、分布等药动学因素；全身给药，但药物不能有效透过血-乳屏障等原因，都可以使得抗菌药物不能在患处达到有效的药物浓度。

(4) 抗生素残留　　现今，抗生素治疗是奶牛防治的常用药物，长期大量的使用使得耐药菌株增多，一些疾病又变得难以用常规方法治愈，使得治疗中加大抗生素剂量，形成不健康循环。我国食品卫生规定，在应用抗生素治疗时及停药后 5 天内的牛奶不可以食用。更重要的是，耐药性细菌还可通过乳汁进入人体，而牛奶中残留的抗生素也会诱发人体发生超敏反应（项开合等，2007）。

因此，在使用抗生素治疗奶牛乳腺炎还存在很多问题，解决关键问题可以通过增强抗生素的靶向性和能效，从而减少抗生素使用量，来解决抗生素残留问题和病原菌耐药性带来的危害。兽医临床实践中治疗奶牛乳腺炎仍然主要使用抗生素药物，但是其已经不能作为治疗奶牛乳腺炎的有效策略，因此对治疗乳腺炎更为安全行之有效的方法已成为研究者主要的研究热点。

二、中草药治疗

中草药制剂的组成主要以动物、植物和矿物质，它们都是动物体易吸收的天然物质，不仅含有生物活性物质，如生物碱、多糖、皂苷、挥发油、蒽类和有机酸等，而且还含有矿物质、维生素及未知的营养因子，兼有预防、治疗和营养的多重作用（张少华，2009）。中草药制剂具有抑菌杀菌和增强动物机体免疫的功能，因此，中草药制剂既可治疗奶牛隐性乳腺炎，也可提高机体的抵抗力（魏成威，2010）。中草药制剂奶牛乳腺炎的特点主要是标本兼治、无残留、毒副作用小、不易产生耐药性（夏祖和张淼，2006）。治疗剂型散剂、煎剂、注射剂、灌注剂、透皮剂等，大多为复方剂，给药途径主要有拌料食用、灌服、乳房灌注、注射、药浴等（高海慧，2014）。

中草药含有的活性成分多，大多数有抑菌杀菌的作用。例如，黄连、穿心莲、大蒜、大青叶、板蓝根、金银花等，有抗病毒作用和广谱抗菌作用。王忠红等（2004）将川芎、当归等 10 味中药组方后加水稍煎，进行药液灌服、药渣拌料饲喂，结果显示，奶牛乳腺炎治愈率达 63.6%，总有效率为 97.0%；张振国等（2010）用中药灌注剂对临床型乳腺炎进行治疗，其治疗的有效率为 100%，治愈率为 90.91%；冯士彬等（2011）研究的中药复方透皮贴剂对奶牛临床型乳腺炎有较好的治疗效果，能够改善患病奶牛微循环、抑制炎症反应、降低血液黏度和炎性细胞因子的含量。

(一) 中药防治乳腺炎的中医理论

奶牛乳腺炎症状表现为乳房实质、间质的炎症。而中医将乳腺炎定义为乳痈,所谓乳痈是指乳房红肿疼痛,乳汁排出不畅,以致结脓成痈的急性化脓性病症。中医理论认为,在引起乳腺炎的原因中,除外伤外,由厥阴肝经郁滞、任冲二脉失调,以及阳明经实热引起的乳房脉络不畅,乳汁积滞,发热腐败,进而发炎是发病的主要原因(张振国,2009)。乳痈可分为肝郁气滞型、热毒壅盛型,以及气血瘀滞型(孟庆娟,2011)。导致乳房生痈的主要原因是湿热毒气上蒸,或乳汁蓄留不能及时排除,导致乳房胀满,使得经络阻塞,气血运行不畅,进而导致乳房发生硬块。

兽医临床中,将乳腺炎主要分为临床乳腺炎和隐性乳腺炎,而中兽医中所定义的"乳痈",所指的是临床型的乳腺炎。其致病机制是由于内分泌的失调,导致邪毒入侵乳房,使得乳房中的经络循环受阻,导致乳汁"蓄积不去"或"蓄结并与气血相搏",最终导致"恶汁于内",从而形成乳痈。而正是因为邪毒的蕴结,使得乳络不畅,进而导致乳汁凝滞,使乳房出现炎症的基本特征,红、肿、热、痛。而乳汁方面,由于炎症的产生,使得乳汁分泌减少,出现腐坏,引起的全身性的症状则导致精神萎靡,体温升高,不思饮食。

对于隐性乳腺炎,其主要症状是产奶量下降。中兽医则将其病因归结为气血两虚导致。中兽医认为,在孕畜的分娩过程中,分娩所致的消耗及出血情况,以及产后可能产生的持续性出血,导致阴血骤虚和阳气易浮,这些都是导致其体质变差,血虚的原因。而从乳汁方面出发,由于任冲气血所化,导致血虚使得泌乳量减少,而由于气虚无法推动血液的循环,导致了血瘀进而影响乳汁的分泌。对于乳痈的治疗,《内经》云,"诸痈疮疖,皆属于火";《医宗金鉴》说,"乳痈由肝气郁结,胃热壅滞而成",因而,乳痈的治疗原则应以疏肝气、散郁结,清胃热、清肿毒、通乳络为主,也就是说,其总的治疗原则为:清热解毒、消肿散结、散瘀止痛、通经活血、活络通乳。补益气血的药物也常常被用来防治奶牛隐性乳腺炎(郭旭东等,2012)。

中兽医学在长期的临床治疗中,形成了大量的中药古方,经过研究者对这些组方的分析发现,这些组方有着共同的特点,即所有组方均遵循乳痈的治疗原则,以清热解毒药和活血化瘀药为主。其中,清热解毒药多以蒲公英、连翘、金银花、鱼腥草、紫花地丁为主,以牛蒡子、麻黄为辅助的解表药,从而辅助增强方剂中清热解毒作用;同时,再辅以木香、陈皮、青皮等理气药,帮助改善胃肠功能,帮助调节机体营养物质摄入并发挥镇痛作用。而活血化瘀药则主要以桃仁、没药、赤芍、穿山甲、乳香、王不留行等为主,发挥改善血液循环,除血祛瘀,并促进泌乳,发挥抑菌消炎作用。正是基于这些中草药的功能性质,形成了有显著治疗作用的古方,包括银归汤、仙方活命饮、牛蒡汤、透脓散、瓜蒌散、瓜蒌牛蒡汤、二花消痈汤、藕节汤、逍遥散、麻连甘草汤、内托生肌散等(伞治豪,2014)。

现代中兽医临床中,针对奶牛生产特点,从发病机制及奶牛的生理特性入手,治疗与调理相结合,对传统中兽医组方进行了一定的改进,取得了更好的临床治疗效果,达到标本兼治,治补结合的中医药理论本质。例如,在已经成型的汤剂中,加入有补中益气作用的药物,如党参、熟地黄、白术、甘草、黄芪、当归、阿胶、白芍等,以补充奶牛分娩消耗及产后大量分泌乳汁导致的气血流失,并增强机体免疫力。目前已研制出了乳炎康、消炎增乳散、蜂胶合剂乳炎散、公英瓜蒌汤、乳腺炎消王等中药制剂。

(二) 中草药防治乳腺炎的作用机制

1. 补充营养,增强物质代谢　　许多中草药都含有各种营养物质和活性成分,这些物质不仅可以促进奶牛乳腺上皮细胞增殖,增强乳腺代谢活动,增加产奶量;还可以促进消化吸收功能,改善机体造血机能和血液循环,提高饲料利用率和奶牛生产性能。有报道称,619种常用中草药中均含有一定的常量元素和微量元素(周二顺,2015)。

2. 增强免疫机能　　中草药含有多种有效的免疫活性成分,如有机酸、生物碱、多糖等。有机酸如桂皮酸可使机体内的白细胞含量升高,马兜铃酸可使巨噬细胞的吞噬能力增强;生物碱不但能增强机体的细胞免疫还可加强体液免疫;多糖可通过加强巨噬细胞的作用,增强机体的细胞免疫功能。

甘草浸膏副产品复方添加剂能提高血清蛋白的含量,降低球蛋白的含量,增强奶牛的细胞免疫功能,对泌乳期奶牛的免疫功能有明显影响;中药组方(黄芪、当归、白芍、益母草、蒲公英、王不留行等)能明显提高隐性乳腺炎奶牛的免疫功能;大豆黄酮能显著提高奶牛血清及乳清中特异性抗体水平,表明奶牛的体液免疫功能明显增强;大豆异黄酮能增强乳腺肥大细胞 IL-4 的分泌水平,增强奶牛的免疫功能,从而提高血清和乳清中生长激素和催乳素的含量,提高奶牛泌乳性能。

3. 抗菌作用　　多种中草药具有抑菌作用,如瓜蒌和金银花的有效成分对金黄色葡萄球菌、牛棒状杆菌、嗜热链球菌、产气荚膜梭菌、乳房链球菌、乳杆菌6种致病菌有良好的抑制效果;虎杖和连翘分别对金黄色葡萄球菌和大肠杆菌具有较好的抑制效果;体外抑菌试验显示,黄芩对引起奶牛乳腺炎的4种病原菌有较好的抑制效果;复方丹参液在治疗奶牛隐性乳腺炎时具有较强的抑菌作用等(周二顺,2015)。

4. 抗炎作用　　炎症是奶牛乳腺炎的主要临床症状,也是导致乳腺受损及奶牛产奶量下降的主要原因。因此,消除炎症反应是治疗乳腺炎的首要任务。多种中草药均具有抗炎作用,如金银花和蒲公英能显著的抑制肉芽组织增生;连翘和黄芩可明显抑制炎性渗出和水肿;栀子苷在体内体外均可抑制脂多糖诱导的炎症反应。

(三) 中草药治疗奶牛乳腺炎的效果

根据乳腺炎的发病机制和中药药理学理论,对于采用中草药治疗乳腺炎应以清热解毒、抗菌消炎、通经活血、消肿止痛、活络通乳为原则,从而保持机体内环境的动态平衡,达到防治目的,中草药可直接杀死奶牛乳腺炎病原菌,也可通过提高机体的免疫力间接抑制或吞噬病原菌,同时中草药又具有解热、镇痛、抗炎等作用,对于奶牛乳腺炎的发生可达到标本兼治的效果(王志刚等,2006)。已有研究证实,中草药所含有的有效免疫活性成分可以增强动物机体的非特异性免疫和巨噬细胞的吞噬功能,并且能提高淋巴细胞的转化率。同时一些微生态制剂与中草药结合使用能提高奶牛对饲料的消化率,促进营养物质的消化和吸收,清除氧自由基及修复被损伤的乳腺细胞等(项开合等,2007)。目前,研究人员已经研制出多种中草药复方制剂及中草药的提取物用于奶牛乳腺炎的预防和治疗。已报道证实，芦荟、鱼腥草等中草药对治疗奶牛乳腺炎均能起到良好的效果（高瑞峰，2014）。李德鑫（2003）用"乳炎消"注射液治疗奶牛急慢性乳腺炎，有独特的疗效，特别对久治不愈的顽固性乳腺炎疗效确切，且不易复发。金凤等（2005）报道治疗奶牛隐性乳腺炎时用党参、黄芪和当归等中草药制剂的药效明显。杜健等（2007）实验结果表明，在奶牛日粮中添加中草药制剂能够显著提高其产奶量。王雄清等（2006）研究表明，黄芩、黄柏、黄连、丹参、石榴皮5种中药对奶牛乳腺炎6种病原菌均有明显的抑菌作用，并研制成功相应的治疗中草药组方。吴静等（2007）研究中草药对奶牛乳腺炎病原菌的体外抑菌作用，结果表明，选择的7种中草药煎剂及5种复方制剂进行病原菌的体外抑菌试验，野菊花、黄连、大青叶、复方Ⅰ（黄连、黄芩、野菊花）、复方Ⅴ（大青叶、金银花、鱼腥草）组方均具有较强的体外抑菌效果。刘澜等（2009）对常规26种中草药对奶牛隐性乳腺炎病原菌体外抗菌作用的研究表明，对奶牛金黄色葡萄球菌抑菌效果为黄连、紫花地丁、当归、红花、板蓝根最好，对大肠杆菌为黄连、紫花地丁、瓜蒌、板蓝根、黄芩、白芍最好。对于中草药的抗炎作用，王秋芳等给患有隐性乳腺炎的奶牛日粮中添加黄芪、蒲公英和益母草等中草药，可显著提高中性粒细胞的吞噬能力和淋巴细胞刺激指数，并能显著降低隐性乳腺炎的发病率（王秋芳等，2002）。史彦斌等（2000）试验表明，消炎酿对蛋清蛋白引起的急性足趾肿胀和白细胞游走有明显的抑制作用。耿梅英等（2006）研究表明，中草药地榆的提取物对奶牛乳腺炎的治疗效果明显优于头孢氨苄。张振国（2009）的研究证实，使用金银花和蒲公英为主要成分的中药对发病奶牛进行治疗能够起到良好的治疗效果，其效果明显优于头孢唑林钠。因此，我国传统中草药及其有效成分在临床中治疗奶牛乳腺炎具有非常大的优势和潜力。

（四）中草药治疗奶牛乳腺炎的不足

目前中草药使用广泛，在很多方面确实优于抗生素，但也存在着许多的不足。例如，中草药的成分极其复杂，再加上中医理论知识的复杂性，给研究者带来特别大的困难，中草药的很多作用机制尚不清楚。目前中药方剂的加工制作仍然是比较原始的传统工艺，粉碎、搅拌后制成，技术水平落后，制作程序复杂、有效成分含量低，很难适应当前规模化养殖、集约化养殖规模的需求。中草药有一定的毒性，对于其在体内的药代动力学，以及添加到饲料中是否会影响饲料的营养成分还需要进一步研究确认。中草药来源广泛，同一味中药，在不同的地区、不同的时间采摘，其药物成分也不同，质量很难保证。到目前为止中草药配方尚无固定标准，研究者多根据自行研究的项目自行配伍。因此现阶段应该在现有研究的基础上，以中兽医理论为依据，对中草药配伍制定相应的标准，并在实验中完善，作为日后研究者的实验设计依据。在治疗和控制奶牛乳腺炎方面，使用中草药日益受到人们的重视，虽然目前还有许多困难，但随着越来越多的研究者对中草药研究的深入，中草药在防治奶牛乳腺炎方面会发挥重要的作用。

（五）中草药治疗奶牛乳腺炎的前景展望

目前，一些中草药制剂已经用于防治奶牛乳腺炎，治疗效果不错。有些研究人员针对中草药治疗乳腺炎的作用机制做了相关研究，认为中草药治疗奶牛乳腺炎主要是通过抗菌消炎和提高机体非特异性免疫功能来实现的。因此，应用中草药防治乳腺炎不仅可以使畜牧业可持续发展，还可以为人们提供绿色食品。

利用中草药治疗乳腺炎同样也存在一些问题，如作用机制尚未完全调查清楚、剂型不够合理等。因此，今后我们应该：①制定合理规范的中草药制剂质量标准；②加强开展中草药剂型研究；③加强中草药作用机制研究。总之，以中兽医理论为参考依据，利用现代医学的科研成果，建立高效的中草药活性成分的分离方法，以科学数据阐述作用机制，研制新剂型，开发出高效率的、稳定性强的、安全质量可控的纯中药制剂。随着人们对中草药日益重视以及深入研究，相信中草药将来会在防治奶牛乳腺炎方面产生巨大作用。

三、中西药结合治疗

朱志达和韩张兴（2001）用"中药乳炎消"与青霉素、链霉素联合用药治疗慢性顽固性乳腺炎，可以将治愈率从17%提高到75%，治疗亚急性乳腺炎的治愈率能达到100%，对急性乳腺炎的治愈率达到75%。李玉文和秦建华（2010）在研究几种治疗方法对奶牛临床型乳腺炎的治疗效果时发现，中西药结合组对奶牛乳腺炎治愈率较高。

中草药与抗生素组成复方药物的治疗方法，既利用了抗生素抑制病原菌生长的作用，同时又发挥中草药清热解毒、消炎止痛、增强机体抵抗力的作用，可以缓解临床症状，控制炎症恶化，还可减少抗生素的药物残留，缩短中草药治疗的疗程。孙凌志等（2007）采用复方中草药结合青霉素、普鲁卡因的组方，用乳房封闭法治疗奶牛乳腺炎，有急性乳腺炎102例，隐性乳腺炎12例，慢性乳腺炎76例，都取得了满意的治疗效果。

四、物理疗法

对患病奶牛挤奶时，按摩10~15 min，共2~3次，可以缓解炎症症状，促进乳腺内的血液循环，保持乳腺导管畅通，使患病奶牛排出乳汁。向50℃的温水中加入适量的乙醇，并把毛巾浸湿，轻轻擦拭乳房，每次30~40 min，每日2~3次。同时还可以利用激光辐射治疗奶牛乳腺炎，而且无公害，操作简单，性能可靠。例如，利用以超高频磁场为基础研制出的一种可以在挤奶同时治疗乳腺炎的手提式超高频磁场挤奶器，以1500 Hz的频率，每天对患病区进行1 min的辐射，3~4次就可以痊愈（付云贺，2015）；另有报道其对隐性乳腺炎的治愈率为85%~90%，对临床型乳腺炎的治愈率为61.5%，可以大大减轻兽医人员的劳动量（丁丹丹，2014）。

五、细菌素治疗

细菌素被认为是可以选择的方法，其较传统的抗生素疗法具有一些优势。我们要加强对人类健康的关注，首先是致病菌抗药性的出现，再就是发展可供选择的抗感染药物也是必需的。细菌素发挥作用的基础是可以靶向敏感菌株的表面受体，从而起到抵抗特殊细菌菌株的作用。在诊断乳腺炎时，要对菌株进行有效的鉴定，这样才能对特殊致病菌采取有效的治疗措施。细菌素通过溶菌素可以快速杀死易感的微生物。这种快速的反应可以确保病原菌不产生耐药性。使用的抗生素通常是广谱的，可以杀死革兰氏阳性菌或者革兰氏阴性菌。细菌素具有特异性靶向的优势。如果需要更广谱的活性的话，2～3 种细菌素联合使用可以在治疗中确保杀死多种病原菌。细菌素的最低抑菌浓度（MIC）已确定，从而可减少细菌素在治疗中的使用剂量（高瑞峰，2014）。

乳腺炎治疗策略中给药方式是很重要的，通过联合使用抗菌药物，抑制剂存在于乳头管中，靶向在乳头开放时的病原菌，可以阻断细菌在乳腺组织中繁殖。局部用制剂也可以使用，由于病原菌侵染力的降低使得局部用制剂更容易作为药物传递的方法。在乳头皮肤上，细菌素的持久性和稳定性是至关重要的。目前，细菌素基础上的产品实验已经成功。乳酸菌肽已经广泛用于泌乳期，乳链球菌素已经被应用于奶牛干奶期的治疗，而且乳球菌肽也成为商品化的乳头消毒剂（Cotter et al.，2005）。奶牛乳头管的环境和给药途径是在确定治疗方法时的一个重要问题。LAB 产出的细菌素普遍认为是安全的，与其他抗生素相比，其更容易被接受。泌乳期的抗生素疗法要求有停药期，由于生产时间的损失而导致经济损失。乳汁中细菌素的残留更多是通过消化酶分解成多肽。因此，如果细菌素疗法可以替代抗生素疗法，那么停药期可以显著减少。

乳腺炎乳汁在奶制品生产中会带来严重的经济损失，所以，研究有效的、安全的、可选择的方法是必要的。病原菌耐药性的出现影响着养殖业的健康发展。然而利用灵敏的特异的分子诊断技术可以做到对病原菌的快速鉴定。依据临床和环境数据建立抗菌谱也是非常重要的。细菌素可以作为一种治疗解决方法，来增强其他管理措施而减少在乳腺炎治疗中抗生素的使用剂量（高瑞峰，2014）。

六、微生态制剂治疗

微生态制剂又称微生态调节剂，是根据微生态理论制备的制剂，其有效成分是活菌、死菌以及其代谢产物，可以有效地提高机体的免疫能力。史文君（2006）观察微生态制剂治疗奶牛乳腺炎的疗效，发现使用微生态制剂治疗单一乳房的临床乳腺炎有明显效果，治愈率为 90%，治疗的有效率达到 100%。

七、细胞因子治疗

细胞因子（cytokine，CK）是指由免疫细胞和某些非免疫细胞合成和分泌的一类高活性多功能蛋白质多肽分子。在控制奶牛乳腺炎上，该因子可调节乳腺中各种免疫细胞的功能，增强乳腺的免疫机能，提高乳腺免疫细胞对入侵病原体的抵抗力和杀伤力，可单独也可与抗生素协同应用，同时可作为乳腺炎疫苗的佐剂。目前，应用最多的细胞因子主要包括：白细胞介素-2（IL-2）、γ-干扰素（IFN-γ）、集落刺激因子（CSF）。

白细胞介素（IL）是免疫系统分泌的主要在白细胞间起免疫调节作用的蛋白，由于人细胞因子对牛细胞存在一定的交叉反应，因此，早在 1994 年，Shuster 等利用重组人白介素 1 受体拮抗剂对牛的革兰氏阴性菌性乳腺炎进行治疗。目前，在控制奶牛乳腺炎上具有代表性的是 IL-2，其功能是促进 T 细胞增殖分化和细胞因子生成；增强 TC 细胞、NK 细胞和 LAK 细胞的活性；促进 B 细胞增殖和抗体的生成。有报道称，在治疗金黄色葡萄球菌性乳腺炎时，将重组牛白细胞介素-2 与头孢菌素联合用药，其效果比单用抗生素疗效平均提高了 20%～30%（张颖，2014）。

干扰素（IFN）是 1957 年最早发现的细胞因子，因其能干扰病毒感染而得名。IFN-γ 是由抗原刺激 T 细胞产生，具有增强 NK 细胞的活性，抗体介导的 T 细胞和细胞毒性 T 细胞的作用以及提高巨噬细胞、中性粒细胞的吞噬作用和细胞毒性作用，调节内毒素脂多糖诱导的炎症反应。IFN-γ 在控制大肠杆菌性乳腺炎上起到一定作用。许金俊（2004）曾利用 IFN-γ 和环丙沙星对隐性乳腺炎或轻度临床型乳腺炎进行对比性治疗。结果表明，单独使用抗生素虽短期内治疗效果较好，但停药后很快复发，单独使用 IFN-γ 短期内没有明显效果，但将二者协同用药可达到有效且稳定的治疗效果，产奶量有所提高，充分证明了 IFN-γ 能增强和改善抗生素的治疗效果。马定坤等（2011）研究 γ 干扰素对金黄色葡萄球菌性的临床型乳腺炎的作用时发现，γ 干扰素配合乳酸环丙沙星治疗该疾病有协同作用。白细胞介素-2 是由辅助型 T 淋巴细胞分泌的细胞因子，在机体免疫系统中发挥着重要作用，白细胞介素-2 还可以诱发其他细胞因子，协同发挥作用。

集落刺激因子（CSF）是一组促进造血细胞增殖、分化和成熟的因子，具有刺激骨髓多能干细胞增殖，使之向中性粒细胞、巨噬细胞和肥大细胞等分化的功能。经粒细胞-巨噬细胞集落刺激因子（GM-CSF）作用后的中性粒细胞有明显的趋化性和噬菌性。重组牛巨噬细胞集落刺激因子（rboGM-CSF）能明显提高血液和乳腺中中性粒细胞的胞内杀伤活性。Takahashi 等（2004）研究表明，重组牛集落刺激因子（rboGM-CSF）有治疗由金黄色葡萄球菌引起的隐性乳腺炎的潜力。王玲等（2011）研究表明，rboGM-CSF 能有效地保护乳腺免受金黄色葡萄球菌引起的新感染，可使其下降 47%，对于已患有金黄色葡萄球菌性乳腺炎的奶牛乳区注入 rboGM-CSF，治疗率达 75%，治愈率达 22%。

目前，利用细胞因子治疗奶牛乳腺炎仍存在一些潜在的毒副作用和不良反应。例如，在干奶期高剂量使用重组牛白介素-2 会引起奶牛流产，重组牛干扰素 γ 只有在感染早期给予低剂量能有效治疗奶牛乳腺炎等。

八、溶菌酶疗法

溶菌酶（lysozyme）又称细胞壁质酶或 N-乙酰胞壁质聚糖水解酶，是一种碱性水解酶，广泛存在于自然界的动物、植物及微生物中，能催化水解细胞壁中的 N-乙酰胞壁酸和 N-乙酰氨基葡萄糖之间的 β-1,4 糖苷键，使细胞壁不溶性多糖分解成可溶性糖肽，从而细菌内容物逸出，细胞壁溶解（刘纹芳等，2006）。溶菌酶具有抗菌消炎、抗病毒、增加免疫力、促进组织修复等功效（张璐莹等，2012）。由于人类和动物的细胞无细胞壁，因此无毒、无害。重组溶菌酶对金黄色葡萄球菌、无乳链球菌、停乳链球菌、大肠杆菌等具有较强的溶菌杀菌作用，特别是耐药菌株有很好的抑菌作用，逐渐可以替代抗生素对奶牛乳腺炎的防治，安全、无残留、无不良反应，具有良好的发展前景。孙怀昌（2005）研制出用人溶菌酶重组质粒来治疗奶牛乳腺炎，用药一周后，乳汁中的细菌数明显下降，产奶量有了不同程度的增加，尤其对经抗生素治疗无效的乳腺炎，治疗有效率达 83.3%，且不易产生耐药性。由上海高科联合生物技术研发有限公司研制的重组溶葡萄球菌酶对引起奶牛乳腺炎的金黄色葡萄球菌、表皮葡萄球菌及耐药菌株具有良好的抑杀作用，对链球菌、牛多杀性巴氏杆菌具有一定的抑杀效果（贾敏等，2007）。戴晏等（2009）对上海市光明荷斯坦金山种奶牛场采集患病奶牛的奶样，注射人溶菌酶的重组质粒 pcDNAKLYZ 并以注射抗生素作对照，结果显示，重组质粒的抑菌效果显著高于抗生素的抑菌效果。沈诚等（2011）将人溶菌酶重组质粒pcDNAKLYZ注射到患有隐性乳腺炎的奶牛，其乳样中的细菌数明显下降。目前，国内对溶菌酶大多仍集中于基础研究，对于其分离、纯化等方法仍需进一步的改进，以便于大规模的生产。

九、抗菌肽疗法

抗菌肽是生物体内经诱导产生的一种具有生物活性的小分子多肽，广泛存在于植物、昆虫、人等生物机体内，是机体免疫防御机制的重要组成部分之一，对细菌、真菌、原虫、病毒、癌细胞均有不同程度的杀伤力，能作用于其细胞膜，破坏细胞膜的完整性，并产生穿孔现象，造成细胞内容物溢出胞外而死亡，但对高等动物的正常细胞无害。抗菌肽与传统抗生素联用，可提高二者的疗效，甚至拓宽传统抗生素的抗菌谱。王思贤（2006）曾利用抗菌肽基因*ABPS1*具有广谱抗菌活性的特点，构建抗菌肽乳腺特异性表达载体，用于奶牛乳腺炎的防治有明显的治疗效果。马晓艳（2007）利用新疆家蚕抗菌肽对隐性乳腺炎的致病菌进行药敏试验，结果显示，该物质对分离到的革兰氏阳性球菌、革兰氏阳性杆菌的抑菌作用强于革兰氏阴性杆菌，对葡萄球菌、链球菌、粪肠球菌有抑制作用，而对大肠杆菌没有抑制作用。宋丽华等（2011）曾对患隐性乳腺炎的奶牛饲喂抗菌肽制剂进行效果试验，共分对照组、高剂量组和低剂量组，其中高剂量组痊愈率为65%，体细胞降低了67.74%，低剂量组痊愈率为77.5%，体细胞率降低了69.53%。但是目前对于抗菌肽在体内外的毒性数据很少，其在体内的稳定性有待研究，大规模生产的费用较高。

十、基因治疗法

一些研究报道，奶牛乳腺炎的发生与遗传有一定关系，在基因水平防治乳腺炎有广阔的发展前景。有人研究出一种抗菌肽-金黄色葡萄A70和A53，二者是由葡萄球菌分泌的，A70和A53不仅可以抑制葡萄球菌活性，还可以抑制链球菌活性。孙怀昌（2005）证实，人溶菌酶基因的重组质粒对临床型和隐性乳腺炎有一定治疗作用，还发现重组质粒的治疗疗效有剂量和次数依赖性。Wang等（2012）发现在患乳腺炎的奶牛乳腺组织中有两个转录变异体（X-CLU和Y-CLU），同时研究证明中国斯坦奶牛的丛生基因在正常乳腺和患乳腺炎的乳腺组织均有两个不同的片段，大小分别为60 kDa和80 kDa。

十一、病原微生物药敏试验研究进展

随着养殖业的深入发展，抗菌药物的大量盲目的使用，导致很多致病菌株产生耐药性，使得抗菌药物对病原微生物控制效果越来越差，给养殖业造成重大的经济损失，同时药物残留问题也给人类健康带来了严重后果。抗生素药敏试验已成为指导临床合理科学用药不可缺少的环节。药敏检测的方法完善是快速获取科学准确的病原微生物耐药谱的前提条件，是实现有效的临床治疗的保障（王娟，2014）。

（一）兽医临床药敏检测方法

现阶段，由于药敏试验要求较高，一些养殖场或是基层兽医门诊进行药敏试验条件不够。不能有效地对病原微生物的药敏性作出准确的判断。药敏试验方法的简便、快速化有利于扩大基层兽医防治的应用的广度，已成为广大基层兽医工作者的迫切需求。基层兽医临床中的药敏试验方法有以下几种。

1. 药敏纸片法 参照美国临床试验室标准化委员会（NCCLS），采用 K-B 琼脂纸片法，此法已是国际公认的成熟技术，以其简单实用的特点，一直作为药敏试验的金标准，用来校正其他的药敏试验方法的准确性（王娟，2014）。现阶段据调查东北地区基层防控部门药敏试验方法，多采用药敏纸片法，但是由于需要进行致病菌的分离与培养计数，至少需要 3 天时间，且在操作过程中会出现技术不准、数据分析不详细、判断不准确等问题，这需要兽医人员熟练的技术。市场上药敏纸片种类有限，且多以人医用药为主，自制的纸片又存在诸多的不确定因素。另外，整个操作过程烦琐，工作量大，设备材料成本较高，大大降低了检测效率。这些都给基层兽医的推广增加障碍。

2. 试管梯度稀释法 该方法包括的抗生素浓度为 2 倍稀释的，如 2^0 μg/mL、2^1 μg/mL、2^2 μg/mL、2^3 μg/mL、2^4 μg/mL，分装于试管中。接种细菌浓度为 $1\times10^5 \sim 5\times10^5$ cfu/mL 细菌悬液于试管中，置于 35℃培养过夜，主要用于测定不同抗菌药物对不同细菌的 MIC，检查无细菌生长的浓度即为该细菌 MIC；同时也能间接地比较出抗菌药的抗菌效果，还有对抗菌药物耐药产生快与慢和恢复情况。此类方法几乎不进行测定和应用，缺点是大量的工作，操作过于烦琐，手动操作易出现错误（Jorgensen and Ferraro，2009），结果判定需借助比浊度。但是，目前，标准 96 孔细胞培养板的使用，使试管肉汤稀释试验小型化和机械化，并且更加实用方便，正在逐步得到推广（王娟，2014）。

3. 动物试验法 这种方法是将豚鼠、鸡、兔等动物随机分组，分别接种致病菌使其发病，然后用 1/8 倍、1/4 倍、1/2 倍、1 倍、2 倍、4 倍、8 倍治疗剂量，分别对各组进行治疗，每组内动物均采用同一倍数剂量，每天给药 2 次，观察并记录治疗效果。少数基层兽医工作者会对新抗菌药物进行试治试验，最主要还停留在治疗性诊断初级阶段。

（二）Etest 法

Etest 法是扩散法的一种变式，是扩散法和稀释法结合体，用非渗透性塑料条作为检测条的载体，检测条的一端至另一端包被着不同浓度呈指数函数连续变化的抗生素药物，在测试条的背面标注经校准过的 MIC 数值。将检测条贴在琼脂平板表面，35℃培养过夜，由于药物经琼脂扩散会形成一个连续变化的浓度梯度，琼脂平板上会根据待检菌株药敏性差异，形成大小不同的椭圆形抑菌圈，在抑菌圈与检测条的交界处，所对应的数值即为该菌对此药物的 MIC 值（Jorgensen and Ferraro，2009）。

（三）自动化药敏仪

基于肉汤微量稀释法原理设计，以测定药物的 MIC 值。目前在美国通过 FDA 批准使用的 4 种自动化系统有 Phoenix、Sensititre ARIS 2X、Vitek 1 和 2、WalkAway（Jorgensen and Ferraro，2009）。它们可以通过检测培养物的浊度、荧光度等来判读结果，如此可以更规范准确地读数，能检测细菌生长的微妙变化。仪器的使用具有快速敏感，分析功能强大的优点，但是价格昂贵，药物选择不灵活，难以在临床基层推广。

（四）基于辅助显色试剂的快速药敏检测

传统药敏试验检测在终点判断时，可以利用一些显色剂来辅助观察，实现更直观、快速、准确地判断。细菌的存在与增长过程，能产生 ATP、水解酶以及氧化还原酶。我们可以利用 ATP 生物荧光法、氧化还原指示剂法如刃天青、四甲基偶氮唑盐（MTT）等（王娟，2014）。

荧光显色剂要有专门的荧光激发装置，配套使用，氧化还原显色剂可通过肉眼观察，定性判定；也可用分光光度计定量检测。此方法简便、廉价、实用性强，可对不同病原菌进行药敏测定，可以丰富药敏检测方法，以满足各种类型病原菌的药敏检测需求。尤其是畜牧水产行业，可以缩短检测时间，有效用药，达到有效治疗的目的。目前，MTT 显色剂多用于病原微生物药敏性试验研究，但是，MTT 指示剂有一定的毒性，需避光使用和保存，给临床应用带来一定的不便。刃天青是一种具有安全性、稳定性、对细胞无毒性作用且易溶于水的新型染料。本实验选用刃天青作为显色剂，进行致奶牛乳腺炎大肠杆菌药敏试验。

十二、耐药性问题

近些年，由于畜主常在不确定病原的情况下，大量滥用抗生素，使得耐药菌株不断增多且耐药性越来越强，抗生素的疗效明显下降或无效，在治疗过程中一度陷入僵局，而牛奶中严重的抗生素残留，直接危害了人类的健康。有资料表明，细菌耐药性可通过质粒在不同种类的细菌间传播，同时耐药菌株还可在动物和人类之间传播，对人类健康和公共卫生安全带来威胁（赵红波等，2010）。因此，了解细菌的耐药机制，摸清耐药性产生的趋势，势必对治疗奶牛乳腺炎有一定的帮助。

（一）耐药性的发生机制

1. 灭活酶或钝化酶的产生　　细菌产生灭活酶是引起耐药性的最重要机制，该酶可破坏抗生素，使其在作用菌体前即被破坏或失活，不能发挥抗菌作用。例如，β-内酰胺酶，在革兰氏阳性菌中，以金黄色葡萄球菌产生的青霉素酶最为重要。钝化酶主要是氨基苷类钝化酶，它既可使药物不易进入细菌体内，又可使药物不易与细菌的内靶位（核糖体 30S 亚基）结合，从而丧失抑制蛋白质合成的能力（买尔旦·马合木提，2003）。

2. 抗生素的渗透障碍　　细菌外膜上的某种特异多孔蛋白发生变异，使其通透性降低，从而阻碍抗生素进入细胞内膜靶位，如不动杆菌、铜绿假单胞菌等。

3. 靶位的改变　　细菌通过改变靶位酶或改变靶位的生理重要性，使其不能被抗生素所作用，从而导致耐药性的产生。

4. 主动外排系统　　在能量支持下，细菌细胞膜上的一类蛋白质能特异地将进入细胞内的抗生素主动排出细胞外，从而导致耐药性的产生，目前已发现细菌中普遍存在该系统，且该系统与细菌的外膜障碍或灭活酶或靶位改变共同发挥耐药的功能，如四环素类、喹诺酮类等（张颖，2014）。

（二）耐药的基因机制

根据细菌耐药的基因机制，细菌可通过不同机制产生遗传变异和对抗菌药物的耐药性。据遗传特性，将细菌耐药性分为两类。

1. 自身基因突变　某些细菌的染色体上本身带有耐药基因，由于遗传基因 DNA 自发变化，具有典型的种属特异性，能够代代相传，但这种耐药性比较稳定，且与是否接触药物无关。

2. 获得外源基因　细菌耐药基因常从附近其他细菌细胞的染色体或质粒、转座子中获得。通过转化、转导、接合、转座进行传播。其中最常见的耐药质粒是对抗生素耐药性编码的 R 质粒（耐药因子）（张颖，2014）。

（三）我国奶牛乳腺炎耐药情况

王正兵等（2011）曾对我国临床型奶牛乳腺炎中分离的无乳链球菌、停乳链球菌、金黄色葡萄球菌、大肠杆菌进行耐药性检测，得知金黄色葡萄球菌对青霉素 G、丁胺卡那、复方新诺明的耐药率达 30%～70%，无乳链球菌对青霉素 G、氨苄西林、链霉素、恩诺沙星、阿莫西林、复方新诺明的耐药率达 50%～100%，停乳链球菌对青霉素 G、氨苄西林、链霉素、多黏菌素 B 的耐药率达 50%～90%，大肠杆菌对青霉素 G、链霉素、四环素、红霉素、林可霉素、万古霉素及复方新诺明的耐药率达 50%～100%。张静（2012）对乳腺炎中金黄色葡萄球菌的耐药情况进行分析，对红霉素和甲氧苄啶的耐药率分别为 100%和 71.93%，对头孢西丁、氯霉素、庆大霉素和环丙沙星的耐药率在 19%～30%。燕霞等（2012）曾在浙江省主要奶牛养殖区分离到的 70 株金黄色葡萄球菌进行耐药性检测，主要对青霉素 G、庆大霉素、红霉素和四环素耐药，表型耐药率分别为 90%、67.1%、50%和 60%。李宏胜等（2012）曾对我国部分地区奶牛场采集的临床型奶牛乳腺炎的乳汁中的分离出的无乳链球菌进行耐药情况的测定，对青霉素 G、链霉素、恩诺沙星、复方新诺明等耐药率达 61.74%～85.22%。徐继英等（2012）调查大肠杆菌性奶牛乳腺炎的耐药状况，发现 95 株大肠杆菌对 16 种抗菌药物中的 8 种耐药率超过 50%，青霉素的耐药率甚至达到 100%，耐药 6 种以上的菌株占到 51.58%。

因此，在奶牛乳腺炎的治疗中，我们应找准病因及主要致病菌，结合实验室检测，筛选出最适宜的药物，严格掌握药物的适应证、剂量、疗程，从而达到最佳的治疗效果。

第二节　中草药制剂对奶牛乳腺炎致病菌抗菌效果分析

奶牛乳腺炎是危害奶牛养殖业最常见的疾病之一，同时也是导致奶牛淘汰的第三大疾病（Biffa et al.，2005）。使用抗生素仍然是防治乳腺炎的重要手段，特别是对急性、多发性和亚急性乳腺炎具有重要应用价值。随着抗菌药物的广泛应用，耐药菌株的增多，细菌对常用抗菌药物的耐药性越来越强。另外，乳腺炎乳汁中的大量致病菌及其产生的毒素或者治疗后残留的大量抗生素，除了能引起致病菌产生耐药性外，还会使某些饮用者产生过敏反应，可直接危害人类健康（李成应，2008）。随着人们对食品安全要求的不断提高和国家对乳制品质量监管的更加严格，临床治疗奶牛乳腺炎常用的抗生素将逐渐被限用或禁用。所以开发抗菌作用强、毒副作用小、无药物残留的中草药制剂来替代传统抗菌药物疗法，成为近年来研究的热点。中草药是天然药物，包括动物、植物和矿物质，不仅含有生物碱、多糖、皂苷、挥发油、蒽类和有机酸等有效生物活性物质，而且含有矿物质、维生素及未知的营养因子，兼有预防、治疗和营养的多重作用（张少华，2009）。中草药既可抑菌杀菌、增强动物机体的免疫功能，又可作为特种生物活性添加剂调节机体的新陈代谢，促进营养物质的吸收。中草药毒性低，对动物及人类毒副作用小，且不易产生抗药性，药残留相对低（丁月云，2004）。

因国家地区和年份季节的不同，奶牛乳腺炎的致病菌有一定的差别，一般以细菌感染为主，其中葡萄球菌和大肠杆菌比较常见（关红等，2010）。前期实验结果表明规模化奶牛场奶牛乳腺炎致病菌主要是金黄色葡萄球菌和大肠杆菌。本实验根据奶牛乳腺炎的发病机制和中草药的药性，选取具有清热解毒、抗菌消炎、可提高机体免疫力的中草药，采用水煎提取法，对奶牛乳腺炎金黄色葡萄球菌和大肠杆菌进行抑菌试验，筛选出具有体外抑菌强度大、最低杀菌浓度小的中草药，并通过正交设计配制复方中草药，为研制治疗奶牛乳腺炎的方剂提供理论依据以及为临床应用奠定试验基础。

一、材料与方法

（一）实验材料

1. 实验菌株、药物和试剂

菌株：从奶牛乳腺炎分离、纯化、鉴定得到的金黄色葡萄球菌和大肠杆菌各一株。

药物：板蓝根、蒲公英、紫花地丁、鱼腥草、连翘、秦皮、诃子、金银花、乌梅、黄芩、当归和黄连12味中草药均购自某药房，并按照2005年《中国兽药典》（第2版）的方法进行鉴定，符合生药质量标准。

营养琼脂、营养肉汤均购自某生物公司。

2. 实验仪器 生物显微镜，高压灭菌器，超净工作台，电子可调电炉，电子分析天平，干燥箱，恒温培养箱，冰箱。

（二）实验方法

1. 药液制备 12味中草药各取40 g，以400 mL去离子水浸泡2 h后煎煮，煮沸后文火煎煮30 min，4层纱布过滤，再加入200 mL去离子水煎煮，煮沸后文火煎煮30 min，纱布过滤，合并所得药液，2000 r/min离心10 min，弃去沉淀，文火加热药液浓缩至40 mL，使其最终成为生药浓度为1 g/mL的水煎剂，分装后100℃灭菌30 min，4℃保存备用（张少华，2009；白峰，2010）。

2. 菌悬液的制备 首先将金黄色葡萄球菌和大肠杆菌在营养琼脂平板上接种，37℃培养16~18 h复壮，在平板上挑取1个典型的菌落接种于10 mL普通肉汤培养基37℃培养16~18 h，然后取100 μL该培养液加到10 mL普通肉汤培养基37℃培养16~18 h，将肉汤培养扩增的金黄色葡萄球菌、大肠杆菌用肉汤培养基稀释至0.5麦氏标准比浊管（相当于1×10^8 cfu/mL），备用。

3. 单味中草药的体外抑菌试验 采用平板挖洞灌药法进行体外抑菌试验（刘建文，2003）。取稀释后的金黄色葡萄球菌和大肠杆菌菌悬液各200 μL，分别均匀涂于营养琼脂平板表面，然后在平板上用6 mm打孔器在涂菌的琼脂平板上均匀打4个孔。封底方法：打好的孔在酒精灯火焰上略微加热，使底部局部培养基熔化后重新凝固，与平皿底壁紧密结合，防止加入的药液从底部的缝隙流出。

将 12 种制备好的中草药水煎剂用移液枪分别加入接种有金黄色葡萄球菌和大肠杆菌菌种的琼脂平板中，每孔以注满为准，注意将平皿平放防止药液溢出影响试验结果准确性。每种药重复两孔作为平行样，对照孔加灭菌生理盐水。37℃培养 24 h，记录抑菌圈直径大小，每孔取不同方向测量 2 次，求 4 次测量数据的平均值，同时设空白对照。

抑菌效果判定标准参照王桂英（2010），即≥20 mm 为极敏，15~19 mm 为高敏，10~14 mm 为中敏，<10 mm 为低敏（耐药）。

4. 中草药对金黄色葡萄球菌和大肠杆菌最低抑菌浓度（MIC）和最低杀菌浓度（MBC）的测定 采用试管二倍稀释法测定 MIC 和 MBC（刘建文，2003）。根据琼脂扩散法体外抑菌试验结果，在无菌条件下，取无菌试管 10 支（15 mm×100 mm），每管加入营养肉汤 2 mL，在第 1 管加入药物原液 2 mL，充分混匀后吸取 2 mL 加入第 2 管中，以此二倍稀释至第 9 管，最后从第 9 管中吸取 2 mL 弃去，第 10 管不加药液为 2 mL 纯培养基，作为对照。第 1~10 管药物浓度分别为 500 mg/mL、250 mg/mL、125 mg/mL、63 mg/mL、31 mg/mL、16 mg/mL、8 mg/mL、4 mg/mL、2 mg/mL、0 mg/mL。再在每个试管中加入制备好的各种菌悬液 0.1 mL，混匀，置于 37℃恒温培养箱培养 18~24 h。肉眼观察，以培养物透明、无细菌生长的最低药物浓度作为该药的 MIC。再从透明管取培养物 50 μL 接种到营养琼脂平板，用灭菌"L"形玻璃棒涂布均匀，并设定阳性对照和阴性对照各一个营养琼脂平板，37℃恒温培养箱培养 18~24 h。计菌落数，菌落数不超过 3 个的最低提取物浓度即为该提取物的 MBC。试验重复 2 次。根据 MIC 和 MBC 值评价中草药的体外抗菌活性。

5. 抗菌中草药组方的筛选 采用正交试验设计，将对金黄色葡萄球菌和大肠杆菌均敏感的药物按照 L_8（2^7）正交表（表 5-1）配制成相应的 8 种中草药制剂。采用试管二倍稀释法测定每个中草药组方对金黄色葡萄球菌和大肠杆菌的 MIC（刘建文，2003）。根据琼脂扩散法体外抑菌试验结果，在无菌条件下，取无菌试管 10 支（15 mm×100 mm），每管加入营养肉汤 2 mL，在第 1 管加入药物原液 2 mL，充分混匀后吸取 2 mL 加入第 2 管中，以此二倍稀释至第 9 管，最后从第 9 管中吸取 2 mL 弃去，第 10 管不加药液为 2 mL 纯培养基，作为对照。第 1~10 管药物浓度分别为 500 mg/mL、250 mg/mL、125 mg/mL、63 mg/mL、31 mg/mL、16 mg/mL、8 mg/mL、4 mg/mL、2 mg/mL、0 mg/mL。再在每个试管中加入制备好的各种菌悬液 0.1 mL，混匀，置于 37℃恒温培养箱培养 18~24 h。肉眼观察，以培养物透明、无细菌生长的最低药物浓度作为该药的 MIC。

表 5-1　L_8（2^7）正交表
Table 5-1　The orthogonal table L_8（2^7）

	A	B	C	D	E	F	G
组方 1	1	1	1	1	1	1	1
组方 2	1	1	1	0	0	0	0
组方 3	1	0	0	1	1	0	0
组方 4	1	0	0	0	0	1	1

续表

	A	B	C	D	E	F	G
组方 5	0	1	0	1	0	1	0
组方 6	0	1	0	0	1	0	1
组方 7	0	0	1	1	0	0	1
组方 8	0	0	1	0	1	1	0

注：A、B、C、D、E、F 和 G 代表 7 个不同的因素（中草药），1 表示有该味中草药，0 表示没有该味中草药

二、实验结果

（一）药物的敏感性测定结果

平板挖洞灌药法测定 12 味中草药的水煎剂对奶牛乳腺炎两种主要致病菌的体外敏感性，结果见表 5-2。

表 5-2 12 味中草药对金黄色葡萄球菌和大肠杆菌的抑菌效果

Table 5-2 The results of antibiotic susceptivity of 12 Chinese herbs on *S. aureus* and *E. coli*

中草药名称	抑菌圈直径/mm	
	金黄色葡萄球菌	大肠杆菌
板蓝根	15.5	9.3
蒲公英	16.8	12.5
紫花地丁	16	10
鱼腥草	9.3	—
连翘	24.3	7.3
秦皮	18	12.3
诃子	19	10.8
金银花	15.8	13.3
乌梅	17.3	18.5
黄芩	16.5	13.5
当归	16	8
黄连	22.8	12.3
对照（灭菌生理盐水）	—	—

注：表中 "—" 代表未产生抑菌圈

由表 5-2 可知，金黄色葡萄球菌对连翘和黄连极度敏感，对板蓝根、蒲公英、紫花地丁、秦皮、诃子、金银花、乌梅、黄芩和当归高度敏感，对鱼腥草耐药；大肠杆菌对乌梅高度敏感，对板蓝根、鱼腥草、连翘、和当归均耐药。对金黄色葡萄球菌和大肠杆菌均敏感的中草药是蒲公英、紫花地丁、秦皮、诃子、金银花、乌梅、黄芩和黄连。

（二）中草药对金黄色葡萄球菌和大肠杆菌的 MIC 和 MBC

12 味中草药的水煎剂对奶牛乳腺炎两种主要致病菌的 MIC 和 MBC 测定结果见表 5-3。

表5-3 12味中草药对金黄色葡萄球菌和大肠杆菌的MIC和MBC测定结果
Table 5-3 The MBC and MIC of 12 Chinese herbs on *S. aureus* and *E. coli*

中草药名称	金黄色葡萄球菌		大肠杆菌	
	MIC/（mg/mL）	MBC/（mg/mL）	MIC/（mg/mL）	MBC/（mg/mL）
板蓝根	16	32	125	250
蒲公英	32	32	125	250
紫花地丁	63	125	250	500
鱼腥草	250	>500	>500	>500
连翘	16	16	125	250
秦皮	125	125	125	250
诃子	8	32	63	125
金银花	63	250	125	250
乌梅	16	32	16	16
黄芩	16	32	250	500
当归	125	250	250	500
黄连	8	16	16	32

由表5-3可以看出，大多数中草药对金黄色葡萄球菌的MIC和MBC均比较低，其中诃子、黄连的抑制作用最强，MIC达到8 mg/mL，其次是板蓝根、连翘、乌梅和黄芩，MIC是16 mg/mL，蒲公英、紫花地丁和金银花也有较强的抑制作用，MIC是63 mg/mL，而鱼腥草的抑制作用较弱。大多数受试中草药对大肠杆菌的抑菌和杀菌作用比较差，其中抑制作用最强的是乌梅和黄连，MIC达到16 mg/mL，其次是诃子，MIC是63 mg/mL，而鱼腥草的MIC大于500 mg/mL，基本上对大肠杆菌没有抑制作用。

（三）抗菌中草药组方筛选结果

试管二倍稀释法中草药组方筛选结果见表5-4。

表5-4 $L_8（2^7）$正交法中草药抑菌结果
Table 5-4 Bacteriostatic results of the Chinese herbs in orthogonal test on $L_8（2^7）$

		A	B	C	D	E	F	G	对金黄色葡萄球菌的抑菌浓度/（mg/mL）	对大肠杆菌的抑菌浓度/（mg/mL）
	组方1	1	1	1	1	1	1	1	8	16
	组方2	1	1	1	0	0	0	0	31	63
	组方3	1	0	0	1	1	0	0	31	63
	组方4	1	0	0	0	0	1	1	16	16
	组方5	0	1	0	1	0	1	0	16	31
	组方6	0	1	0	0	1	0	1	16	31
	组方7	0	0	1	1	0	0	1	8	16
	组方8	0	0	1	0	1	1	0	31	31
金黄色葡萄球菌	K_1	23	24	24	25	23	24	26		
	K_0	24	23	23	22	24	23	21		
	R	−1	1	1	3	−1	1	5		

		A	B	C	D	E	F	G	对金黄色葡萄球菌的抑菌浓度/（mg/mL）	对大肠杆菌的抑菌浓度/（mg/mL）
大肠杆菌	K_1	20	20	21	21	20	22	23		
	K_0	21	21	20	20	21	19	18		
	R	−1	−1	1	1	−1	3	5		

注：A. 板蓝根，B. 蒲公英，C. 连翘，D. 诃子，E. 金银花，F. 乌梅，G. 黄连；1 表示有该味中草药，0 表示没有该味中草药；K_1 表示有该味中草药组方的中草药制剂中培养物透明、无细菌生长的试管数，K_0 表示没有该味中草药组方的中草药制剂中培养物透明、无细菌生长的试管数；$R=K_1-K_0$

试验结果分析：在表 5-4 中，由于 1 表示有该味中草药，0 表示没有该味中草药，$R=K_1-K_0$，采用直观分析方法，R 值为正值时，表明此中草药在该组方中起到提高抑制细菌生长的作用，R 值越大说明作用越明显，R 值为负值时表明此味中草药在该组方中不起疗效。例如，金黄色葡萄球菌，因子 A：$K_1=7+5+5+6=23$，$K_0=6+6+7+5=24$，极差 $R=23-24=-1$；同样方法计算因子 B、C、D、E、F、G 的 K_1、K_0 和级差 R。大肠杆菌也是如此计算。综合金黄色葡萄球菌和大肠杆菌的抑菌结果，可以确定因子 C（连翘）、D（诃子）、F（乌梅）、G（黄连）均能明显提高对金黄色葡萄球菌和大肠杆菌的抑制作用，因此作为组方的主要成分。该组方对金黄色葡萄球菌和大肠杆菌的 MIC 分别是 8 mg/mL 和 16 mg/mL。

三、讨论

（一）选药依据分析

中兽医学有着悠久的历史和丰富的应用经验，在抗生素发明和使用以前，对保障家畜健康起着重要作用。中兽医传统理论认为乳腺炎即乳痈，是痰、湿、气、血郁结不散，化而为炎，与肝胃两经有关；或因饲养管理不善，久卧湿热之地，湿热毒气上蒸，侵害乳房；或因胃热壅盛、肝郁气滞、乳络失畅以致乳房气血凝滞、瘀结而生痈肿；或因牛犊吸乳时咬伤乳头，邪毒入侵引起痈肿；或因患牛拒绝挤奶，使乳汁停滞，乳房胀满等而导致本病（付秀花和王恬，2001；刘俊杰，2010）。此外，兽医技术人员应用激素治疗奶牛生殖系统疾病时，因其正常激素分泌被扰乱，也可导致奶牛乳腺炎的发生。因此，中药方剂的组方应以清热解毒、通经活血、消肿散结、活络通乳为原则。

中草药抑制细菌生长的有效成分很多，主要成分有生物碱、有机酸、皂苷及黄酮等物质，大多数可溶于水。由于中草药成分复杂，有些有效成分还有可能出现相互间"助溶"作用，使本来不溶于水或难溶于水的成分在用水作浸出溶剂时被浸出；另外，水具有溶解范围广、极性大、经济易得、无药理作用等特点，所以本实验选去离子水作为浸出溶剂。

本试验挑选出的中草药中，板蓝根清热解毒，有提高机体免疫功能，根部主要成分是靛苷、靛红、靛蓝等，还有抗革兰氏阳性和阴性细菌的抑菌物质，如酮类、β-谷甾醇等。蒲公英味甘微苦、性平，可清热解毒、消肿散结，含蒲公英甾醇、蒲公英赛醇、蒲公英苦素、咖啡酸、胆碱、有机酸、菊糖等成分，可使细菌发生菌体细胞膨大，细胞壁增厚等改变，从而具有抑菌抗菌作用。有资料表明蒲公英有显著的催乳作用，治疗乳腺炎十分有效，无论煎汁口服，还是捣泥外敷，皆有效验。紫花地丁性寒味微苦，清热解毒，凉血消肿，解毒力强，主要成分是苷类、黄酮类等。鱼腥草味苦、性微寒，具有清热解毒、排脓消痈之功效，全草含挥发性油，其中有效成分为癸酰乙醛（即鱼腥草素）、月桂醛、芳樟醇等，能够提高机体的免疫力，对感染性疾病的治疗有着重要的意义。连翘清热解毒，清心火而解毒消痈，泻胆热而解郁散结，主要成分是连翘酚、甾醇化合物、皂苷（无溶血性）及黄酮醇苷类、马苔树脂醇苷等，抗微生物的主要有效成分是连翘脂苷和连翘苷。秦皮味苦、性寒，具有清热燥湿之功效，主要有效成分是马栗树皮素、秦皮苷、丁香苷、东莨菪素等，具有抗菌、消炎、镇痛的作用。诃子味苦、酸、涩，性平，主要含三萜酸类、诃子素、诃子酸、诃黎勒酸、鞣料云实精、诃子鞣质等，对常见的人体致病菌均有明显的抗菌活性，表现出广谱抗菌作用，如绿脓杆菌、白喉杆菌、金黄色葡萄球菌、大肠杆菌、肺炎球菌、伤寒杆菌、变形杆菌、溶血性链球菌等（庄玉坚，2009）。药理作用研究表明，金银花具有清热解毒、凉散风热、抑菌、抗病毒、抗炎、抗氧化、增强免疫功能等作用，赵彦杰研究发现，采用75%乙醇提取的金银花叶提取液对大肠杆菌、金黄色葡萄球菌、枯草芽孢杆菌、鼠伤寒沙门氏菌、青霉、黑曲霉、黄曲霉和酿酒酵母等均有一定的抑制作用（赵彦杰，2007），其主要成分是绿原酸类、苷类、黄酮类、挥发油类，其中绿原酸被认为是金银花中抗菌解毒、消炎利胆的主要有效成分。乌梅味酸涩，性平，主要化学成分是枸橼酸、苹果酸、草酸、琥珀酸和延胡索酸等，具有抗菌、解毒、收敛止血和去腐生新的作用，体外筛选发现，乌梅对多种致病菌有抑制作用，如痢疾杆菌、大肠杆菌、伤寒杆菌、副伤寒杆菌、百日咳杆菌、脑膜炎双球菌等，宋波等（2010）研究发现，乌梅对大肠杆菌和白假丝酵母菌都有较强的抑菌作用，MIC值分别是1.25 mg/mL、2.5 mg/mL和5 mg/mL。黄芩味苦，性寒，主要含黄芩苷、黄芩素等成分，抗菌谱较广，对多种细菌、皮肤真菌、钩端螺旋体等都有抑制作用。即使对青霉素等抗生素产生抗药性的金黄色葡萄球菌，对黄芩仍然很敏感。王雨玲（2010）研究发现，黄芩对体外幽门螺旋杆菌的MIC是3.9 mg/mL，效果非常明显，宋波等（2010）研究发现黄芩的MIC是2.5 mg/mL，具有较强的抑菌效果，此外，对Ⅰ型变态反应（过敏反应）作用显著。当归味甘、辛，性温，含藁本内酯、正丁烯酰内酯等挥发油，对体外大肠杆菌、白喉杆菌、霍乱弧菌及α、β溶血性链球菌等均有抗菌消炎作用。黄连味苦，性寒，主要成分是小檗碱（黄连素）、黄连碱等，其主要有效成分是小檗碱。小檗碱是一种广谱抗菌药，在临床上已应用多年，疗效确切，对多种革兰氏阳性、阴性菌，以及真菌、霉菌、病毒、原虫、线虫具有抑制杀灭作用（Čerňáková and Košťálová，2002）。

（二）中草药单药的敏感性测定

采用平板挖洞灌药法检测12味中草药对金黄色葡萄球菌和大肠杆菌的体外抑菌敏感性，研究发现金黄色葡萄球菌对连翘和黄连极度敏感，对板蓝根、蒲公英、紫花地丁、秦皮、诃子、金银花、乌梅、黄芩和当归高度敏感，对鱼腥草耐药；大肠杆菌对乌梅高度敏感，对板蓝根、鱼腥草、连翘和当归均耐药。对金黄色葡萄球菌和大肠杆菌均敏感的中草药是蒲公英、紫花地丁、秦皮、诃子、金银花、乌梅、黄芩和黄连。

采用试管二倍稀释法测定所选中草药对金黄色葡萄球菌和大肠杆菌的MIC和MBC，结果发现，诃子和黄连对金黄色葡萄球菌的MIC最低，达到8 mg/mL，板蓝根、连翘、乌梅和黄芩次之，MIC 是 16 mg/mL；乌梅和黄连对大肠杆菌的 MIC 最低，达到 16 mg/mL，诃子次之，MIC 是 63 mg/mL，板蓝根、蒲公英、连翘、秦皮和金银花再次之，MIC 是 125 mg/mL。Sato 等研究发现从诃子果实提取物中分离纯化的没食子酸及其乙酯对耐甲氧西林的金黄色葡萄球菌具有很强的抑制作用，其有效抑菌浓度是 1∶128（试管法，相当于 8 mg/mL 生药浓度）。刘东梅等（2008）研究发现，黄连对产 β-内酰胺酶的大肠杆菌等革兰阴性杆菌具有较强的抑菌作用，同时与已报道的研究结果基本一致（吴国娟等，2003；丁月云，2004；赵彦杰，2007；张少华等，2009；白峰等，2010；关红等，2010；宋波等，2010）。

对于抑菌作用较强的中草药，用平板挖洞灌药法检测其对致病菌的抑菌效果时可能被掩盖，用试管二倍稀释法比较精确。但是有些中草药水煎剂色泽比较深、浑浊度比较大，细菌在试管内培养后不易详细观察细菌的生长情况。本试验采用平板挖洞灌药法和试管二倍稀释法两种实验方法进行中草药对致病菌的抑菌效果，克服了上述两种实验方法的不足，既定性又定量地体现了各种中草药水煎剂对金黄色葡萄球菌和大肠杆菌的抑菌作用，可以比较准确地测定各药物的抑制作用，两种实验方法试验结果基本一致。

仔细观察并对比两种实验方法中每种中草药对致病菌的抑菌强度，稍微有些差异，其原因可能有以下两个方面：一方面，有些中草药经过稀释后（用营养肉汤稀释），抑菌效果减弱得快，而有些减弱得慢；另一方面，中草药水煎剂的成分种类繁多，发挥药效作用的化学成分不稳定。如鱼腥草鲜汁对金黄色葡萄球菌有抑制作用，加热后作用减弱，鱼腥草的主要抗菌有效成分癸酰乙醛对多种细菌、抗酸杆菌及真菌等均有较明显的抗菌作用，但该种化合物性质不稳定。

（三）抗菌中草药组方筛选

正交试验设计是针对多因素多水平实验组合中寻求最优水平组合的一种高效的科学试验设计方法，它是利用一套规范化的正交表合理安排试验的因素和水平，并分析试验结果，从而得出最佳优化条件，具有"均匀分散、整齐可比、计算简便"的优点，是目前中草药制备工艺中多因素优选中最常用的设计方法，非常适合中草药组方的研究。中草药的组方非常复杂，通常单独一种中草药就具有多种药理作用，中药的组方研究应采用多指标综合评价，评价药物组合中各种中草药之间的交互作用。另外，可以避免大量的拆方试验，大大缩短中草药组方研究开发的时间。因此，本试验采用了正交试验设计，以药物体外抗菌活性为指标，筛选出抗奶牛乳腺炎金黄色葡萄球菌和大肠杆菌的中草药最佳组方，采用正交设计中药组方，有计算简单、结论可靠的优点，用该法进行中草药配伍试验是一种值得推广的科学方法。

本试验根据正交试验设计原理，评价了板蓝根、蒲公英、连翘、诃子、金银花、乌梅和黄连 7 种中草药的最佳组合，确立了 $L_8(2^7)$ 正交试验，共进行了 8 组试验，根据正交试验结果，最终确立了中草药组方。该组方是由连翘、诃子、乌梅和黄连组成，对金黄色葡萄球菌和大肠杆菌的 MIC 分别是 8 mg/mL 和 16 mg/mL。为临床治疗奶牛乳腺炎提供一定的基础和依据，同时，该优选组方可用于奶牛乳腺炎的临床研究。

本试验所用菌株直接从临床乳腺炎奶牛乳汁中分离并鉴定得到，用中草药水煎剂对金黄色葡萄球菌和大肠杆菌的抗菌活性进行测定，其结果更有说服力。中草药复方制剂用于治疗奶牛乳腺炎时，不仅毒性低，不易产生抗药性，药残留相对低等特点。宋华容等（2009）认为，中草药复方制剂可以增强奶牛机体的免疫功能，对 T 淋巴细胞介导的免疫反应有明显的促进作用，提高细胞免疫功能。使用中草药治疗奶牛乳腺炎在临床上已经是屡见不鲜，所用的中草药种类也很多，绝大部分是在体外药敏试验的基础上选定的，另外一些是根据临床经验选定的。由于体内各种酶、激素等化学成分对药物处理的影响，体内外作用效果有一定差异，因此对于体内试验还需进一步研究。中草药治疗奶牛乳腺炎的药理作用主要表现为抑制病原微生物生长或杀菌、抗病毒、解热、抗炎、增强机体的免疫力、抗氧化、清除自由基等，所以深入研究中草药对致病菌的抑菌和杀菌作用，对控制和治疗耐药菌感染所引起的疾病具有非常重要的实际意义。

四、小结

（1）奶牛乳腺炎金黄色葡萄球菌对连翘和黄连极度敏感，乳腺炎大肠杆菌对乌梅高度敏感。

（2）正交设计显示连翘、诃子、乌梅和黄连均能明显提高对奶牛乳腺炎金黄色葡萄球菌和大肠杆菌的抑制作用，可以作为组方的主要成分，该组方对金黄色葡萄球菌和大肠杆菌的 MIC 分别是 8 mg/mL 和 16 mg/mL。

第六章　奶牛乳腺炎疫苗研究

第一节　奶牛乳腺炎疫苗研究进展

目前国内外防治奶牛乳腺炎主要靠加强牛场管理和抗生素治疗，由于抗生素疗法在公共卫生上的局限性和耐药菌株的产生，研究者进行了大量的疫苗接种实验，结果证明，利用疫苗预防和控制奶牛乳腺炎发生有一定的效果。同时也从理论上说明，奶牛乳腺炎是乳腺组织受病原微生物感染，具有一定的传染性，而且以金黄色葡萄球菌、大肠杆菌和链球菌为主，占乳腺炎发生的一半以上，因此，开发针对这些病原微生物主要抗原蛋白的菌苗十分必要（尹荣兰，2009）。采用接种疫苗预防和控制奶牛乳腺炎还有许多优点，如有效时间持久、乳汁无药物残留、能够降低乳腺感染的严重程度，而且操作简便、费用低廉（林锋强等，2002）。当然不同地区引起奶牛乳腺炎的主要病原菌不尽相同，再者乳腺的生理解剖特征及其防御机制等原因不同，使疫苗在田间实验的效果不够理想，有学者认为，筛选疫苗佐剂和研制多价苗、多联苗，是研究乳腺炎疫苗的重要方向（林锋强等，2002）。

乳腺的特殊生理和组织结构，使得在利用疫苗预防奶牛乳腺炎上存在许多独特性问题。首先，乳腺炎是乳腺抵抗侵入病原菌的一种免疫反应，特异性增强免疫反应也可能使疾病恶化；其次，乳房内储存有大量的乳汁，使乳腺内抗感染的免疫成分被稀释，也是致病菌良好的培养基，乳汁中的脂类和酪蛋白极大地降低了乳腺和乳汁中抗原呈递细胞的吞噬和杀菌活性；最后，乳腺的自身位置使其即使在最好的管理和卫生条件下也能接触泥浆、粪尿，这使乳腺经常处于环境性乳腺炎病原的威胁下。由于这些内在的困难，研制安全有效的疫苗变得非常困难（Shinefield，2006），同时也很难判定一种乳腺炎疫苗是成功还是失败的。成功的疫苗是否要具有降低乳腺炎的严重程度和发病率，预防感染的发生，消除已存在的感染，还是三者兼有。然而，结果远非像其他成功的疫苗所料的那样好，因为大多乳腺炎疫苗难以发挥应有的作用。基于以上乳腺免疫的特殊性，开发可靠有效的佐剂也成为未来乳腺炎疫苗研究的一个关键问题。尽管乳腺免疫接种存在困难，但在过去的十来年里，奶牛乳腺炎疫苗的研制也取得了显著进展。

奶牛乳腺炎应用疫苗防治有很多优点，如无药物残留、降低感染程度以及防控亚临床型乳腺炎，有效控制奶牛乳腺炎的发病率，降低奶产业的损失。用于临床实验的主要有金黄色葡萄球菌疫苗、大肠杆菌疫苗和链球菌疫苗等。

一、金黄色葡萄球菌疫苗

近年来由于抗生素的大量使用，金黄色葡萄球菌的耐药性越来越强，甚至出现耐药性极强的"超级细菌"，使得抗生素治疗效果越来越差，这给预防及治疗乳腺炎都提出新的严峻挑战。疫苗被认为是预防金黄色葡萄球菌感染最具前景的方法。Watson 等利用金黄色葡萄球菌弱毒株皮下接种奶牛，结果显示，该疫苗具有中等程度的免疫保护，同时，也证实了机体 IgG2 在对金黄色葡萄球菌免疫中的重要性。鉴于此，开发出由金黄色葡萄球菌和类毒素为抗原，矿物油为佐剂的疫苗，该疫苗能显著降低金黄色葡萄球菌引起的乳腺炎发生率。可是在大规模田间实验中，该疫苗仅起到了一定的保护作用，但是与对照组相比，统计学无差异显著性（武泽轩，2010）。Fattom 等（2004）将荚膜多糖 5 和 8 偶联到载体蛋白上，在小鼠体内诱导抗荚膜多糖的抗体反应，证实所诱导的抗体具有调理作用。Zhou 等（2006）融合了 Cap 和 Fnbp 作为蛋白疫苗免疫小鼠，发现该疫苗能对小鼠起到免疫保护作用。

随着医学分子生物学实验技术的发展，人们利用基因重组技术将编码某种抗原蛋白的外源基因与真核表达载体重组后，直接导入动物细胞内，并通过宿主细胞的转录系统合成抗原蛋白，诱导宿主产生对该抗原蛋白的免疫应答，以达到预防和治疗疾病的目的，这种疫苗称为核酸疫苗（Kimman et al.，2007）。核酸疫苗的出现，促使国内外学者针对金黄色葡萄球菌乳腺炎疫苗加深了疫苗的研究深度。1999 年，Brouillette 等研制了针对金黄色葡萄球菌 ClfA 的核酸疫苗，小鼠接种该疫苗后，产生了高效价的特异性抗体，但是腹腔攻毒结果显示其保护率不够理想。2004 年，Shkreta 等设计了针对金黄色葡萄球菌 ClfA 和 Fnbp 的二价核酸疫苗，接种奶牛并用重组蛋白加强免疫了一次，结果发现能引起奶牛明显的细胞免疫反应和体液免疫反应，对奶牛乳腺炎的发生起到了一定的保护作用。2007 年，Gaudreau 等（2007）构建的 Clfa 和 FnbpA 等多基因核酸疫苗，可使 55% 的免疫小鼠在攻毒后受到保护，而对照组仅有 15% 的免疫小鼠得到保护。100 多年来，国内外针对金黄色葡萄球菌疫苗的研发经历了全菌灭活苗、亚单位疫苗及 DNA 疫苗几个阶段。

（一）全菌灭活苗

全菌灭活苗又称死疫苗，经化学或物理方法杀灭病原体制得，需多次接种才能激发有效的免疫，且主要是体液免疫反应。其对胞外感染的病菌保护效果较好，但对寄生在奶牛乳腺细胞中的细菌、寄生虫或病毒则保护效果很差甚至没有效果，还存在毒力逆转、一些活病原体或有害成分被无意保留等缺陷，在使用中会出现副作用大、效果差、维持时间不长等问题。Wright 于 1902 年将体外培养的金黄色葡萄球菌全菌灭活制备成灭活苗免疫牛，免疫效果不理想。Pankey 等利用多种噬菌体型金黄色葡萄球菌裂解物制成疫苗，该疫苗有助于乳腺炎自愈，但不能抵御新的感染。Watson 等用弱毒株 W79 皮下免疫奶牛，结果显示具有中等保护力，开发出由类毒素、矿物油及含荚膜金黄色葡萄球菌组成的疫苗，能降低金黄色葡萄球菌乳腺炎发生率，但大规模临床实验中其保护力与对照组无统计学差异。尚佑军等利用奶牛乳腺炎多联苗对泌乳母鼠全身免疫，显示能增强局部的抗感染能力，能延缓乳腺组织病变进程，但介于各地流行菌株存在较大差异，使得广泛推广受到限制（宋战辉，2014）。

（二）亚单位疫苗

亚单位疫苗指以微生物的某些主要抗原制备的不含有核酸且能诱发机体抗体产生的疫苗。由于它没有遗传物质、宿主以及培养基的成分，显示出较高的免疫作用。目前市场上已有商品化亚单位疫苗，如 Somato-Staph、Lysigin 及大肠杆菌 J5 菌苗。Lysigin 是一种多价苗，由 4 种噬菌体型 *S. aureus*（Ⅰ、Ⅱ、Ⅲ、Ⅳ）菌体裂解物构成，能提高自愈率，但不能抵御新发生的感染。黏附素、荚膜多糖、酶和毒素作为影响 S. aureus 致病性的主要因子，是目前疫苗研究的主要探索对象，且不断有新的研究成果呈现。

肽聚糖能锚定宿主细胞壁，并使 S. aureus 定植于宿主靶细胞外基质的黏附素，被认为是 S. aureus 感染早期最重要的致病因子（宋战辉，2014）。

凝集因子 A（clumping factorA，ClfA）是 S. aureus 黏附宿主细胞的重要因子之一，存在于几乎所有金黄色葡萄球菌中。ClfA 是与纤维蛋白原（Fg）结合最主要的蛋白，其与 Fg 的结合力决定 S. aureus 与纤维蛋白的结合能力。ClfA 的纤维蛋白原结合作用位于 A 区，其具有免疫原性。以 ClfA 的 A 区免疫奶牛，可以产生特异性抗体应答。Brouillette 等的研究表明，ClfA DNA 疫苗诱导的特异性抗体应答经小鼠抗血清孵育，其对 Fg 的黏附力下降显著，且这些经孵育的 S. aureus 在体外更易被巨噬细胞吞噬。ClfA 的 A 区、FnBPA 及 FnBPB 制备的联合抗体，可提高对感染主动脉斑块小鼠的免疫保护作用。研究表明，利用油佐剂进行乳化的 rClfA-A 对感染 S. aureus 的小鼠乳腺炎有非常强的免疫保护效果（宋战辉，2014）。

纤连蛋白结合蛋白（fibronectin-binding protein，FnBP）是 S. aureus 另一个非常重要的黏附素，介导 S. aureus 与细胞表面的纤连蛋白（fibronectin，Fn）的结合，使得病原菌可以黏附在宿主细胞的表面，继而入侵宿主组织。FnBPA 和 FnBPB 由两个相似且相当保守的基因分别编码，其黏附作用基本相同。Lammers 等研究结果显示，缺乏 FnBPA 和 FnBPB 表达的 S. aureus 不能凝集细胞，且黏附、侵入奶牛乳腺细胞的能力有显著降低。95%以上 S. aureus 都含有 FnBPA，Shinji 等将 fnbA 突变体、fnbB 突变体及 fnbA、fnbB 双突变体对组织分别侵染，发现 fnbA 突变体和 fnbA、fnbB 双突变体均使脾细胞中白介素-6 和核因子 κB 显著降低，而 fnbB 突变体侵染后无显著下降，说明 FnBPA 在体内外侵染中作用更为重要。Castagliuolo 等用编码 4 种黏附因子（纤连蛋白结合蛋白 A、纤连蛋白结合蛋白、凝集胶原和凝聚因子）的 pDNA 混合物进行鼻腔内接种，体外实验显示可以抑制 S. aureus 黏附到牛乳腺上皮细胞上，血清抗体检测水平高，免疫小鼠对由金黄色葡萄球菌引发的乳腺感染产生有效抵御作用。另外，研究表明，ClfA 和 FnBPA 的联合抗体对 S. aureus 小鼠主动脉补片感染的保护有增强作用。

荚膜多糖（capsular polysaccharide，CP）是 S. aureus 在宿主及环境得以生存的重要表面结构成分。它聚集在感染部位的中性粒细胞上，通过干扰多形核白细胞（PMN）吞噬作用促进 S. aureus 侵袭，还能阻止中性粒细胞对抗 S. aureus 细胞壁成分抗体的识别，帮助病菌逃避宿主免疫细胞的吞噬作用。S. aureus 不论是活的还是灭活的，均能通过厚的荚膜逃避调理作用，因此，荚膜多糖成为传统疫苗免疫效果差的原因之一，以 CP 作为抗原进行疫苗研制也成为研究的热点。

Guidry 等将从美国 178 个奶牛场分离的 S. aureus 进行血清分型，T5（18%）、T8（18%）及 T336（59%）为优势血清型，而对欧洲样本荚膜多糖进行血清分型结果显示 T5（34%）、T8（34%）、T336（30%）、未分型（2%），表明血清型有地域差别。

荚膜多糖是小分子，属于 T1（T-cell independent）抗原，纯化的 CP 免疫原性低，当多糖共价偶联载体蛋白质后免疫动物，可增加其 T 细胞依赖性免疫应答，使机体产生荚膜多糖抗体，二次免疫可使抗体水平提高，实现对机体保护的作用。Johnson 等把 CP5、

CP8 与铜绿假单胞菌外毒素 A 进行偶联后制成疫苗，取得较好的免疫效果。Fattom 等将 CP5 和 CP8 偶联到绿脓假单胞菌无毒的重组外毒素 A 上并进行小鼠免疫，结果显示产生的荚膜多糖抗体有一定的调理作用。Eldridge 等将纯化的 CP5、CP8 及 CP336 与载体蛋白和可降解的微生物球结合，显示荚膜多糖抗体效价提高，免疫的时间相对较长，使中性粒细胞吞噬作用增强，对阻止 S. aureus 与乳腺上皮细胞的黏附作用明显。Nanraa 等将 ClfA 和 CP 在动物体内进行表达，显示 ClfA 和 CP 两种抗原的表达不同，且 ClfA 产生的抗体能增强 CP 抗体的作用，将二者进行连接作为靶抗原进行疫苗研制可能更有效。Hanh R 等研制的针对 α-毒素、黏附素和荚膜多糖的亚单位疫苗，免疫家兔后保护效果良好。Naidu 等将荚膜多糖和纤维结合素在体外进行偶联制成疫苗，结果显示，S. aureus 临床乳腺炎发生率有所降低。研究表明，纯化的蛋白与荚膜多糖偶联制得的疫苗能诱导调理牛体内 IgG 2，对预防 S. aureus 乳腺炎有效（宋战辉，2014）。

运用基因工程的方法制备的亚单位疫苗抗原产量大、纯度高，还可对有特殊性质病原菌开展疫苗研究，这是常规方法生产无法相比的，另外，亚单位疫苗良好的安全性也为其发展提供了保障。

（三）DNA 疫苗

DNA 疫苗又称核酸疫苗，利用基因重组技术将抗原蛋白编码基因与真核表达载体进行重组，在动物机体细胞内，利用宿主细胞的转录系统合成抗原蛋白，以诱导宿主对该抗原蛋白产生免疫应答，达到防治疾病的目的。该类疫苗抗原无需纯化，重组体用量相对较少，接种后重组体在机体内大量繁殖所产生的抗原刺激机体产生免疫保护作用，载体本身还有佐剂作用以增强免疫效果。目前常用载体包括痘病毒、腺病毒、腺相关病毒、脊髓灰质炎病毒、枯草杆菌、乳酸菌、卡介苗等。用编码病菌抗原的质粒 DNA 进行免疫，成为防治乳腺炎新的思路。

Brouillette 等研制的可表达 ClfA 蛋白的质粒，免疫小鼠时有效价很高的特异性抗体产生，但腹腔攻毒时不能产生很好的保护。Nour 等研制的针对金黄色葡萄球菌 ClfA 蛋白的核酸疫苗对奶牛进行免疫，反应强烈。Castagliuolo 等分别构建的 FnBPA、Can、ClfA 及 Efb 疫苗，小鼠滴鼻免疫时产生的特异性抗体水平较高。Gaudreau 等构建的 FnBPA、ClfA 及 Srt 多基因核酸疫苗，免疫后对小鼠攻毒，55%产生免疫保护作用，对照组仅 15%小鼠得到保护。Shkreta 等研制的 ClfA（221～550 个氨基酸）和 FnBP（D121-34；D320-33）二价核酸苗免疫怀孕 7 个月奶牛，加强免疫采用相应的重组蛋白，结果显示该二价苗与亚单位蛋白苗联合使用时能引起明显的体液及细胞免疫反应，对乳腺炎起到保护作用。Nour 等将 CTLA-4 和 ClfA 重组 DNA 疫苗免疫奶牛，奶牛有较强的体液及细胞免疫反应产生（宋战辉，2014）。

载体作为一个活的大分子，可能会诱导机体产生对载体本身的免疫，导致该类疫苗只能用于未感染过该类载体微生物或者未接种过的动物，另外，活载体疫苗不易进行加强免疫，这在一定程度上限制了其发展。再者，DNA 疫苗刺激机体产生的免疫反应比自然感染时弱，且外源 DNA 被导入机体后有被整合的危险，因此，国内外目前还没有一种核酸疫苗被批准使用。

二、大肠杆菌疫苗

目前，E. coli（O111：B4）J5 菌苗是关于大肠杆菌研究较多的疫苗，它为大肠杆菌不完全 O-多糖的突变株，其特征是裸露展现的脂多糖核心区和类质 A 区，该核心区抗原可以刺激机体产生抗革兰氏阴性菌核心抗原 IgG，能有效抵抗其他血清型的交叉感染，从而产生抗其他血清型抗原的抗体。2000 年 4 月，我国与美国加州大学合作开发 J5 菌苗，现已有成品上市应用。到目前为止，美国已研制出 4 种 "奶牛乳腺炎疫苗"，其中有 3 种是使用大肠杆菌 J5，这 4 种疫苗都能刺激奶牛对肠道细菌和大多数革兰氏阴性菌的产生抗体，并与多种粗糙型和光滑型革兰氏阴性菌发生交叉反应。

J5 疫苗由 5 mL 弗氏不完全佐剂与 5 mL 菌液（10^9 cfu/mL）组成。通常采用皮下三次免疫接种，效果较好，该菌苗免疫后可以提高血清和乳汁中抗大肠杆菌 J5 核心抗原的抗体效价。大量数据表明 J5 疫苗显著降低大肠杆菌性乳腺炎临床症状的严重程度。田间试验表明，实验组奶牛乳腺炎的发病率比对照组减少了近 70%，此外使用疫苗可降低再感染率，免疫效果与疫苗注射次数有关，注射一次的奶牛，临床乳腺炎的发病率减少 10%，注射两次的减少 30%~40%，注射三次的减少近 80%（赵建荣，2009）。

三、链球菌疫苗

无乳链球菌、停乳链球菌和乳房链球菌能引起奶牛乳腺炎，但不会产生显著的免疫力，所以现在没有可靠的多价疫苗。对链球菌疫苗的研究较少。总体认为对灭活的链球菌苗对奶牛无保护力。Giraudo 等的一个田间试验中采用乳房链球菌和停乳链球菌多联灭活苗，接种疫苗并不能显著降低奶牛由链球菌引起的乳腺炎的发生。到现在为止仍然没有预防链球菌乳腺炎商业化疫苗。Michae 等使用血纤维蛋白溶酶受体蛋白（GapC 蛋白）基因编码的乳房细胞表面相关的无乳链球菌蛋白 GapC 蛋白，但试验证明，基因工程疫苗 GapC 蛋白对乳房链球菌炎症有保护，同时对无乳链球菌乳腺炎无保护性，说明 GapC 基因工程疫苗缺乏种间交叉保护性。GapC 和 Mig 重组疫苗免疫的奶牛，检测体细胞的水平，两者都可以使牛奶体细胞的含量降低，但 GapC 蛋白苗可使得 SDG8 显著下降，Mig 反而没有减少。该方面机制还不是很清楚，还有很大的研究空间（樊杰，2014）。

四、奶牛乳腺炎疫苗佐剂研究现状

佐剂是在疫苗配制过程中与疫苗混合的混合物，属于非特异性免疫增强剂，用于增强机体对抗原的免疫应答量级或者改变免疫应答类型。随着生物技术的不断发展，疫苗研究取得飞速发展的同时，疫苗佐剂也逐渐走入人们的研究领域，筛选出安全有效的疫苗佐剂成为疫苗研究中的重要课题。目前兽用疫苗佐剂主要包括铝胶盐、蜂胶、转移因子、油水乳剂、核酸及其类似物、细胞因子、脂质体等。这些佐剂通过增强抗原的免疫原性，延缓抗原释放、降解和排除，刺激淋巴细胞增殖分化，促进相关细胞因子的表达等方式来发挥作用（侯伟杰，2014）。

（一）铝胶盐

铝胶盐混合物的主要成分是氢氧化铝，具有提升免疫力的能力，因其安全有效性已经被作为疫苗佐剂广泛应用，包括白喉-破伤风-百日咳疫苗、人的乳头状瘤病毒疫苗、流感疫苗以及肝炎疫苗等（Philippa et al.，2009）。铝胶盐佐剂能促进 TH2 免疫应答，并且动物试验显示，铝胶盐佐剂能诱导产生强烈的非特异性的 IgE 应答。研究显示，铝胶盐佐剂能通过诱导 NLRP3 炎性小体的活化来促进 IL-18、IL-33 和 PGE2 的释放，释放的这些细胞因子会促进 TH2 型细胞因子（如 IL-4、IL-5 和 IL-13）的产生，从而促进 TH2 型免疫应答（Terhune and Deth，2013）。然而铝胶盐在诱导细胞毒性 T 细胞和 TH1 免疫应答方面的作用非常弱。

（二）蜂胶

蜂胶是一种天然树脂混合物，由蜜蜂从植物、芽孢及渗出物中采集加工而来。蜂胶具有多种生物学活性，包括抗细菌、抗真菌、抗原生动物、抗氧化、抗癌、抗炎活性，同时还具有免疫调节作用。

蜂胶的免疫调节作用首先表现在能刺激抗体的产生，增强机体的体液免疫。Fischer 等（2007）将蜂胶作为佐剂，用于制备灭活的疱疹病毒疫苗，研究结果显示，小鼠接种加入蜂胶的疫苗后抗体水平明显增加。蜂胶作为鲤鱼气单胞菌疫苗的佐剂时也非常有效。同未加佐剂的疫苗组相比，佐剂疫苗组鲤鱼吞噬细胞的活性明显增强，并且其血清抗体水平也明显增加。蜂胶还可增强先天应答。蜂胶作用于小鼠后，通过上调巨噬细胞和脾细胞 TLR-2、TLR-4 的表达及促炎性细胞因子（IL-1 和 IL-6）的产生，活化了先天免疫应答的早期阶段（Orsatti et al.，2010a）。然而，蜂胶对脾脏淋巴细胞的增殖具有抑制作用，作用于小鼠 3 天后，能抑制脾细胞 IFN-γ 的产生（Orsatti et al.，2010b）。

（三）转移因子

转移因子（transfer factor，TF）是一种可透析的淋巴细胞提取物，包括许多分子，来自于机体免疫系统核心的 TF 储存了免疫系统对于细菌和病毒的记忆信息。

自从 40 年前 Sherwood Lawrence 发现了转移因子之后，许多临床报道证明 TF 能够作为疾病的一种免疫调节剂，用于增强细胞免疫。一些研究指出，TF 可以被作为一种预防手段将免疫力转移给感染者，另外，TF 还具有修饰免疫系统的能力。研究者将猪 TF 用作猪细小病毒灭活疫苗的佐剂，试验结果显示，同未加入佐剂的疫苗组相比，加入 TF 的疫苗组的细胞免疫应答水平增强，产生了高水平的 IFN-γ 和强烈的 T 细胞应答；然而两组间在抗体滴度上没有明显差异（$P>0.05$）（Wang et al.，2012）。表明 TF 能增强细胞免疫应答能力，可被作为灭活疫苗的免疫增强剂。

第二节　奶牛乳腺炎金黄色葡萄球菌疫苗的初步研制

奶牛乳腺炎的广泛流行，是危害奶牛养殖业发展最严重的疾病之一。据资料报道，一头奶牛发生一次临床型乳腺炎后,本胎次减少产奶量 500 kg 左右,约占总产奶量的 20%,同时影响其终身产奶水平,乳汁质量也受影响,乳脂率下降 0.3 个百分点,乳体细胞数上升到 100 万～150 万/mL（吴东桃等,2001）。虽然现代化奶牛场控制奶牛乳腺炎发生的方法不断改良,但仍不能有效控制奶牛乳腺炎的流行和发生。目前食品安全问题日益受到关注,用大量抗生素治疗和控制奶牛乳腺炎的方法受到限制。利用疫苗防治乳腺炎具有乳汁无药物残留、操作简单和费用低廉的优点（林锋强等,2002）。如果能研制开发出有效的乳腺炎疫苗,将是预防和控制奶牛乳腺炎最经济可行的措施。

酶联免疫吸附测定（enzyme linked immunosorbent assay,ELISA）是以酶联免疫吸附试验为基础的测定技术,根据抗原抗体反应的特异性和酶促反应的高敏感性而建立的免疫分析技术。该试验创始于 1971 年,瑞典学者 Engvall 等以及荷兰学者 Van Weerman 和 Schuurs 分别报道将免疫技术发展为检测体液中微量物质的固相免疫测定方法,定为酶联免疫吸附试验。1974 年,Voller 等又将固相支持物改为聚苯乙烯反应板,使 ELISA 技术得以推广应用。ELISA 的基本原理是:先将已知的抗体或抗原结合在某种固相载体上,并保持其免疫活性,测定时将待检标本和酶标记的抗原或抗体按一定的步骤与固相载体表面吸附的抗原或抗体发生反应。然后用洗涤的方法分离抗原抗体复合物和游离的成分。最后加入酶作用底物催化显色,进行定性或定量测定。

本实验的目的是使用从奶牛临床乳腺炎分离鉴定的金黄色葡萄球菌研制出奶牛乳腺炎金黄色葡萄球菌疫苗,通过实验室一系列评价,并采用间接 ELISA 方法评价奶牛乳腺炎疫苗的抗体效价水平。

一、材料与方法

（一）实验材料

1. 实验菌株、动物和试剂

实验菌株：从奶牛临床乳腺炎分离、纯化、鉴定得到的金黄色葡萄球菌一株。

实验动物：昆明系小白鼠,购自广西中医学院实验动物中心,体重 18～22 g；新西兰大白兔,购自广西大学农业科学实验实习基地,体重 1.5～2 kg；荷斯坦奶牛：广西农垦金光乳业有限公司奶牛场,2～3 胎次的泌乳奶牛。

药品试剂：法国 Seppic（赛比克）公司生产的两种兽医专用佐剂 MONTANIDE ISA 206VG 和 MONTANIDE ISA 50V2（由法国赛比克公司上海代表处郑菁惠赠）,兔抗牛 IgG-HRP 购自北京奥博星生物技术有限责任公司,牛脑心浸粉（Brain Heart Infusion）培养基购自美国 BD 公司,脱脂奶粉,甲醛,邻苯二胺（OPD）,犊牛血清白蛋白（BSA）。

包被液（0.05 mol/L pH9.6 碳酸盐缓冲液）：Na_2CO_3 0.16 g,$NaHCO_3$ 0.29 g,加水至 100 mL。

洗涤液（0.01 mol/L pH7.4 磷酸盐缓冲液）：NaCl 8 g,KH_2PO_4 0.2 g,$Na_2HPO_4 \cdot 12H_2O$ 2.9 g,KCl 0.2 g,吐温-20 0.5 mL,加水至 1000 mL。

稀释液（100 mL）：90 mL 洗涤液,10 mL 小牛血清,也可用封闭液代替。

封闭液：5%脱脂奶粉,用洗涤液配制。

底物缓冲液（现用现配）：0.1 mol/L Na_2HPO_4 5.14 mL,0.05 mol/L 柠檬酸 4.86 mL,

OPD 4 mg，溶解后加 3% H_2O_2 0.05 mL。

终止液：2 mol/L H_2SO_4。

2. 实验仪器 生物显微镜，高速组织捣碎机，高压灭菌器，超净工作台，酶标仪，超声波细胞破碎仪，电子分析天平，干燥箱，恒温培养箱，冰箱，96 孔可拆酶标板，移液器。

(二) 实验方法

1. 灭活疫苗抗原的研制

（1）金黄色葡萄球菌的培养 挑取在鲜血琼脂平皿上 37℃培养 18 h 的金黄色葡萄球菌菌落接种 100 mL 牛脑心浸粉培养基，37℃振荡培养 18 h，然后取 25 mL，接种到 500 mL 的牛脑心浸粉培养基，37℃振荡培养 18 h，取样涂片纯检。

（2）水相的制备 向 500 mL 金黄色葡萄球菌培养液加入 4 mL 甲醛，37℃振荡灭活 24 h，然后 4000 r/min 离心 20 min，弃去上清液，加 PBS 缓冲液悬浮细菌，再离心，重复数次至上清液完全透明，将沉淀的细菌悬浮于少量 PBS 缓冲液中，进行细菌计数，根据计数结果，稀释细菌悬浮液，使金黄色葡萄球菌浓度为 2×10^8 cfu/mL（Tollersrud et al.，2002）。

将盛放有金黄色葡萄球菌细菌悬浮液的烧杯置于冰浴中，用超声波细胞破碎仪将细菌菌体超声裂解，35 W 超声 20 min，取少量悬浮液涂片镜检，若是超过 20%以上的细菌菌体完整，再次超声裂解，直至完整的细菌菌体低于 20%，即为制作疫苗的水相。

（3）水相的无菌检验 将制作疫苗的水相接种血琼脂平皿培养基，37℃培养 24 h，观察细菌的生长情况。并且取 100 μL 水相加入到含有 2%犊牛血清的营养肉汤培养基中，37℃振荡培养 16~20 h，观察细菌生长情况。

（4）疫苗的配制

疫苗 1：油相 1（MONTANIDE ISA 206 VG）与水相质量比为 1∶1（体积比为 5.4∶4.6），用高速组织捣碎机，8000 r/min 搅拌下逐渐将水相加入到油相 1 中，搅拌 20 min，分装，4℃保存。

疫苗 2：油相 2（MONTANIDE ISA 50 V2）与水相体积比为 1∶1（质量比为 5.4∶4.6），用高速组织捣碎机，12 000 r/min 搅拌下逐渐将水相加入到油相 2 中，搅拌 8 min，分装，4℃保存。

（5）疫苗乳液类型的检测 将两只 250 mL 的大口烧杯分别装入约 200 mL 蒸馏水，分别将一滴疫苗 1 和疫苗 2 乳液滴在烧杯水面上，不要搅动，观察液滴在水面的稀释状态，判定标准如下。

水包油包水型：液滴部分自我稀释，并使烧杯内蒸馏水呈现出乳白色外观。

油包水型：液滴浮在水面，并且用手轻轻摇动烧杯后，蒸馏水依然澄清。

2. 疫苗的稳定性检验和安全性检验

稳定性检验：两种疫苗均取少量密封后分别放置在 4℃和室温，连续观察 4℃下两年内和室温下一个月内破乳情况。

安全性检验：两种疫苗各接种 5 只家兔，腹股沟内侧肌内注射 1 mL，连续测定直肠体温，并观察精神、饮欲和食欲 7 天。

3. 白鼠攻毒保护实验　　将两种疫苗和生理盐水分别肌内注射免疫小白鼠（0.1 mL/只）各 10 只，接种后 21 天，以最小致死量的金黄色葡萄球菌攻毒（分别约为 7×10^8 cfu），观察攻毒后小白鼠的精神、饮欲和食欲变化情况，记录死亡时间，并对死亡小白鼠进行剖检。

4. 间接 ELISA 抗原的制备及浓度测定　　将培养的金黄色葡萄球菌 4000 r/min 离心 20 min，弃去上清液，加 PBS 缓冲液悬浮细菌，再离心，重复数次至上清液完全透明。将沉淀的细菌悬浮于少量 PBS 缓冲液中，细菌悬液用超声波细胞粉碎仪 35 W 冰浴粉碎 30 min，然后 4000 r/min 离心 10 min 除去大的细菌碎片，即为金黄色葡萄球菌包被抗原，–20℃保存备用。

抗原浓度的测定是采用紫外扫描法进行测定，具体方法为：将金黄色葡萄球菌抗原用紫外分光光度计法测定其在 260 nm 和 280 nm 波长的 OD 值（刘箭，2004），通过下面公式计算抗原蛋白的含量：

$$抗原蛋白浓度（mg/mL）= 1.45A_{280}-0.74A_{260}$$

5. 阳性血清和阴性血清的制备　　选 6 头产后一个月的泌乳荷斯坦奶牛，用本实验制作的金黄色葡萄球菌水相和生理盐水分别接种免疫 2 头，臀部肌内注射 5 mL，免疫后 14 天颈静脉采血，分离血清，EP 分装，每管 0.5 mL，于–20℃冰箱保存。同时琼脂双向免疫扩散法初步测定血清效价。

6. 琼脂双向免疫扩散法初步测定阳性血清效价　　取 1 g 琼脂糖加入 100 mL PBS 液，加热充分溶解，待冷至 45～50℃时倒入洁净灭菌的平皿中，每个平皿倒入 18～20 mL，置超净工作台平面上，加盖凝固后将平皿倒置，放入 4℃冰箱备用。反应孔现用现打，用孔径为 4 mm 的金属打孔器在 1%琼脂糖凝胶上打成 7 孔梅花样，封底方法是在酒精灯上均匀加热底部，以免渗漏。完全冷却后在中心孔加入金黄色葡萄球菌抗原 6，外周 6 孔加入不同稀释比例的阳性血清（1∶2、1∶4、1∶8、1∶16、1∶32、1∶64），每个孔均加入 100 μL。加样完毕后，室温静置 10 min，然后将平皿放入湿盒内置 37℃恒温培养箱中反应（注意放平），分别于 24 h 和 48 h 观察结果（吴继尧，2001）。结果判定：将平皿置于日光灯下观察，若抗体孔（外周孔）与抗原孔（中心孔）之间出现白色的沉淀线，则可判为阳性。

7. 间接 ELISA 测定抗体方法的建立

（1）抗体测定方法的建立参照间接 ELISA 操作程序　　间接 ELISA 操作步骤（朱立平，2000）：

A. 用制备的抗原做适当稀释后包被 96 孔聚苯乙烯微量反应板，100 μL/孔，置湿盒内 4℃过夜；

B. 洗涤，每孔加洗涤液（pH 7.4，含 0.05% Tween-20）200 μL，1 min 后拍干，洗涤 3 次；

C. 加入封闭液，100 μL/孔，37℃湿盒内 40 min；

D. 洗涤，同 B；

E. 加入稀释的阳性血清，100 μL/孔，37℃湿盒内 1 h；

F. 洗涤，同 B；

G. 加入 1∶750 稀释的兔抗牛 IgG-HRP，100 μL/孔，37℃湿盒内 1 h；

H. 洗涤，同 B；

I. 加入 OPD 底物缓冲液 100 μL/孔，37℃湿盒内避光反应 30 min；

J. 加入 2 mol/L H$_2$SO$_4$ 50 μL/孔，终止反应。于 492 nm 波长处测 OD 值，并记录结果。

K. OD 值以阳性血清的 OD 值/阴性血清的 OD 值（P/N）呈现最大者的反应条件作为确定 ELISA 最佳反应条件的判定依据。

（2）间接 ELISA 反应条件的选择和优化　　依照参考文献对间接 ELISA 进行条件的选择和优化。

A. 抗原最佳包被浓度的确定。将金黄色葡萄球菌抗原用包被缓冲液分别稀释 1∶5 倍、1∶10 倍、1∶20 倍、1∶40 倍、1∶60 倍、1∶80 倍、1∶100 倍、1∶120 倍、1∶140 倍、1∶160 倍、1∶180 倍和 1∶200 倍，分别加入抗原，4℃过夜，洗涤后加封闭液，37℃ 40 min，洗涤后加 1∶200 倍稀释的阳性血清，37℃孵育 1 h，洗涤后加 1∶750 倍（推荐稀释倍数 1∶500 倍至 1∶1000 倍）稀释的酶标二抗，37℃孵育 1 h，洗涤后加入底物缓冲液，37℃避光显色 30 min，最后加终止液，空白孔调零，492 nm 波长处测 OD 值，以 P/N 值呈现最大者为抗原最佳包被浓度。

B. 血清最佳稀释度的确定。以最佳包被浓度的抗原包被，4℃过夜，洗涤后加封闭液，37℃ 40 min，洗涤后分别加入 1∶10 倍、1∶20 倍、1∶40 倍、1∶60 倍、1∶80 倍、1∶100 倍、1∶120 倍、1∶140 倍、1∶160 倍、1∶180 倍、1∶200 倍、1∶240 倍、1∶280 倍、1∶320 倍、1∶360 倍、1∶400 倍、1∶450 倍、1∶500 倍、1∶600 倍、1∶800 倍、1∶1000 倍、1∶1200 倍、1∶1400 倍和 1∶1600 倍稀释的血清，37℃孵育 1 h，洗涤后加 1∶750 倍（推荐稀释倍数 1∶500 倍至 1∶1000 倍）稀释的酶标二抗，37℃孵育 1 h，洗涤后加入底物缓冲液，37℃避光显色 30 min，最后加终止液，空白孔调零，492 nm 波长处测 OD 值，以 P/N 值呈现最大者为血清最佳稀释度。

C. 抗原最佳包被条件的选择。以最佳包被浓度的抗原包被，然后分别置于 4℃ 24 h、4℃ 18 h、4℃ 12 h、4℃ 8 h、4℃ 6 h、37℃ 4 h、37℃ 3 h、37℃ 2 h、37℃ 1.5 h、37℃ 1 h 和 37℃ 0.5 h 条件下处理，洗涤后加封闭液，37℃ 40 min，洗涤后加入最佳稀释度的血清，37℃孵育 1 h，洗涤后加 1∶750 倍（推荐稀释倍数 1∶500 至 1∶1000）稀释的酶标二抗，37℃孵育 1 h，洗涤后加入底物缓冲液，37℃避光显色 30 min，最后加终止液，空白孔调零，492 nm 波长处测 OD 值，以 P/N 值呈现最大者为抗原最佳包被条件。

D. 最佳封闭条件的确定。以最佳包被浓度的抗原包被，最佳包被条件处理，洗涤后分别不封闭、0.1%犊牛血清白蛋白、0.3%犊牛血清白蛋白、5%犊牛血清、10%犊牛血清、5%兔血清和 5%奶粉封闭，再分别置于 37℃下 20 min、30 min、40 min 和 60 min 处理，洗涤后加入最佳稀释度的血清，37℃孵育 1 h，洗涤后加 1∶750 倍（推荐稀释倍数 1∶500 至 1∶1000）稀释的酶标二抗，37℃孵育 1 h，洗涤后加入底物缓冲液，37℃避光显色 30 min，最后加终止液，空白孔调零，492 nm 波长处测 OD 值，以 P/N 值呈现最大者为最佳封闭条件。

E. 酶标二抗最佳稀释度的确定。以最佳包被浓度的抗原包被，最佳包被条件处理，最佳封闭条件封闭，洗涤后加入最佳稀释度的血清，37℃孵育 1 h，洗涤后分别加入 1∶400 倍、1∶500 倍、1∶600 倍、1∶650 倍、1∶700 倍、1∶750 倍、1∶800 倍、1∶850 倍、1∶900 倍、1∶1000 倍、1∶1100 倍和 1∶1200 倍稀释的酶标二抗，37℃孵育 1 h，洗涤后加入底物缓冲液，37℃避光显色 30 min，最后加终止液，空白孔调零，492 nm 波长处测 OD 值，以 P/N 值呈现最大者为酶标二抗最佳稀释度。

F. 抗原抗体最佳作用时间的确定。以最佳包被浓度的抗原包被，最佳包被条件处理，最佳封闭条件封闭，洗涤后加入最佳稀释度的血清，37℃分别孵育 5 min、10 min、15 min、20 min、30 min、40 min、60 min、80 min、100 min 和 120 min，洗涤后加入最佳稀释度的酶标二抗，37℃孵育 1 h，洗涤后加入底物缓冲液，37℃避光显色 30 min，最后加终止液，空白孔调零，492 nm 波长处测 OD 值，以 P/N 值呈现最大者为抗原抗体最佳作用时间。

G. 酶标二抗最佳作用时间的确定。以最佳包被浓度的抗原包被，最佳包被条件处理，最佳封闭条件封闭，洗涤后加入最佳稀释度的血清，最佳作用条件反应，洗涤后加入最佳稀释度的酶标二抗，37℃分别孵育 5 min、10 min、15 min、20 min、30 min、40 min、60 min、80 min、100 min 和 120 min，洗涤后加入底物缓冲液，37℃避光显色 30 min，最后加终止液，空白孔调零，492 nm 波长处测 OD 值，以 P/N 值呈现最大者为酶标二抗最佳作用时间。

H. 底物最佳显色时间的确定。以最佳包被浓度的抗原包被，最佳包被条件处理，最佳封闭条件封闭，洗涤后加入最佳稀释度的血清，最佳作用条件反应，洗涤后加入最佳稀释度的酶标二抗，最佳作用条件反应，洗涤后加入底物缓冲液，37℃分别避光显色 5 min、10 min、15 min、20 min、30 min、40 min 和 60 min，最后加终止液，空白孔调零，492 nm 波长处测 OD 值，以 P/N 值呈现最大者为底物最佳显色时间。

（3）间接 ELISA 阴阳临界值的确定　按照上面已经确定好的各步骤的条件用间接 ELISA 方法检测 9 份阴性血清，每份血清设两个重复，测定 OD_{492}，计算平均值及标准差。根据公式计算阴阳临界值：

$$阴阳临界=阴性血清样品平均 OD 值+3×标准差$$

（4）建立的间接 ELISA 特异性阻断实验　最佳稀释浓度的阳性血清在反应前与最佳包被浓度的金黄色葡萄球菌抗原混合后 37℃作用 30 min，作为阻断品，与未加抗原处理（未阻断）的阳性血清同时作间接 ELISA 测定，比较阻断前后血清 OD 值的变化。并计算（未阻断孔 OD 值-阻断孔 OD 值）/未阻断孔 OD 值，若此值大于 0.5，则判断为阻断阳性（秦春香，2007），分别设 4 个样品。

（5）标准曲线及回归方程的建立　用奶牛免疫后第 7 天、14 天、21 天、35 天、50 天、70 天和 97 天的血清各 3 份共 21 份，1∶100 稀释后，再倍比稀释，至 1∶6400 倍，按照已经建立的 ELISA 方法检测，每份血清的效价定为 OD 值大于规定吸收值（即阴阳临界值）的血清最高稀释度（崔焕忠，2004）。根据每份血清 1∶100 倍稀释时的 OD 值和血清效价的关系，进行统计学分析其相关性。以每份血清 1∶100 倍稀释的 OD 值为横坐标，以相应血清的 ELISA 效价倒数的以 2 为底的对数为纵坐标，做回归曲线，经证实二者为直线关系（崔焕忠，2004；程安春等，2004）。每份血清效价可以用 1∶100 稀释的 OD 值从回归曲线上查出或解回归方程求出。

为验证回归方程对血清的预测效价与实际效价的吻合性，实测 5 份血清的效价，然后将这 5 份血清 1∶100 倍稀释时 OD 值测出，代入回归方程，计算预测值，将实测效价与预测效价比较。

（6）奶牛免疫反应及抗体消长规律的检测

组别：疫苗1组、疫苗2组和对照组，各5头泌乳奶牛，免疫时全部健康状况良好。

免疫程序：每头泌乳奶牛在臀部肌内注射5 mL，对照组奶牛以生理盐水替代。

不良反应：记录每一头实验奶牛在实验阶段内的不良反应，如体温、采食量、注射部位的肿胀反应。

采血：分别于接种前第3天和接种后第0天、7天、14天、21天、35天、50天、70天、97天、120天、150天颈静脉采血并及时分离血清，−20℃保存。

将所有血清1∶100稀释，应用本实验建立的间接ELISA方法，检测每份血清的OD值，求出同一个时间点血清样品的平均OD值，作出抗体消长变化曲线。

将OD值代入回归方程，计算出每份血清的ELISA效价。

二、实验结果

（一）疫苗合格性检查

1. 水相的无菌检验　经过对制作疫苗的水相接种血琼脂平皿培养基，37℃培养24 h未发现菌落生长；接种2%犊牛血清的营养肉汤培养基培养后，37℃培养24 h培养基澄清，染色镜检未发现细菌，水相无菌检验合格。

2. 疫苗的乳液类型　疫苗1滴到水面后部分乳液自我稀释，使烧杯内的蒸馏水呈现出乳白色外观，判定为水包油包水型乳液；疫苗2滴到水面后浮在水面，用手轻轻摇动烧杯，蒸馏水依然澄清判定为油包水型乳液。同时经显微镜检查和液滴法检测疫苗1为水包油包水型乳液，与该佐剂的设计要求相符（图6-1）；经显微镜检查和液滴法检测疫苗2为油包水型乳液，与该佐剂的设计要求相符（图6-2）。

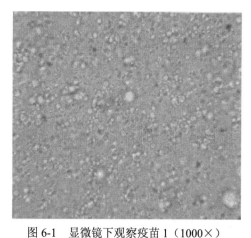 

图6-1　显微镜下观察疫苗1（1000×）　　　图6-2　显微镜下观察疫苗2（1000×）

Figure 6-1　Vaccine 1（1000×）　　　　　Figure 6-2　Vaccine 2（1000×）

3. 疫苗稳定性检验　稳定性检验结果显示两种疫苗均在室温下一个月内未出现佐剂（油相）析出，疫苗1在4℃下三年半未出现佐剂（油相）析出，疫苗2在室温下两个月内出现少量佐剂（油相）析出（图6-3），在4℃下两年未出现佐剂（油相）析出，查看佐剂（MONTANIDE ISA 206 VG 和 MONTANIDE ISA 50 V2）的说明书，未达到破乳并影响疫苗使用的情况，因此两种疫苗的稳定性均合格。

4. 疫苗安全性检验 安全性检验结果表明，仅疫苗 2 的一只大白兔在注射后次日出现食欲下降、精神轻度沉郁现象、注射部位有轻微红肿，无化脓现象，第三天即恢复正常，其他家兔注射疫苗后均未出现异常现象，说明该疫苗安全可靠。

（二）小白鼠攻毒保护实验

两种疫苗对小白鼠都有一定的保护率，详见表 6-1。攻毒后小白鼠表现出不同程度的精神沉郁、不愿走动、饮欲和食欲较少，5 天后均恢复正常。

图 6-3 疫苗 2 在室温下存放两个月后的表现

Figure 6-3 Vaccine 2 stored at room temperature for two months

表 6-1 奶牛乳腺炎疫苗对小白鼠的免疫保护实验结果

Table 6-1 The protection rate of mouse after challenged which are vaccinated with mastitis vaccines

组别	小白鼠数量/只	攻毒后观察天数及小白鼠存活数量							保护率/%
		1 天	2 天	3 天	4 天	5 天	6 天	7 天	
疫苗 1 组	10	8	7	7	7	7	7	7	70
疫苗 2 组	10	8	6	6	6	6	6	6	60
对照组	10	3	1	0	0	0	0	0	0

（三）间接 ELISA 可溶性包被抗原蛋白浓度的测定

紫外分光光度计法测定抗原的蛋白质浓度是 113.4 μg/mL。

（四）琼脂双向免疫扩散法初步测定阳性血清效价

经琼脂双向免疫扩散法初步测定阳性血清中金黄色葡萄球菌抗体，结果表明，金黄色葡萄球菌的抗体效价是 1∶16。

（五）间接 ELISA 方法的建立

1. 抗原最佳包被浓度的确定 将制备的金黄色葡萄球菌抗原用包被缓冲液分别稀释 1∶5 倍、1∶10 倍、1∶20 倍、1∶40 倍、1∶60 倍、1∶80 倍、1∶100 倍、1∶120 倍、1∶140 倍、1∶160 倍、1∶180 倍和 1∶200 倍，包被抗原，进行 ELISA 试验，结果显示，抗原稀释比例为 1∶10 倍时，P/N 值最大（3.21），因此将抗原的最佳包被比例确定为 1∶10 倍，此时的浓度为 11.34 μg/mL（113.4 μg/mL/10）。结果见图 6-4。

2. 血清最佳稀释度的确定 以 1∶10 倍稀释金黄色葡萄球菌抗原并包被，4℃过夜，洗涤后加封闭液，37℃ 40 min，洗涤后分别加入 1∶10 倍、1∶20 倍、1∶40 倍、1∶60 倍、1∶80 倍、1∶100 倍、1∶120 倍、1∶140 倍、1∶160 倍、1∶180 倍、1∶200 倍、1∶240 倍、1∶280 倍、1∶320 倍、1∶360 倍、1∶400 倍、1∶450 倍、1∶500 倍、1∶600 倍、

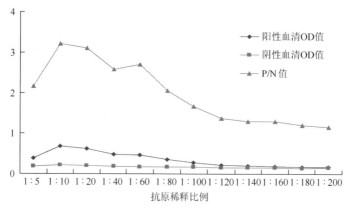

图 6-4　金黄色葡萄球菌抗原最佳包被浓度

Fig 6-4　The best coating concentration of *S. aureus* antigen

1∶800 倍、1∶1000 倍、1∶1200 倍、1∶1400 倍和 1∶1600 倍稀释的血清，进行 ELISA 试验，结果显示，血清稀释比例为 1∶200 倍时，P/N 值最大（3.43），因此将血清的最佳稀释比例确定为 1∶200 倍。结果见图 6-5。

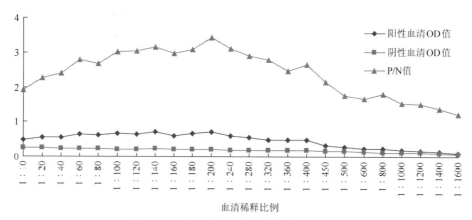

图 6-5　金黄色葡萄球菌血清最佳稀释比例

Figure 6-5　The best dilution of serum in *S. aureus* indirect ELISA

3. 抗原最佳包被条件的选择　以 1∶10 倍稀释金黄色葡萄球菌抗原并包被，然后分别置于 4℃ 24 h、4℃ 18 h、4℃ 12 h、4℃ 8 h、4℃ 6 h、37℃ 4 h、37℃ 3 h、37℃ 2 h、37℃ 1.5 h、37℃ 1 h 和 37℃ 0.5 h 条件下处理，进行 ELISA 试验，结果显示，包被抗原后 37℃ 1.5 h 的处理 P/N 值最大（3.24），因此将金黄色葡萄球菌抗原最佳包被条件确定为 37℃ 1.5 h。结果见图 6-6。

4. 最佳封闭条件的确定　以 1∶10 倍稀释金黄色葡萄球菌抗原并包被，放置于 37℃ 处理 1.5 h，洗涤后分别不封闭、0.1%犊牛血清白蛋白、0.3%犊牛血清白蛋白、5%犊牛血清、10%犊牛血清、5%兔血清和 5%奶粉封闭，再分别置于 37℃下 20 min、30 min、40 min 和 60 min 处理，进行 ELISA 试验，结果显示，5%犊牛血清 37℃ 30 min 封闭时 P/N 值最大（3.45），因此将 5%犊牛血清 37℃封闭 30 min 确定为金黄色葡萄球菌抗原最佳封闭条件。结果见图 6-7。

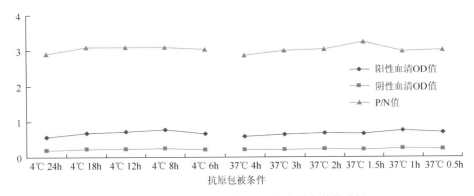

图 6-6　金黄色葡萄球菌抗原最佳包被条件的选择

Figure 6-6　The best coating condition of *S. aureus* antigen

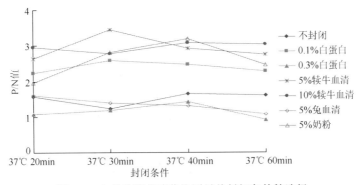

图 6-7　金黄色葡萄球菌抗原最佳封闭条件的选择

Figure 6-7　The best sealing condition in *S. aureus* indirect ELISA

5. 酶标二抗最佳稀释度的确定　　以 1∶10 倍稀释金黄色葡萄球菌抗原并包被,放置于 37℃处理 1.5 h,用 5%犊牛血清 37℃封闭 30 min,洗涤后加入 1∶200 倍稀释的血清,37℃孵育 1 h,洗涤后分别加入 1∶400 倍、1∶500 倍、1∶600 倍、1∶650 倍、1∶700 倍、1∶750 倍、1∶800 倍、1∶850 倍、1∶900 倍、1∶1000 倍、1∶1100 倍和 1∶1200 倍稀释的酶标二抗,进行 ELISA 试验,结果显示,酶标二抗稀释 1∶800 倍时 P/N 值最大（3.37）,因此将 1∶800 倍确定为酶标二抗的最佳稀释倍数。结果见图 6-8。

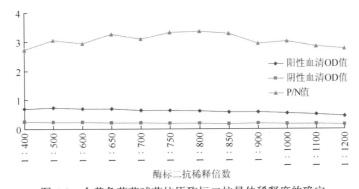

图 6-8　金黄色葡萄球菌抗原酶标二抗最佳稀释度的确定

Figure 6-8　The best dilution of HRP secondary antibody in *S. aureus* indirect ELISA

6. 抗原抗体最佳作用时间的确定　　以 1∶10 倍稀释金黄色葡萄球菌抗原并包被，放置于 37℃处理 1.5 h，用 5%犊牛血清 37℃封闭 30 min，洗涤后加入 1∶200 倍稀释的血清，37℃分别孵育 5 min、10 min、15 min、20 min、30 min、40 min、60 min、80 min、100 min 和 120 min，进行 ELISA 试验，结果显示，抗原抗体反应 20 min 时的 P/N 值最大（3.51），因此将抗原抗体最佳作用时间确定为 20 min。结果见图 6-9。

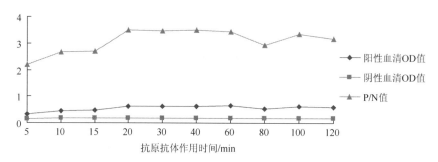

图 6-9　金黄色葡萄球菌抗原抗体最佳作用时间的确定

Figure 6-9　The best time of antigen and antibody response in *S. aureus* indirect ELISA

7. 酶标二抗最佳作用时间的确定　　以 1∶10 倍稀释金黄色葡萄球菌抗原并包被，放置于 37℃处理 1.5 h，用 5%犊牛血清 37℃封闭 30 min，洗涤后加入 1∶200 倍稀释的血清，37℃反应 20 min，洗涤后加入 1∶800 倍稀释的酶标二抗，37℃分别孵育 5 min、10 min、15 min、20 min、30 min、40 min、60 min、80 min、100 min 和 120 min，进行 ELISA 试验，结果显示，酶标二抗作用 15 min 时的 P/N 值最大（3.47），因此将酶标二抗最佳作用时间确定为 15 min。结果见图 6-10。

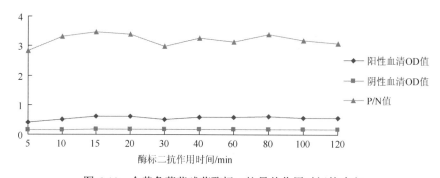

图 6-10　金黄色葡萄球菌酶标二抗最佳作用时间的确定

Figure 6-10　The best time of antigen and HRP secondary antibody response in *S. aureus* indirect ELISA

8. 底物最佳显色时间的确定　　以 1∶10 倍稀释金黄色葡萄球菌抗原并包被，放置于 37℃处理 1.5 h，用 5%犊牛血清 37℃封闭 30 min，洗涤后加入 1∶200 倍稀释的血清，37℃反应 20 min，洗涤后加入 1∶800 倍稀释的酶标二抗，37℃反应 15 min，洗涤后加入底物缓冲液，37℃分别避光显色 5 min、10 min、15 min、20 min、30 min、40 min 和 60 min，进行 ELISA 试验，结果显示，底物显色 30 min 时的 P/N 值最大（3.54），因此将底物最佳显色时间确定为 30 min。结果见图 6-11。

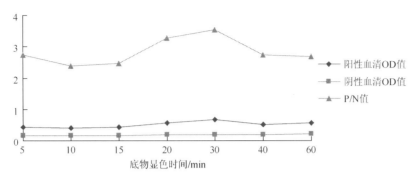

图 6-11 金黄色葡萄球菌底物最佳显色时间的确定

Figure 6-11　The best coloration time of substrate in *S. aureus* indirect ELISA

（六）间接 ELISA 阴阳临界值的确定

分别以已经建立的间接 ELISA 方法检测 9 份阴性血清，每份血清设两个重复，计算 OD_{492} 的平均值及标准差。样品阴阳性临界值=阴性样本 OD_{492} 平均值+3×标准差。测定结果见表 6-2。

表 6-2　间接 ELISA 阴阳临界值的确定

Table 6-2　Determination of optimal cutoff value of indirect ELISA

	平均 OD 值	标准差	阴阳临界值
阴性血清	0.199	0.027	0.28

（七）建立的间接 ELISA 特异性阻断实验

将金黄色葡萄球菌抗原 1∶10 倍稀释与 1∶200 倍稀释的阳性血清混合，37℃作用 30 min，进行 ELISA 实验，结果见表 6-3。

表 6-3　间接 ELISA 特异性阻断实验

Table 6-3　The specificity blocking of indirect ELISA

		未阻断孔 OD 值	阻断孔 OD 值	(N−P)/N	阻断情况
金黄色葡萄球菌	1	0.577	0.184	2.14	是
	2	0.611	0.218	1.8	是
	3	0.505	0.243	1.08	是
	4	0.521	0.237	1.2	是

（八）标准曲线及回归方程的建立

根据每份血清 1∶100 倍稀释时的 OD 值和血清效价的关系，以每份血清 1∶100 倍稀释的 OD 值为纵坐标，以相应血清的 ELISA 效价倒数的以 2 为底的对数为横坐标，做回归曲线，求出间接 ELISA 回归方程 $y = 2.945x + 6.377$（$R^2=0.911$），见图 6-12。表 6-4 是实测 5 份血清 1∶100 倍稀释时的 OD 值，分别代入回归方程即得预测效价，同时 5

份血清 1∶100 倍稀释再倍比稀释后测定 OD 值,大于阴阳临界值的稀释度即该血清的实际效价。

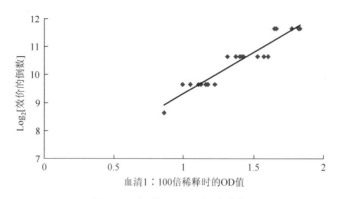

图 6-12 间接 ELISA 标准曲线

Figure 6-12 Indirect ELISA Standard curve

表 6-4 预测效价与实际效价的比较

Table 6-4 The predictive value compared with the actual potency

		血清编号				
		1	2	3	4	5
金黄色葡萄球菌	预测效价	1∶800	1∶400	1∶1600	1∶3200	1∶1600
	实际效价	1∶911	1∶492	1∶1915	1∶2794	1∶1493

(九)奶牛对疫苗的免疫反应

相比对照组奶牛,疫苗 1 组和疫苗 2 组奶牛在接种后第二天早晨平均体温升高幅度最大,为 0.4℃;疫苗 1 组奶牛在接种后第一天下午平均体温升高幅度最大,为 0.3℃;疫苗 2 组奶牛在接种后第一天下午平均体温升高幅度最大,为 0.5℃;未经治疗疫苗 1 组和疫苗 2 组奶牛于接种后第 4 天即恢复正常体温(图 6-13 和图 6-14)。下午采食量出现轻微影响,未经治疗均于注射疫苗 3~5 天后恢复。

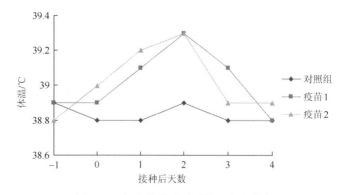

图 6-13 奶牛接种后早晨体温变化曲线

Figure 6-13 Cows morning body temperature curve after inoculation

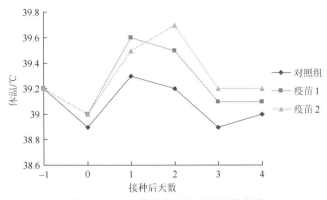

图 6-14 奶牛接种后下午体温变化曲线

Figure 6-14 Cow afternoon body temperature curve after inoculation

（十）抗体水平的检测

应用本实验建立的间接 ELISA 方法，将血清 1∶100 倍稀释，分别检测疫苗 1 组奶牛和疫苗 2 组奶牛血清中金黄色葡萄球菌特异性抗体的 OD 值，求各组奶牛同一时间点的平均 OD 值，绘制抗体消长变化曲线。结果表明，免疫前各实验组之间、实验组与对照组之间特异性抗体水平均无明显差异，抗体水平均较低，平均 OD 值为 0.154，接种后 7～14 天特异性抗体效价逐渐升高，疫苗 1 组奶牛于接种后 14 天达到最高峰（OD 值 1.756，相当于抗体效价 1∶2995），随后逐渐缓慢下降；疫苗 2 组奶牛于接种后 21 天达到最高峰（OD 值 1.573，相当于抗体效价 1∶2062），随后逐渐缓慢下降。接种疫苗后第 150 天，两个实验组奶牛血清的特异性抗体效价仍然高于阴阳临界值。各组奶牛血清特异性抗体效价及其差异性见图 6-15。

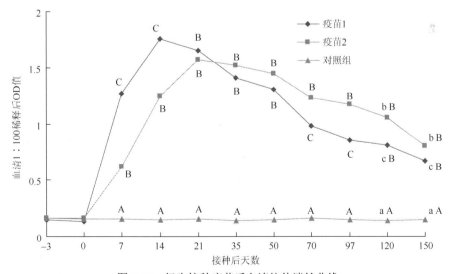

图 6-15 奶牛接种疫苗后血清抗体消长曲线

注：同时间点上不同大写字母表示差异极显著（$P<0.01$），同时间点上不同小写字母表示差异显著（$P<0.05$）

Figure 6-15 The curves of serum antibodies after vaccination in dairy cows

Different capital letters on the same time point indicate highly significant differences ($P<0.01$), different lowercase letters on the same time point indicate significant differences ($P<0.05$)

三、讨论

（一）关于疫苗的研制

疫苗免疫是预防乳腺炎的最佳方法，预防和控制奶牛乳腺炎的发生是在不影响奶牛健康和乳汁生产的前提下实施的，通过接种疫苗预防奶牛乳腺炎发生的前提是不能影响奶牛的健康和正常的泌乳。常用佐剂对动物都有一定的刺激性，如蜂胶、铝胶，可能会影响到奶牛的局部炎症反应，引起精神沉郁、采食量下降等，产奶量随之下降，从而影响奶牛场的经济效益，因此疫苗佐剂的选择必须慎重。法国Seppic（赛比克）公司生产的MONTANIDE ISA 206 VG是一种含有十八碳烯酸和无水甘露醇等酯类的矿物油佐剂，制备水包油包水乳液；Montanide ISA 50 V是一种基于甘露醇酯的矿物油佐剂，制备油包水乳液。这两种佐剂不会导致严重的不良反应，并且黏稠度很低（Iyer et al., 2000）。也有报道称MONTANIDE ISA 206能有效增强疫苗的免疫原性，刺激性小，安全无毒，易于乳化，工艺简单（钟辉，2005）。本实验疫苗的制作严格按照说明书操作，从而也制备了合格的乳液类型，对灭活菌液无菌检验、成品疫苗的理化性质检验均符合要求。稳定性也很高，4℃至少可以保存两年时间，以及室温下可以保存一个月。其对新西兰大白兔的安全性比较高，与佐剂的选择和制备有很大关系。

田海燕（2009）研究认为，MONTANIDE ISA 206 VG油佐剂疫苗组比铝胶佐剂疫苗组能够更好地提高机体的抗体水平，产生较强的免疫力，且能够更好地维持较长时间的高抗体水平。MONTANIDE ISA 206 VG油佐剂苗相比不完全弗氏佐剂苗，稳定性好，刺激性小，在免疫接种时也能产生较好的免疫效力，具有较好的应用优势。开发和研制安全、高效、无毒副作用的新型免疫佐剂以增强疫苗的免疫效力并减少疫苗的毒副作用将是未来的发展趋势（吴超和邹全明，2005）。新型免疫佐剂的安全性、有效性、选择性和可控性的有机结合将成为其发展的必然趋势，也是佐剂最终通过临床试验考核，最终全面应用于生产实践所应具备的基本条件。

（二）小白鼠攻毒保护实验

研制的两种疫苗接种小白鼠后第21天，用最小致死量的金黄色葡萄球菌攻毒，达到了60%以上的保护率，说明这两种疫苗免疫原性好，对金黄色葡萄球菌表现出较强的抵抗力和较好的保护力。同时也可以推断免疫奶牛后也可有效刺激牛体产生高水平的抗金黄色葡萄球菌血清抗体，并能够通过各种机制向乳汁中转移，从而增强了奶牛乳腺组织中抗金黄色葡萄球菌抗体水平，达到预防或清除金黄色葡萄球菌感染的目的。

（三）间接ELISA条件的选择和优化

自从1971年瑞典学者Engvall等建立ELISA诊断方法以来，以其简便快速、特异性强、敏感性高、安全和可自动化等优点被广泛应用于多种病毒或细菌的抗原或抗体的检测（Juliarena et al., 2007；Leyva et al., 2007）。同时，由于ELISA实验中变异因素多，如抗原、待检样品及酶标抗体的浓度与抗原、抗体纯度等，都直接影响ELISA方法的特异性和敏感性。间接ELISA用于临床诊断检测时，必须进行一系列实验条件的选择和优化，主要优化指标由抗原的包被条件和工作浓度、封闭条件、待检样品和酶标二抗的工作浓度，以及与被检样品、酶标二抗和底物的作用时间组成。降低间接 ELISA 方法的非特异性是建立 ELISA 方法成功与否的关键所在。间接 ELISA 检测奶牛乳腺炎疫苗抗体过程中，由于泌乳奶牛本身含有一定的相应抗体，势必造成本底色问题，严重影响检测结果。产生非特异性的因素是多方面的，通过反应体系的优化，酶标板、试剂及显色系统的筛选比较等，很好地解决此问题。

本实验选用最常用的 96 孔可拆聚苯乙烯酶标板作为固相载体，用超声裂解的金黄色葡萄球菌抗原包被酶标板，最佳包被浓度是 11.34 μg/mL，高于冯万新（2008）建立的金黄色葡萄球菌抗原的包被浓度（4.24 μg/mL）。

ELISA 反应封闭物的选择中，根据测得的 P/N 值，确定 5%犊牛血清作为封闭物。5%脱脂奶粉的封闭效果也较好，但由于奶粉的成分比较复杂，而且封闭后的载体不易长期保存，因此未选用。由封闭结果来看，封闭物并非浓度越高越好，如不得不用犊牛血清，浓度大时反而 P/N 值低，可能是浓度过大时，封闭物未能完全洗脱，覆盖在包被物上，使 ELISA 的特异性反应降低。本实验采用的是国产酶标二抗，按推荐浓度 1：1000 稀释使用，在试验中尝试将酶标二抗从 1：400 稀释到 1：1200，结果显示 1：800 时的 P/N 值最高。进口酶标二抗的稀释度可以达到 1：5000，甚至更高，但是价格昂贵，故未选用。

1：200 倍稀释的阳性血清在反应前与 1：10 倍稀释的金黄色葡萄球菌抗原混合，测定其 OD 值，实验数据显示，金黄色葡萄球菌抗原特异性阻断了血清中金黄色葡萄球菌抗体。因此本实验建立的间接 ELISA 方法具有显著的特异性阻断作用，可以用来检测血清中的金黄色葡萄球菌抗体。

针对研制的两种奶牛乳腺炎疫苗，通过反应体系的筛选与优化，最终初步建立了间接 ELISA 检测奶牛血清中金黄色葡萄球菌抗体的方法。

在间接 ELISA 结果判定标准中，有两种方法，一种是用测定阴性样品的 OD 值和标准差来确定阴阳临界值。另一种是计算被检样品 OD 值与阴性样品 OD 值的比值，比值高于 2，即判为阳性，比值小于 2 而大于 1.5 为可疑，小于 1.5 为阴性。本研究采用第一种判定标准，根据大量的阴性样品的 OD 值，采用统计学方法确定阴阳临界值。依据的原理是样品的 OD 值大于阴性样本 OD 值的平均数加上三倍标准差时，可以在 99.9%的水平上判为阳性（朱立平，2000）。标准曲线和回归方程的建立是在血清 1：100 倍稀释的基础上再次倍比稀释，因此 Log_2[效价的倒数]会使一部分数据出现在同一个水平上，但是吸光度有一定的变化幅度范围，使得 R^2 值偏小。

（四）奶牛血清抗体水平消长规律

疫苗 1 组奶牛和疫苗 2 组奶牛接种后体温稍微有所升高，属于疫苗接种后的正常反应，下午对照组奶牛体温升高的原因是亚热带地区气温炎热，致使奶牛下午体温比早晨体温较高。免疫组奶牛与对照组奶牛相比，接种部位没有出现明显的临异常反应，也没有影响奶牛生产性能。

本实验接种奶牛采用一次免疫，从图 6-15 可以看出，奶牛接种疫苗后即产生抗体并逐渐升高，疫苗 1 在接种后 14 天达到最高值，疫苗 2 在接种后 21 天达到最高值，然后逐渐下降。从而也证实了奶牛接种这两种疫苗后确实都能激发奶牛产生高水平的抗金黄色葡萄球菌抗体。在接种后 150 天，血清中金黄色葡萄球菌抗体还维持在一定的水平，因此根据季节性乳腺炎发生长短，应当在奶牛乳腺炎高发季节到来之前半个月接种疫苗 1 或疫苗 2 免疫。李宏胜等（2007）也证明了在奶牛乳腺炎高发季节到来之前接种疫苗可明显地降低乳腺炎发病率。

水边油包水型的疫苗 1 产生抗体的速度比油包水型的疫苗 2 产生抗体的速度要快，并且下降速度较快。与赵建荣（2009）和李宏胜等（2007）研制的铝胶疫苗相比，MONTANIDE 系列佐剂制作的疫苗抗体效价产生较快（抗体峰值出现早且高于铝胶疫苗），血清中抗体效价也较高，也可能与制作疫苗的菌株有关。

奶牛乳腺炎疫苗对乳腺的保护作用主要是增强乳腺内的免疫力，包括体液免疫和细胞免疫。疫苗免疫效果的优劣除了与抗原性、免疫佐剂有关外，还与免疫途径有很大的关系。McDowell 等研究表明，在乳房淋巴结近处皮下注射疫苗会提高乳汁中的抗体水平。Nordhaug 等用含有灭活的金黄色葡萄球菌、荚膜、毒素和矿物油为原料制成的疫苗，在奶牛乳房淋巴结处注射了 2 次，使整个泌乳期血清和乳汁中均产生了相当高的抗体。李宏胜等（2007）的研究结果表明，在乳牛后海穴注射疫苗优于肌内注射，不仅可提高疫苗保护率而且可减少疫苗用量，减少注射部位副反应。

本实验研制的奶牛乳腺炎疫苗尚处于初级阶段，根据上述分析和讨论，本实验研制的疫苗在奶牛上的免疫途径、剂量和攻毒试验方面还需做进一步深入研究。

四、小结

（1）研制的奶牛乳腺炎金黄色葡萄球菌疫苗 1 和疫苗 2 的乳液类型分别是水包油包水型乳液和油包水型乳液，实验室检查合格。

（2）疫苗 1 和疫苗 2 对小白鼠的攻毒保护率分别是 70% 和 60%。

（3）根据常规间接 ELISA 方法，通过反应体系的优化，确定了最佳反应条件，初步建立了间接 ELISA 检测奶牛乳腺炎金黄色葡萄球菌抗体的方法。

（4）疫苗 1 组奶牛于接种后 14 天达到最高峰（1∶2995），疫苗 2 组奶牛于接种后 21 天达到最高峰（1∶2062），直至接种疫苗后第 150 天，两个实验组奶牛血清的特异性抗体效价仍然高于阴阳临界值。

第七章 奶牛乳腺炎的危害

奶牛乳腺炎主要是由病原微生物侵入乳腺，或受环境及应激等因素影响，引起乳腺组织实质和间质发生炎症反应，从而影响泌乳功能，相对于奶牛其他疾病，乳腺炎造成的经济损失最为严重（赵中利，2012）。与乳腺炎相关损失包括：临床型乳腺炎的治疗费用，瞎乳后终身丧失泌乳能力所造成的经济损失，治愈后产奶量不能恢复所致的饲料浪费，患病时期牛奶的废弃，患病母牛的过早淘汰，隐性乳腺炎引起的产奶量下降；抗生素治疗后，降低牛奶品质造成的损失，动物寿命的降低，降低动物的利用价值，降低牛奶的质量，以及更高的健康护理和治疗费用。乳腺炎最严重的影响是减少了奶制品的产量，占总经济损失的69%～80%。相比未感染的牛，患乳腺炎的牛平均每天产奶量减少0.5 kg。Janzen等报道在感染牛群中，每头每天减产0.35～2.7 kg。估计每头牛每个泌乳期损失100～200美元。

第一节 奶牛乳腺炎对生产性能的危害

一、产奶量下降

感染乳腺炎的奶牛为了清除机体内的病原微生物和尽快修复损伤的乳腺组织，会产生大量的白细胞并向受损伤乳区迁移，大量的白细胞聚集在一起，使部分乳腺管道被堵塞，分泌的乳汁无法及时排出，导致泌乳细胞的泌乳量降低甚至停止泌乳，最后还有可能萎缩，丧失泌乳功能。患有临床型乳腺炎的奶牛与健康奶牛相比，患病奶牛每天的产奶量减少0.5 kg，占乳腺炎总损失的69%～80%。据报道显示，我国的奶牛每头每天因隐性乳腺炎产奶量减少约3.7 kg，在世界范围内，每年的牛奶生产因乳腺炎损失约400万吨（周二顺，2015）。

二、牛奶质量降低

影响乳成分的因素有很多，如遗传、胎次、产犊季节、305天产奶量、泌乳时期、季节变化、品种、营养状况、疾病等，乳成分之间也存在着不同程度的关联（王小龙，2014）。

在正常情况下，奶牛机体内的乳汁与血液具有相同的渗透压。当乳汁中的乳糖减少时，为了保持它们之间的渗透压平衡，血液中的某些成分会及时进入乳汁中进行补充（王会珍，2007）。发生乳腺炎时病原菌侵入奶牛乳腺之后，刺激巨噬细胞和乳腺上皮细胞释放趋化因子，促使白细胞（主要是中性粒细胞）从血液向乳腺迁移，使原乳中白细胞的比例从5%～25%上升到90%左右（Leitner et al.，2000）。由于牛奶中含有大量的体细胞，致使乳汁成分发生以下变化：①乳清蛋白含量增加，而酪蛋白含量减少；乳糖和乳脂量降低，pH升高。②磷、钙、钾含量均略有减少，氯化物增多；碱反应增强，而热稳定性下降。这些变化使牛奶中的营养成分明显降低，严重影响牛奶质量。正常情况下，牛奶中的体细胞数为2万～20万/mL。在某些发达国家中，如果牛奶中的体细胞数量超过30万/mL，奶价就要有一定的折扣，如果超过 50 万/mL 则会被拒收（周二顺，2015）。

当奶牛患乳腺炎时，牛乳的成分会发生很大的变化，酪蛋白和乳球蛋白含量降低，血清蛋白和免疫球蛋白增高，而且中性粒细胞的含量也大幅升高。体细胞含量是评价牛乳质量的一个重要指标，健康奶牛所产牛奶中体细胞的含量为 2 万～20 万/mL，而患乳腺炎奶牛所产的牛奶中体细胞的含量可达 80 万～100 万/mL。体细胞含量高的牛奶中 pH 可达 7.2，严重影响牛奶质量，而且缩短其保质期（郑国卫和潘鸿飞，2006）。

患乳腺炎的牛奶乳腺上皮细胞合成的 α-乳白蛋白、β-乳白蛋白、酪蛋白含量会降低，但来自血液的免疫球蛋白和血清蛋白含量上升。当奶牛患乳腺炎时，乳汁中的体细胞数会显著上升，体细胞线性分值越大，乳脂率、乳蛋白率、乳糖率越高（刘琴等，2012）。SCC 在 50 万～100 万/mL，牛奶中总固体和非脂乳固体含量最高，但当 SCC 大于 100 万/mL 时，总固体含量即下降（宋维政等，2010）。陈建坡等（2013）研究结果表明：①当 SCC 小于 40 万/mL 时，原料奶的成分没有显著的差异（$P>0.05$）；②随着 SCC 的增大，原料奶的脂肪水解程度增强，但 SCC 小于 40 万/mL 的两组原料奶的脂肪水解程度没有显著差异（$P>0.05$）；③不同 SCC 原料奶的酪蛋白构成表现不同，其蛋白质水解程度随 SCC 的增大而增加，但 SCC 在 40 万/mL 以下的两组原料奶的蛋白水解程度没有显著差异（$P>0.05$）。

常玲玲等（2011）采样测定 478 头南方荷斯坦奶牛自然年度的乳成分和体细胞数，结果表明，乳脂率与体细胞数呈极显著负相关，而乳蛋白率与体细胞数呈极显著正相关。SCC 在泌乳末期最高，泌乳初期水平较低且稳定，SCC 极显著影响奶牛的产奶量，对乳脂率和乳蛋白有显著影响，与产奶量有极显著的负相关，与乳脂率有极显著的正相关，与乳蛋白率无明显的相关性。WilUam 等研究发现，SCC 为 1 万的原料奶和为 10 万的原料奶相比，酪蛋白含量减少，乳清蛋白含量升高，酪蛋白组成成分中 β-酪蛋白含量减少，γ 酪蛋白含量上升。Santos 等将 4 组不同 SCC 的牛奶（2.6 万、37.6 万、72.5 万和 1113 万）进行巴氏杀菌，然后在同一温度下贮藏，发现 SCC 高的牛奶即便储存温度较低，在贮藏时间一段时间后也出现了异味，说明牛奶中蛋白和脂肪酶解速度也较快。其研究数据显示，SCC 为 75 万/mL 比 10 万/mL 的贮藏奶，酪蛋白含量减少 20%，奶粉的得减少 4%，如果 SCC 在 1 万/mL 以上，酪蛋白含量将减少 25%，奶粉得率减少 5%。Ma 等（2011）研究发现，高 SCC 牛奶中含有的脂肪氧化酶比低 SCC 牛奶中多，游离脂肪酸的含量变化是低 SCC 牛奶的 2～3 倍。

患隐性乳腺炎的牛，其牛奶中中长链脂肪酸和自由脂肪酸较多，但碳原子数在 16～18 的长链脂肪较少，脂肪酸总量也较少。但有研究结果表明，和健康奶牛相比，患隐性乳腺炎的牛脂肪酸组成没有显著差异，但患临床乳腺炎的奶牛乳汁中的乳脂肪和多不饱和脂肪酸含量显著增加（$P<0.05$）。毛永江等（2011）将隐性乳腺炎与健康奶牛的奶中脂肪酸进行对比，发现隐性乳腺炎牛奶中脂肪酸总量、饱和脂肪酸和不饱和脂肪酸的绝对含量都显著低于健康奶牛，但是饱和脂肪酸与单不饱和脂肪酸的相对含量高于健康奶牛，多不饱和脂肪酸绝对含量低于健康奶牛（$P<0.05$）。常玲玲等（2011）研究发现，奶牛患隐性乳腺炎后，牛奶中脂肪酸的总量降低，乳汁中的饱和脂肪酸和单不饱和脂肪酸的百分比高于正常乳，多不饱和脂肪酸的百分比则低于正常乳。隐性乳腺炎导致总脂肪酸含量下降，

却存在大量游离脂肪酸。

目前我国的牛奶质量整体较差,严重制约着我国奶牛业的发展。2008年的调查发现,我国的奶牛乳汁中乳蛋白率和乳脂率分别为2.8%和3.1%,而国外一些发达国家的奶乳蛋白率和乳脂率分别为3.2%和3.5%(王加启和赵圣国,2009),说明我国的牛乳质量与发达国家相比还有很大的差距。

三、经济损失严重

奶牛乳腺炎对我国乃至世界奶牛养殖业造成了巨大的经济损失,主要有以下几个方面。①产奶量减少:无论是临床型的乳腺炎还是隐性的乳腺炎,都会大幅降低奶牛产奶量。有研究发现,隐性乳腺炎痊愈后,产奶量并没有得到提高。②牛奶废弃:每100 kg废弃的牛奶所造成的经济损失比降低100 kg产奶量还多。③奶牛过早淘汰:奶牛乳腺炎不仅能够降低产奶量和牛奶质量,还能够使乳房受损,甚至导致泌乳停止,最终使奶牛被淘汰。④医疗费用:兽医服务及治疗乳腺炎的药物花销也是奶牛乳腺炎所造成经济损失的一部分(付云贺,2015)。

感染乳腺炎的奶牛所产的牛奶在采用抗生素治疗期间和休药期内不能食用,而且患病牛的乳汁发生了理化性质的变化,乳品质明显降低,有报道称,全世界每年因乳腺炎而废弃的牛奶约达400万吨(王会珍,2007)。另外,由于乳腺炎使奶牛的产奶量降低,而且20%以上的奶牛因再次感染乳腺炎导致乳池萎缩,使大量正值产奶高峰期的奶牛被淘汰,从而给奶牛养殖场带来了巨大的经济损失。感染乳腺炎的奶牛在兽医诊疗费和医药费以及劳动量增加等方面所造成的损失同样不可轻视。在20世纪90年代,美国理事会通过调查发现,美国奶牛场每年对一头患乳腺炎奶牛的治疗费用约为182美元;加拿大的患该病奶牛每头每年花费140~300美元;部分北欧国家每年在一头患该病奶牛上的费用约为225美元(周二顺,2015)。近年来,我国的奶牛养殖业在农牧业中所占比值逐年上升,正处于奶牛业高速发展时期,但相关技术和设施还不配套和完善,因此,乳腺炎所造成的损失就更加严重,每年高达上百亿元,沉重打击了我国奶牛业的发展。

四、危及消费者健康

目前,针对奶牛乳腺炎的治疗药物主要是抗生素。应用抗生素治疗虽然有一定的效果,但是长期大量的使用抗生素也存在一些问题,如长时间并大剂量地使用抗生素易导致乳汁中含有残留的抗生素、耐药菌株产生等,这些劣质的原奶严重影响乳制品的安全与卫生,严重影响人类健康。另外,当奶牛患乳腺炎时,乳汁中的病原微生物及其毒素大量增加,食用该类牛乳后可导致人出现不适症状,如腹泻、呕吐及过敏等,严重的还会导致过敏性休克。长期饮用这种乳制品还会使机体内的细菌产生耐药性,严重影响人类疾病的医治。

奶牛乳腺炎大多数是由病原微生物的感染诱发的,可引起患病牛奶中的细菌总数和病原菌数会有不同程度地升高。因而,当消费者进食含细菌较多的奶制品就会严重地影响人体健康,甚至还会危害公共卫生安全等。因此也就增加了通过牛奶携带病原菌的危险性和潜在危害,并且,有的乳腺炎病原菌通过乳或乳制品被进食后可影响消费者的健康,甚至污染乳汁中的链球菌、大肠杆菌、金黄色葡萄球菌等病原菌,均能使人发生食物中毒(付云贺,2015)。

乳腺中的病原菌如金黄色葡萄球菌、化脓性链球菌和结核杆菌等，它们不但能引起乳腺炎，还能使人患病。金黄色葡萄球菌能产生并释放毒素，引起食物中毒，使人产生呕吐、发烧、腹泻和脱水等症状；化脓性链球菌多感染操作不当的挤奶员，使其患扁桃体炎、咽喉炎或猩红热等；结核杆菌容易使人患结核病。

五、奶牛的产奶年限缩短

乳腺炎对奶牛的乳腺组织可以造成十分严重的损伤，主要表现在，奶牛的乳腺组织损伤或疤痕组织生成、发生脓肿性纤维化时，可形成乳腺组织的永久性损坏而使其丧失泌乳能力，最终被淘汰。另外，必须淘汰那些久治不愈或持续感染的慢性乳腺炎患牛。当发生大肠杆菌性临床型乳腺炎时，倘若不及时治疗，就会危及奶牛生命，死亡率在10%以上。据Huszenicza等报道，当乳腺炎发生在产后15～28天时，这将推迟恢复奶牛卵巢正常活动，同时乳腺炎周期性发生的严重病例会诱使溶解不成熟的黄体或延长卵泡成熟。据报道，泌乳早期必须淘汰的奶牛占乳腺炎的7%（樊杰，2014）。非临床型乳腺炎和临床型乳腺炎都对奶牛繁殖性能产生消极影响。

六、降低繁殖性能

尽管近些年在奶牛乳房保健方面有了很大的改善，奶牛乳腺炎仍然是危害当今世界规模化奶牛养殖场最常见的疾病之一。泌乳早期乳腺炎的发生极其普遍，有报道称奶牛临床乳腺炎发病率高达37.1%，隐性乳腺炎奶牛阳性率在22.3%～62.9%（Kerro Dego and Tareke，2003；Sori et al.，2005；Getahun et al.，2008），我国部分地区奶牛临床型乳腺炎奶牛发病率为17.5%（孔雪旺和陈功义，2005）和23.1%（邢玫等，2007），隐性乳腺炎奶牛阳性率在40%～64.75%（朱贤龙等，2006；邢玫等，2007；李建军等，2008；寨鸿瑞等，2008）。

普遍认为奶牛乳腺炎对奶牛业的经济损失仅包括降低产奶量和乳汁质量（营养成分降低）、使用抗生素引起的乳汁废弃，以及增加体细胞数（SCC）、治疗费用、奶牛的淘汰率和死亡率等。近几年国内外学者报道奶牛产后至配种前或配种与怀孕之间发生临床乳腺炎和（或）隐性乳腺炎对繁殖性能有着不同程度的负面影响，如增加首次配种天数、配种次数和空怀期天数、改变动情间隔、增加流产率和淘汰率、降低受胎率和妊娠率等（Schrick et al.，2001；Santos et al.，2004；Ahmadzadeh et al.，2009；Nava-Trujillo et al.，2010；杨丰利，2011；Yang et al.，2012），然而当隐性乳腺炎进一步恶化为急性乳腺炎时，对奶牛繁殖性能的影响更加严重和明显（Schrick et al.，2001）。乳腺炎发生的泌乳生理阶段也显得很重要：如果发生在产犊后至首次配种前，会显著增加首次配种的天数和空怀期的天数；然而，如果发生在首次配种至怀孕期间，会增加空怀期天数和配种次数（Schrick et al.，2001；Ahmadzadeh et al.，2009）；怀孕之后发生乳腺炎会增加胚胎死亡率（Santos et al.，2004）。为使大家对奶牛乳腺炎所引起的经济损失有一个全面的认识和了解，现将近年来国内外奶牛乳腺炎对繁殖性能负面影响的研究进展进行如下总结。

（一）临床乳腺炎降低繁殖性能的研究进展

1. 临床乳腺炎增加首次配种天数　　早在1990年，Oltenacu等研究发现临床乳腺炎使首次配种天数增加 1.18 天（$P<0.05$）。Schrick 等（2001）对 752 头奶牛的研究发现在首次配种前发生临床乳腺炎的奶牛首次配种天数增多 9 天（77 天 vs. 68 天），在首次配种和怀孕之间发生临床乳腺炎的奶牛首次配种天数没有显著差异（71 天 vs. 68 天），隐性乳腺炎发展为临床乳腺炎的奶牛首次配种天数增多 26 天（94 天 vs. 68 天）。Nava-Trujillo 等（2010）研究认为，产后至首次配种之间发生临床乳腺炎的奶牛首次配种天数增多 38 天（136 天 vs. 98 天）。杨丰利（2011）对 152 头 2~4 胎次的经产奶牛产后进行临床乳腺炎的发病时间和繁殖指标的调查，结果表明，奶牛怀孕之前发生临床乳腺炎增加了首次配种天数（110 天 vs. 88 天），产后至首次配种之间和首次配种至怀孕之间发生临床乳腺炎的奶牛首次配种天数均极显著多于对照组奶牛（114 天，105 天 vs. 88 天）。但也有报道称产后乳腺炎对首次配种天数没有显著影响，Santos 等（2004）对 1001 头高产奶牛的研究就未发现临床乳腺炎发生在产后至怀孕之间奶牛对首次配种天数有显著影响。

2. 临床乳腺炎降低配种怀孕率和增加配种次数 Kelton 等（2001）对 4555 头奶牛进行跟踪记录，产后 30 天内发生临床乳腺炎的奶牛受孕率是 38%，产后 30 天内未发生临床乳腺炎的奶牛受孕率是 46%，其中首次配种的怀孕率分别是 31% 和 47%。Santos 等（2004）对 1001 头高产奶牛产后进行临床乳腺炎和繁殖参数的记录，发现临床乳腺炎发生在产后至首次配种之间和首次配种至怀孕之间的奶牛首次配种怀孕率显著低于对照组奶牛（分别是 22.1%、10.2% 和 28.7%），在产后 320 天时的怀孕率依然均显著低于对照组奶牛（分别是 72.3%、58.5% 和 85.4%）。Hertl 等（2010）研究发现，临床乳腺炎发生在配种前 14 天至配种后 35 天的奶牛受孕率下降，革兰氏阴性菌引起的乳腺炎受孕率下降 80%。杨丰利（2011）对 152 头 2~4 胎次的经产奶牛产后进行临床乳腺炎的发病时间和繁殖指标的调查，结果表明，怀孕之前发生临床乳腺炎的奶牛首次配种怀孕率显著降低（26% vs. 49%）、产后 70 天内的怀孕率显著降低（12% vs. 22%），产后至首次配种之间和首次配种至怀孕之间发生临床乳腺炎的奶牛首次配种怀孕率显著降低（27%，24% vs. 49%）。然而，Chebel 等（2004）认为，临床乳腺炎发生在首次配种至怀孕之间的奶牛与健康奶牛相比怀孕率没有差异。Nava-Trujillo 等（2010）的研究结果表明，产后至首次配种之间发生临床乳腺炎的奶牛首次配种怀孕率没有显著差异（49.72% vs. 56.1%）。

Schrick 等（2001）对 752 头奶牛的研究结果发现，在首次配种前发生临床乳腺炎的奶牛和在首次配种至怀孕之间发生临床乳腺炎的奶牛配种次数增多（2.1 vs. 1.6，3.0 vs. 1.6），隐性乳腺炎发展为临床乳腺炎的奶牛配种次数显著增多（4.3 vs. 1.6）。Santos 等（2004）对 1001 头高产奶牛产后进行临床乳腺炎和繁殖参数的记录，发现临床乳腺炎发生在首次配种至怀孕之间的奶牛配种次数显著多于对照组奶牛（3.05 vs. 2.59）。Ahmadzadeh 等（2009）对 967 头奶牛进行回顾性分析，泌乳期间仅发生临床乳腺炎的奶牛配种次数显著增多（2.1 vs. 1.6），其他疾病（如卵巢囊肿、胎衣不下、真胃移位、酮病、生产瘫痪和子宫炎）和临床乳腺炎在泌乳期间都有发生的奶牛配种次数显著增多（2.8 vs. 1.6）；产后 56~105 天和 105 天以后发生临床乳房炎的奶牛配种次数均显著多于健康奶牛（2.33，3.11 vs. 1.62）。杨丰利（2011）对 152 头 2~4 胎次的经产奶牛产后进行临床乳腺炎的发病时间和繁殖指标进行调查，结果表明，怀孕之前发生临床乳腺炎的奶牛配种次数显著增多（2.07 vs. 1.72），首次配种至怀孕之间发生临床乳腺炎的奶牛配种次数显著增多（2.24 vs. 1.72）。然而，Nava-Trujillo 等（2010）的研究结果表明产后至首次配种之间发生临床乳腺炎的奶牛配种次数没有显著差异（2.35 *vs.* 2.21）。

3. 临床乳腺炎增加发情间隔天数　　奶牛正常的发情间隔是 18～24 天，低于 18 天或多于 24 天均被认为发情间隔异常。Moore 等在美国加利福尼亚州的两个奶牛场研究临床乳腺炎对发情间隔的影响，第一个奶牛场的奶牛 65%处于第一泌乳期，23%处于第二泌乳期，剩下的 12%的奶牛处于为第三泌乳期及以后，试验期间临床乳腺炎的发病率为 11%，异常发情间隔的奶牛为 56%，结果表明，临床乳腺炎奶牛发情间隔发生异常的风险比例与对照组奶牛基本相似。第二个奶牛场的奶牛 14%处于第一泌乳期，38%处于第二泌乳期，剩下的 48%奶牛处于第三泌乳期及以后，临床乳腺炎的发病率为 14%，异常发情间隔的奶牛为 46%，结果发现，临床乳腺炎奶牛发情间隔异常的风险比例是对照组奶牛的 1.6 倍，经产奶牛发生临床乳腺炎对发情间隔的影响远大于初产奶牛。

4. 临床乳腺炎增加空怀期天数　　Schrick 等（2001）对 752 头奶牛的研究结果发现，临床乳腺炎发生在首次配种之前的奶牛空怀期天数增多 25 天（110 天 *vs.* 85 天），在首次配种和怀孕之间发生临床乳腺炎的奶牛空怀期天数增多 59 天（144 天 *vs.* 85 天），隐性乳腺炎发展为临床乳腺炎的奶牛空怀期天数增多 111 天（196 天 *vs.* 85 天）。Santos 等（2004）对 1001 头高产奶牛进行记录临床乳腺炎和繁殖参数，发现临床乳腺炎发生在产后至首次配种之间和首次配种至怀孕之间的奶牛空怀期显著多于对照组奶牛（165 天 *vs.* 140 天，190 天 *vs.* 140 天），然而临床乳腺炎发生在怀孕之后的奶牛空怀期显著低于对照组奶牛（118 天 *vs.* 140 天）。Ahmadzadeh 等（2009）对 967 头奶牛进行回顾性分析，泌乳期间仅发生临床乳腺炎的奶牛空怀期天数显著增多（140 天 *vs.* 88 天），其他疾病（如卵巢囊肿、胎衣不下、真胃移位、酮病、生产瘫痪和子宫炎）和临床乳腺炎在泌乳期间都有发生的奶牛空怀期天数显著增多（155 天 *vs.* 88 天）；产后 56 天内、56～105 天和 105 天以后发生临床乳房炎的奶牛空怀期天数均显著多于健康奶牛（123 天，141 天，181 天 *vs.* 88 天）。Nava-Trujillo 等（2010）研究结果表明，产后至首次配种之间发生临床乳腺炎的奶牛空怀期显著增多（187 天 *vs.* 144 天）。杨丰利（2011）对 152 头 2～4 胎次的经产奶牛产后进行临床乳腺炎的发病时间和繁殖指标的调查，结果表明怀孕之前发生临床乳腺炎的奶牛空怀期显著增多（158 天 *vs.* 122 天），产后至首次配种之间和首次配种至怀孕之间发生临床乳腺炎的奶牛空怀期显著增多（154 天，164 天 *vs.* 122 天）。

5. 临床乳腺炎增加流产率　　Chebel（2004）认为，临床乳腺炎发生在首次配种至怀孕之间的奶牛增加配种后 31～45 天内的流产率，流产率相当于健康奶牛的 2.8 倍。Santos（2004）对 1001 头高产奶牛进行跟踪记录临床乳腺炎和繁殖参数，发现临床乳腺炎发生在产后至首次配种之间、首次配种至怀孕之间和怀孕之后的奶牛流产率均显著高于对照组奶牛（分别是 11.8%、11.6%、9.7%和 5.8%）。杨丰利（2011）对 152 头 2～4 胎次的经产奶牛产后进行临床乳腺炎的发病时间和繁殖指标的调查，结果表明怀孕之前发生临床乳腺炎的奶牛流产率显著增多（9.3% *vs.* 4.6%）。

（二）隐性乳腺炎降低繁殖性能的研究进展

1. 隐性乳腺炎增加首次配种天数　　Schrick（2001）对 752 头奶牛的研究发现隐性乳

腺炎发生在首次配种之前的奶牛首次配种天数显著多于对照组奶牛（75 天 vs. 68 天），发生在首次配种和怀孕之间的奶牛首次配种天数和对照组奶牛没有显著差异（61 天 vs. 68 天）。Klaas 等（2004）对 1362 头奶牛进行跟踪研究，在产后 42 天内隐性乳腺炎发病率 33.6%，这些奶牛的首次配种天数比对照组奶牛显著增加了 11.7 天，其中首次配种天数最多的奶牛是隐性乳腺炎奶牛，而不是临床乳腺炎奶牛，该研究表明隐性乳腺炎比临床乳腺炎对奶牛繁殖性能的负面影响更大。Pinedo 等（2009）的研究表明，出现至少一次高线性体细胞计数（high linear somatic cell counts，LNSCC）≥4.5 的奶牛首次配种天数显著增多 22 天（108 天 vs. 86 天）。Nguyen 等（2011）认为，乳汁体细胞数（SCC）在 20 万～50 万的奶牛黄体期比乳汁 SCC 在 5 万～10 万的奶牛黄体期长，产后一个月内乳汁 SCC 大于 50 万的奶牛产后首次排卵延迟于 SCC 低于 50 万的奶牛。

2. 隐性乳腺炎增加空怀期天数 Schrick 等（2001）对 752 头奶牛的研究发现隐性乳腺炎发生在首次配种之前的奶牛空怀期天数显著多于对照组奶牛（108 天 vs. 85 天），在首次配种和怀孕之间发生隐性乳腺炎的奶牛空怀期天数没有显著差异（91 天 vs. 85 天）。Pinedo 等（2009）的研究表明，出现至少一次 LNSCC≥4.5 的奶牛空怀期天数增多 48 天（169 天 vs. 121 天）。Nguyen 等（2011）认为，乳汁 SCC 在 20 万～50 万的奶牛空怀期天数多于 SCC 低于 20 万的奶牛。

3. 隐性乳腺炎降低怀孕率和增加配种次数 Nguyen 等（2011）认为，乳汁 SCC 在 20 万～50 万的奶牛怀孕率低于乳汁 SCC 低于 20 万的奶牛。Schrick 等（2001）对 752 头奶牛的研究发现，隐性乳腺炎发生在首次配种之前的奶牛配种次数显著多于对照组奶牛（2.1 vs. 1.6）。Pinedo 等（2009）的研究表明，出现至少一次 LNSCC≥4.5 的奶牛配种次数增加了 0.49 次，流产率是对照组奶牛的 1.22 倍。

（三）乳腺炎对繁殖性能影响的机制研究

1. 体温升高 临床乳腺炎的发生除了乳房和乳汁的变化外，革兰氏阳性菌和革兰氏阴性菌均可引起乳腺感染，导致体温升高。Sartori 等（2003）将奶牛置于热应激条件下观察，发现受精率和高质量的胚胎比例显著降低。试验诱导的乳房链球菌引起的临床乳腺炎使奶牛体温升高（Hockett et al.，2000），体温严重升高，使奶牛的饲料摄入量大大减少，体重急剧下降，最终导致奶牛卵巢周期的紊乱。有研究也证明了热应激的奶牛增加了胚胎的死亡率（Schrick et al.，2001），急性乳腺炎还有可能导致菌血症的发生，但是至今并没有试验证实临床乳腺炎引起的体温升高直接引起胚胎的死亡，因此该作用机制仅限于奶牛全身性临床乳腺炎。

2. 细胞因子 奶牛发生乳腺炎后，会引起机体产生大量的细胞因子，它们会影响卵母细胞、胚胎质量和发育、子宫环境和卵巢功能，这些细胞因子有白细胞介素（IL）-1α、IL-1β、IL-6、IL-10、IL-12 和肿瘤坏死因子-α（TNF-α）（Riollet et al.，2001）。Waller 等（2009）用大肠杆菌的细胞壁成分（脂多糖，LPS）诱导的奶牛乳腺炎乳汁表现为 IL-1β、IL-8 和 TNF-α 浓度增加。Hockett 等（2000）研究结果表明，奶牛发生乳腺炎后乳汁中一氧化氮（NO）和前列腺素 F2α（PGF2α）的浓度增加。

Soto 等（2003）研究表明，一氧化氮阻碍胚胎的发育。隐性乳腺炎奶牛乳汁 IL-6 浓度（30.8 ng/mL）显著高于临床乳腺炎奶牛（18.0 ng/mL）和健康奶牛（5.2 ng/mL）；临床乳腺炎奶牛乳汁溶菌酶浓度（15.6 μmol/L）显著高于隐性乳腺炎奶牛（11.2 μmol/L）和健康奶牛（6.9 μmol/L）；临床乳腺炎奶牛乳汁 NO 浓度（11.5 μmol/L）显著高于隐性乳腺炎奶牛（6.2 μmol/L）和健康奶牛（5.6 μmol/L）（Osman et al.，2010）。隐性乳腺炎奶牛乳汁 NO 浓度（8.89 μmol/L）显著高于健康奶牛（3.96 μmol/L）（Atakisi et al.，2010）。

丙二醛（MDA）是脂质过氧化反应的最终产物，是显示脂质过氧化反应水平的常用指标。GPx 已经被国内外学者们确定是含硒酶，参与催化多类过氧化物的还原反应，保护机体的细胞免受氧化损伤，充足的 GPx 能够保护乳汁的脂类不被氧化。隐性乳腺炎奶牛乳汁中 MDA 浓度显著高于健康奶牛乳汁（28.45 nmol/mL *vs.* 24.37 nmol/mL），而隐性乳腺炎奶牛乳汁中 GPx 的平均活性显著低于健康奶牛乳汁（26.41 IU/mL *vs.* 32.81 IU/mL）（杨丰利，2011；Yang et al.，2012）。

当培养液中含有 TNF-α 时，奶牛受精卵母细胞发育至胚泡期的百分比下降，当培养液中含有 TNF-α、PGF2α 或者 NO 时均会增加胚胎的细胞凋亡数量（Soto et al.，2003）。关于评价乳腺内感染对卵母细胞和胚胎发育影响的研究大多采取体外试验，体外试验意味着所得结果与在动物体内试验情况并不一定完全吻合，但是对于乳腺炎影响繁殖性能的机制提供了重要参考依据。

3. 激素 奶牛的繁殖周期受下丘脑［分泌促性腺激素释放激素（GnRH）］、脑垂体［分泌促卵泡激素（FSH）和促黄体生成素（LH）］和卵巢［分泌雌二醇（E2）和孕酮（P4）］产生的激素的调控。GnRH 刺激 FSH 和 LH 的分泌，FSH 刺激卵泡的起始生长，LH 刺激卵泡的成熟和排卵，通过黄体刺激 P4 的分泌，P4 促使胚胎生长和维持妊娠；E2 由卵泡产生，刺激 LH 峰值的产生引起排卵。因此，以上激素的产生和分泌障碍均会影响繁殖性能。

IFN-β 等细胞因子会抑制 LH 的分泌，革兰氏阴性菌分泌的内毒素引发的乳腺炎使血液中皮质醇的浓度升高，皮质醇抑制 LH 的分泌和峰值的出现。LH 分泌减少或缺乏导致卵泡和卵母细胞发育不良或不能发育，产生不能排卵和黄体功能不全的后果。乳腺炎期间产生的一些细胞因子还直接作用于卵巢，如 IL-6，抑制雌二醇的分泌，使 LH 分泌减少，而 TNF-β 和 IFN-δ 对黄体具有细胞毒性，引起 P4 浓度降低（McCann et al.，2000）。

Hockett 等（2000）认为，奶牛发生乳腺炎会使 PGF2α 合成增加，导致黄体功能降低，最终引起胚胎死亡；在发情周期的黄体期，试验诱导的乳房链球菌乳腺炎奶牛子宫对前列腺素 F2α 的敏感性增加，认为前列腺素 F2α 是通过减少黄体的存在时间或降低胚胎的质量和发育来影响繁殖性能。Schrick 等（2001）也报道奶牛胚胎的质量与子宫腔内前列腺素 F2α 的浓度呈负相关。

（四）展望

综上所述，奶牛产后不管发生临床乳腺炎还是隐性乳腺炎，对繁殖性能都有统计学上明显的负面影响，其主要机制包括乳腺炎引起的机体体温升高、产生的细胞因子、抗氧化能力降低、内分泌激素改变等。繁殖性能降低进而推迟奶牛下一个泌乳期，所引起的经济损失非常大，据笔者所知，尚未有研究人员统计这方面的经济损失。虽然已经有文献报道奶牛发生乳腺炎后引起繁殖性能改变的机制，但是引起奶牛乳腺炎的因素众多，致病菌多种多样，这些报道主要采用体外试验进行研究，因此这些机制的假设需要研究探讨并进一步证实，人们对其认识也将更加清晰，从而阻断中间途径，将奶牛乳腺炎对繁殖性能的影响降到最低。

第二节 奶牛临床乳腺炎对繁殖性能的影响

奶牛乳腺炎引起的经济损失不仅包括产奶量的降低、乳汁的废弃、治疗费用的增加、淘汰率和死亡率升高（Ahmadzadeh et al., 2009），而且在产后首次配种前后发生乳腺炎对繁殖性能也有一定的负面影响。近几年国外学者已经报道在产后首次配种之前和之后发生奶牛临床乳腺炎和（或）隐性乳腺炎对繁殖性能具有一定的负面影响，如增加首次配种天数、增加配种次数和空怀期天数、改变动情间隔、增加流产率和淘汰率、降低受胎率和妊娠率等（Schrick et al., 2001；Santos et al., 2004；Ahmadzadeh et al., 2009；Nava-Trujillo et al., 2010），然而当隐性乳腺炎进一步恶化为急性乳腺炎时，对奶牛繁殖性能的影响更加严重和明显（Schrick et al., 2001）。乳腺炎发生的泌乳生理阶段也显得很重要，如果发生在产犊后至首次配种前，会显著增加首次配种的天数和空怀期的天数，然而，如果发生在首次配种至怀孕期间，会增加空怀期天数和配种次数（Schrick et al., 2001；Ahmadzadeh et al., 2009），怀孕之后发生乳腺炎会增加胚胎死亡率（Santos et al., 2004）。奶牛乳腺炎的发生对繁殖性能影响的机制已有研究，主要是在乳腺炎发生的一开始就释放炎性介质和细菌内毒素，影响机体的内分泌（下丘脑—垂体—卵巢—子宫轴），从而改变繁殖性能，引起乏情、不排卵、不怀孕和其他与激素平衡稳定相关的疾病（Hockett et al., 2000；Suzuki et al., 2001；Hansen et al., 2004）。

提高奶牛的繁殖性能对于产奶量的提高和稳定有着非常重要的意义，大多数关于奶牛乳腺炎和繁殖性能关系的报道是在温带国家的高产奶牛进行调查研究的（Santos et al., 2004；Ahmadzadeh et al., 2009），然而大部分热带和亚热带地区的奶牛产奶量相对较低，对热带环境应激较大，受到多种因素的影响，如季节、公牛的生物刺激和胎次等，繁殖性能方面表现为产后较长的乏情期。更重要的是至今为止国内还没有可参考的文献。因此很有必要对亚热带环境下临床乳腺炎发生的不同阶段对繁殖性能的影响进行展开调查。

一、材料与方法

（一）实验材料

1. 实验动物 规模化奶牛场荷斯坦奶牛群，随机选 2~4 胎次的经产奶牛 152 头，产犊月为 10~11 月，产犊前两个月无任何临床疾病发生。

2. 饲养管理 每栏 90~100 头泌乳奶牛，在半封闭舍饲，每天定时饲喂两次（上午和下午），先喂粗料，后喂精料，每天清槽两次。奶牛产后饲喂的高产料配方为：玉米 45.00%，麸皮 12.00%，豆粕 20.00%，棉籽粕 8.00%，玉米 8.00%，小苏打 1.50%，磷酸氢钙 2.00%，石粉 1.50%，食盐 1.00%，矿物添加剂 8011 1.00%。粗饲料为玉米青贮。

采用鱼骨式自动挤奶器每天挤奶三次，分别是上午、下午和晚上，挤奶时"二次药浴，一次纸巾擦干"，保证一牛一巾。每个挤奶设备定期检查和维护。所有泌乳奶牛的每个乳区在预产期前 8 周停止挤奶,挤奶的最后一次在各个乳区内注射抗生素制剂进行干奶。产犊后将犊牛移至犊牛舍进行饲养管理,不进行犊牛吮乳刺激。

(二)实验方法

1. 临床乳腺炎的确定　　首先由挤奶员在每次挤奶时和饲养员在饲养过程中观察并进行乳腺炎的诊断,临床乳腺炎奶牛具有明显的临床症状,如乳房表现不同程度的红肿、发热和疼痛等炎性症状,或者触诊乳区有肿块,乳汁有絮片、凝块或变色(淡黄色、淡红色不等)。然后由奶牛场兽医技术人员进行乳腺炎的确诊和治疗,治疗方法包括乳腺内注射抗菌药,如乳炎消,以及根据感染的严重程度进行全身治疗,严重的乳腺炎病例也静脉注射抗组胺药物。

2. 记录指标　　实验奶牛每天观察发情情况,并记录人工授精日期,受精后 21 天前后观察是否返情,42~49 天通过直肠检查确定受胎率,并追踪本次妊娠期发生流产情况。繁殖指标的记录包括首次配种天数、配种次数、首次配种怀孕率、空怀期、产后 70 天内怀孕率、流产率;并且记录奶牛产后首次临床乳腺炎发生的时间。

3. 实验分组　　依据实验奶牛临床乳腺炎发生的时间,将所选实验牛分为三组。实验组Ⅰ:产后至首次人工授精之间发生临床乳腺炎的奶牛。实验组Ⅱ:首次人工授精至怀孕之间生在临床乳腺炎的奶牛。对照组:没有发生任何临床疾病的奶牛。

(三)数据统计

用平均数±标准误(SEM)表示各组实验奶牛的首次配种天数、配种次数和空怀期,采用 Office Excel 2003 和 SAS 9.1 数据统计软件对试验数据进行统计分析,t-检验评价各实验组间的差异显著性,用百分数的差异性检验和逻辑回归分析各组间首次配种怀孕率、产后 70 天怀孕率和流产率,以 $P<0.05$ 为差异显著,$P<0.01$ 为差异极显著。

二、实验结果

(一)实验奶牛分组结果

根据实验期间实验奶牛的临床疾病发生的种类和时间可分为 4 个组:实验组Ⅰ、实验组Ⅱ、对照组和其他临床疾病组(该组排除),见表 7-1。实验奶牛在实验期间除了发生临床乳腺炎外,还在产犊后发生胎衣不下、子宫内膜炎。本实验共 152 头奶牛,发生临床乳腺炎的奶牛有 43 头,其中 26 头奶牛发生在产后至首次人工授精之间,为实验组Ⅰ;17 头奶牛发生在首次人工授精至怀孕之间,为实验组Ⅱ;在实验阶段未发生任何临床疾病的奶牛 87 头,为对照组。本实验期间所有实验奶牛的临床乳腺炎发病率为 39.45%。

表 7-1　152 头实验奶牛的分组情况

Table 7-1　The grouping of the 152 dairy cows

组别	头数
实验组Ⅰ	26
实验组Ⅱ	17

续表

组别	头数
对照组	87
发生其他临床疾病	22
总头数	152

（二）奶牛临床乳腺炎对繁殖性能指标的影响

在实验阶段内 152 头实验奶牛中有 43 头奶牛发生临床乳腺炎，87 头临床健康奶牛（表 7-1）。临床乳腺炎组奶牛的首次配种天数和空怀期均极显著多于健康奶牛，即对照组奶牛（110.17±9.56 vs. 87.66±2.25，157.76±14.03 vs. 121.52±7.85）（$P<0.01$）；临床乳腺炎组奶牛的配种次数（2.07±0.27）显著多于对照组奶牛（1.72±0.09）（$P<0.05$）；临床乳腺炎组奶牛的首次配种怀孕率（25.58%）显著低于对照组奶牛（49.43%）（$P<0.05$）；临床乳腺炎组奶牛的产后 70 天内的怀孕率（11.63%）低于对照组奶牛（21.84%）（$P>0.05$）；临床乳腺炎组奶牛的流产率（9.3%）高于对照组奶牛（4.6%）（$P>0.05$）（表 7-2，图 7-1 和图 7-2）。

表 7-2 临床乳腺炎奶牛和健康奶牛繁殖性能指标分析结果
Table 7-2 The analysis of reproductive performance in clinical mastitis and health dairy cows

指标	临床乳腺炎组（$n=43$）	对照组（$n=87$）
首次配种天数	110.17±9.56^A	87.66±2.25^B
配种次数	2.07±0.27^a	1.72±0.09^b
首次配种怀孕率	25.58%^{a}	49.43%^{b}
空怀期	157.76±14.03^A	121.52±7.85^B
产后 70 天内怀孕率	11.63%	21.84%
流产率	9.3%	4.6%

注：同行数据不同上标小写字母表示差异显著，$P<0.05$；同行数据不同上标大写字母表示差异极显著，$P<0.01$

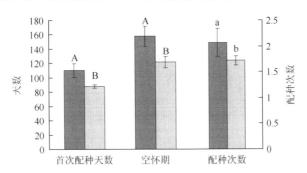

图 7-1 奶牛临床乳腺炎对首次配种天数、空怀期和配种次数的影响

灰色的柱形是产后至怀孕之间发生临床乳腺炎的奶牛，格子的柱形是临床表现健康的奶牛；同一组柱形图内不同小写字母表示差异显著，$P<0.05$；同一组柱形图内不同大写字母表示差异极显著，$P<0.01$

Figure 7-1 The effects of clinical mastitis prior to pregnancy diagnosis on days to first insemination, days open, and services per conception of dairy cows

Bars represent clinical mastitis before pregnancy diagnosis(gray), and mastitis after pregnancy confirmation or uninfected(hatched). Means with different lower case letters within bar groups differ at $P<0.05$, with significant different capital letters within bar groups differ at $P<0.01$

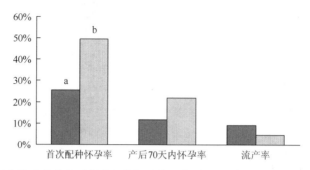

图 7-2 奶牛临床乳腺炎对首次配种怀孕率、产后 70 天内怀孕率和流产率的影响

灰色的柱形是产后至怀孕之间发生临床乳腺炎的奶牛，格子的柱形是临床表现健康的奶牛；同一组柱形图内不同小写字母表示差异显著，$P<0.05$

Figure 7-2 The effects of clinical mastitis prior to pregnancy diagnosis on the conception rate at first postpartum AI, conception rate within 70 days postpartum, and abortion rate of dairy cows

Bars represent clinical mastitis before pregnancy diagnosis (gray), and mastitis after pregnancy confirmation or uninfected (hatched). Means with different lower case letters within bar groups differ at $P<0.05$

（三）奶牛临床乳腺炎发生的时间对繁殖性能指标的影响

在实验阶段内 152 头实验奶牛中有 43 头奶牛发生临床乳腺炎，其中产后至首次人工授精之间发生临床乳腺炎的奶牛有 26 头（实验组Ⅰ），首次人工授精至怀孕之间生在临床乳腺炎的奶牛有 17 头（实验组Ⅱ），87 头临床健康奶牛（对照组）（表 7-1）。实验组Ⅰ奶牛和实验组Ⅱ奶牛的首次配种天数均极显著多于对照组奶牛（113.76±8.45 $vs.$ 87.66±2.25，104.88±8.6 $vs.$ 87.66±2.25）（$P<0.01$）；实验组Ⅰ奶牛和实验组Ⅱ奶牛的空怀期均极显著多于对照组奶牛（153.76±11.69 $vs.$ 121.52±7.85，163.65±14.19 $vs.$ 121.52±7.85）（$P<0.01$）；实验组Ⅱ奶牛的配种次数（2.24±0.3）显著多于对照组奶牛（1.72±0.09）（$P<0.05$）；实验组Ⅰ奶牛和实验组Ⅱ奶牛的首次配种怀孕率均显著低于对照组奶牛（26.92% $vs.$ 49.43%，23.53% $vs.$ 49.43%）（$P<0.05$）（表 7-3，图 7-3 和图 7-4）。

表 7-3 各组奶牛繁殖性能指标分析结果

Table 7-3 The analysis of reproductive performance in each group

指标	实验组Ⅰ	实验组Ⅱ	对照组
首次配种天数	113.76±8.45^A	104.88±8.6^A	87.66±2.25^B
配种次数	1.96±0.2^{ab}	2.24±0.3^a	1.72±0.09^b
首次配种怀孕率	26.92%^{a}	23.53%^{a}	49.43%^{b}
空怀期	153.76±11.69^A	163.65±14.19^A	121.52±7.85^B
产后 70 天内怀孕率	11.54%	11.76%	21.84%
流产率	7.69%	11.76%	4.6%

注：同行数据不同上标小写字母表示差异显著，$P<0.05$；同行数据不同上标大写字母表示差异极显著，$P<0.01$

三、讨论

一直以来，很多学者认为，奶牛临床乳腺炎对奶牛业的经济损失仅包括降低产奶量和乳汁质量、增加体细胞数（SCC）、增加兽医的劳动和治疗费用、由于乳汁成分或抗生素

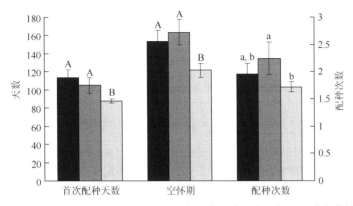

图 7-3 奶牛临床乳腺炎发生的时间对首次配种天数、空怀期和配种次数的影响

黑色柱形表示产后至首次人工授精之间发生临床乳腺炎的奶牛（实验组Ⅰ），灰色柱形表示首次人工授精至怀孕之间生在临床乳腺炎的奶牛（实验组Ⅱ），格子柱形是临床表现健康的奶牛（对照组）；同一组柱形图内不同小写字母表示差异显著，$P<0.05$；同一组柱形图内不同大写字母表示差异极显著，$P<0.01$

Figure 7-3　The effects of timing of first clinical mastitis occurrence on days to first insemination, days open, and services per conception of dairy cows

Bars represent clinical mastitis before first insemination (black, Group Ⅰ), mastitis between first insemination and pregnancy (gray, Group Ⅱ), and mastitis after pregnancy confirmation or uninfected (hatched, Control Group). Means with different lower case letters within bar groups differ at $P<0.05$, with significant different capital letters within bar groups differ at $P<0.01$

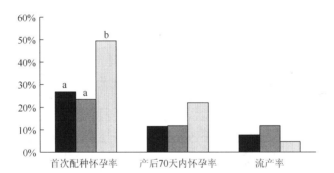

图 7-4　奶牛临床乳腺炎发生的时间对首次配种怀孕率、产后 70 天内怀孕率和流产率的影响

黑色的柱形表示产后至首次人工授精之间发生临床乳腺炎的奶牛（实验组Ⅰ），灰色的柱形表示首次人工授精至怀孕之间生在临床乳腺炎的奶牛（实验组Ⅱ），格子的柱形是临床表现健康的奶牛（对照组）；同一组柱形图内不同小写字母表示差异显著，$P<0.05$

Figure 7-4　The effects of timing of first clinical mastitis occurrence on the conception rate at first postpartum AI, conception rate within 70 days postpartum, and abortion rate of dairy cows

Bars represent clinical mastitis before first insemination (black, Group Ⅰ), mastitis between first insemination and pregnancy (gray, Group Ⅱ), and mastitis after pregnancy confirmation or uninfected (hatched, Control Group). Means with different lower case letters within bar groups differ at $P<0.05$

的使用引起的乳汁废弃，以及增加了奶牛的淘汰率等。最近几年，关于奶牛产后临床乳腺炎对繁殖性能的影响在国外逐渐得到重视（Schrick et al.，2001；Santos et al.，2004；Ahmadzadeh et al.，2009；Nava-Trujillo et al.，2010），如首次人工授精之前和之后发生临床乳腺炎对奶牛的繁殖性能、淘汰率，以及乳汁成分有着显著的影响（Santos et al.，2004）。空怀期超过 90 天，将会引起经济损失（Schrick et al.，2001），包括整个泌乳期产奶量降低、乳汁成分发生改变、治疗费用增加、乳汁的废弃和增加淘汰率。空怀期天数越多，损

失越大，因此久配不孕的奶牛建议及时淘汰，减少经济损失。临床乳腺炎对繁殖性能的影响可能是通过影响下丘脑—垂体—卵巢轴（内分泌机能）和卵泡的发育，但是至今仍没有关于乳腺炎和内分泌变化之间关系的研究文章发表。

本项研究比较了首次人工授精之前发生临床乳腺炎的奶牛和首次人工授精与怀孕之间发生临床乳腺炎的奶牛的繁殖性能，其目的是研究临床乳腺炎的发生与否及其发生时间对奶牛繁殖性能的影响。研究结果表明，首次人工授精之前发生临床乳腺炎显著增加了配种次数、首次配种天数和空怀期，并且显著降低了首次配种怀孕率。与国外报道的基本一致（Schrick et al.，2001；Santos et al.，2004；Ahmadzadeh et al.，2009；Nava-Trujillo et al.，2010）。另外，因为百分数的差异性检验对所有样本的要求是大于等于 5，而实验组 II 的首次配种成功数为 4 头，实验组 I 和实验组 II 产后 70 天内发情数分别是 3 头和 2 头，流产数均是 2 头，因此不能用百分数的差异性检验组间的差异显著性。

临床乳腺炎的发生除了乳房和乳汁的变化外，其他症状还有引起体温升高等，实验诱导的乳房链球菌临床乳腺炎使机体体温升高（Hockett et al.，2000），有研究也证明了热应激的奶牛增加了胚胎的死亡率（Schrick et al.，2001）。急性乳腺炎还有可能导致菌血症的发生，体温严重升高，使奶牛的饲料摄入量大大减少，体重急剧下降。

乳腺内注射大肠杆菌会引起奶牛发生大肠杆菌乳腺炎和内毒素血症，内毒素是革兰氏阴性菌细胞壁脂多糖的组成成分，细菌死亡分解后释放出来，研究表明，内毒素通过释放炎性介质导致黄体溶解和影响受孕和早期胚胎的存活率，在怀孕的第一个月发生黄体溶解会降低受胎率和流产发病率的增加（Lucy，2001）。尤其是内毒素刺激机体生成前列腺素 $F2\alpha$、糖皮质激素、促肾上腺皮质激素和白细胞介素-1，降低促性腺激素释放激素的释放，内毒素还可以导致机体发热反应。

Battaglia 等给绵羊静脉注射内毒素，引起机体的炎症反应，每隔 10 min 取一次垂体门脉的血样和颈静脉血样，观察促性腺激素释放激素、促黄体素、氢化可的松和黄体酮的差异。结果发现，显著降低了促性腺激素释放激素的峰值、降低了促性腺激素释放激素和促黄体素浓度，而氢化可的松和黄体酮的浓度及体温升高了，因此卵泡发育不充分导致雌激素生成减少和不排卵。

McCann 等认为，注射内毒素后，机体释放的细胞因子通过改变氧化亚氮的产生，抑制促性腺激素释放激素，阻止了促黄体素的脉冲式分泌，因此乳腺炎通过改变促黄体素和促卵泡激素的活性或功能，从而影响卵泡的发育和（或）卵母细胞的成熟，能够影响繁殖性能。Peter 等报道子宫内注射内毒素会抑制促黄体素的分泌，导致不能排卵。因此发生乳腺炎的奶牛配种次数和空怀期会因为持久卵泡或卵巢囊肿的形成而增加。

Gilbert 等报道，向荷斯坦奶牛子宫内注射大肠杆菌会导致发情周期缩短，说明前列腺素 $F2\alpha$ 的分泌量增加，并引起黄体发生溶解。奶牛排卵后发生乳腺炎，前列腺素 $F2\alpha$ 的升高会影响胚胎或者子宫。Hockett 等（2000）发现，实验诱导的乳房链球菌乳腺炎奶牛在发情周期的黄体期子宫对前列腺素 $F2\alpha$ 的敏感性增加，认为前列腺素 $F2\alpha$ 是通过减少黄体的存在时间或降低胚胎的质量和发育影响繁殖性能的。Schrick 等也报道，奶牛胚胎的质量与子宫腔内前列腺素 $F2\alpha$ 的浓度呈负相关。

Ahmad 等研究认为，乳腺炎对卵母细胞的影响主要是内分泌状态发生改变，促黄体素峰值的降低（排卵延迟）或者前列腺素 F2α 升高（黄体酮降低）进一步导致持久性卵泡，而持久性卵泡奶牛的胚胎存活率低于成长卵泡奶牛，也更难到达桑葚胚。

国内关于奶牛繁殖性能的研究，如马径军调查发现胎衣顺下奶牛的配种次数为 1.5 次，空怀期为 98 天（马径军，2008），均高于本实验的研究结果（1.72 次，121 天）。从实验结果来看，首次人工授精前发生临床乳腺炎奶牛的首次配种天数多于首次人工授精至怀孕之间发生临床乳腺炎的奶牛，空怀期和配种次数均低于首次人工授精至怀孕之间发生临床乳腺炎的奶牛，但是结果统计分析，不存在差异显著性。首次人工授精前发生临床乳腺炎奶牛和首次人工授精至怀孕之间发生临床乳腺炎奶牛的首次配种天数、空怀期和配种次数均显著多于健康奶牛。首次人工授精前发生临床乳腺炎奶牛的首次配种怀孕率、产后 70 天内怀孕率和流产率与首次人工授精至怀孕之间发生临床乳腺炎奶牛之间没有差异显著性。

首次配种前发生临床乳腺炎的奶牛首次配种天数增加主要是由于卵泡发育不充分，促黄体激素峰值受阻引起的不排卵；或者雌激素生成的降低引起的发情行为不出现。首次配种和怀孕之间发生临床乳腺炎的奶牛配种次数增加、空怀期延长，主要是由于黄体溶解、继而黄体酮降低和早期胚胎的死亡。奶牛的乳腺可以合成前列腺素 F2α，在发生临床乳腺炎的奶牛也发现了前列腺素 F2α 的升高。乳腺内注射肺炎克雷伯菌也可以导致奶牛血浆和乳汁前列腺素 F2α 升高，而前列腺素 F2α 能够引起黄体的退化。

Risco 等研究发现，在怀孕的前 45 天发生临床乳腺炎的奶牛在其后 90 天内发生流产的风险是非临床乳腺炎奶牛的 2.7（95%信赖区间为 1.3～5.6）倍。

鉴于奶牛临床乳腺炎对繁殖性能的负面影响，特别是产后早期乳腺炎，引起的经济损失是巨大且持久的，不仅包括产奶量的损失，还有对奶牛繁殖性能的影响。因此，不论从生产效率方面看，还是繁殖性能方面来看，奶牛场管理者应特别注重乳腺炎的预防和控制。另外，进一步深入研究乳腺炎对繁殖性能的影响机制是很有必要的，从而可以制定相应的预防措施，避免奶牛乳腺炎对繁殖性能的负面影响。

四、小结

（1）产后至怀孕之间发生临床乳腺炎的奶牛首次配种天数和空怀期均极显著多于对照组奶牛，配种次数显著多于对照组奶牛，首次配种怀孕率显著低于对照组奶牛。

（2）实验组 I 奶牛和实验组 II 奶牛的首次配种天数和空怀期均极显著多于对照组奶牛，实验组 II 奶牛的配种次数显著多于对照组奶牛，实验组 I 奶牛和实验组 II 奶牛的首次配种怀孕率均显著低于对照组奶牛。

第八章 奶牛乳腺炎的预防措施

尽管人们对奶牛乳腺炎的预防研究已有 150 多年,取得了一些显著的成绩,在临床上起到了一定作用,但是至今仍没能提出一个彻底控制本病的有效方法。目前对于乳腺炎的治疗,主要还是以抗菌药为主,由于盲目大量使用,耐药菌株不断增多,使得药物疗效下降,同时,乳汁中的抗生素的残留,影响乳品深加工时的发酵过程,并对人类的身心健康产生严重影响。

奶牛乳腺炎的预防主要通过选种、合理饲养及科学管理进行。首先,要选取抗病性强的奶牛,从基因角度出发,通过现代基因工程和遗传育种技术,对奶牛优良基因进行筛选并逐渐培育改进,选择性状优良的奶牛品种(Sordillo,2011)。其次,通过改进制定配方合理的饲料,提高奶牛饲养综合管理水平,能使体细胞数量发生明显的下降,从而提高奶牛的日产奶量(王福慧和杨帆,2011)。除此以外,对于奶牛日粮的合理配比也非常重要,如定期在饲料中合理添加微量元素和维生素等,增加了饲料的营养,提高了奶牛的免疫力,从而使得奶牛乳腺炎的临床症状得到大幅改善(Scaletti and Harmon,2012)。除了饲料营养方面,对养殖场环境的改善及管理水平的提高对降低奶牛乳腺炎的发病率也非常必要(Park et al.,2012)。奶牛乳腺炎应防重于治,做到防微杜渐,防治过程中应该注意考虑到机体应激、免疫压力,以及从流行病学角度调查传染病相关途径等对乳腺炎发生带来的潜在影响(伞治豪,2014)。

一、奶牛乳腺炎的常规预防措施

1. 改善饲养环境 奶牛乳腺炎的发生是由环境、病原菌和牛体三种因素决定的,其中环境因素占有重要的地位。为预防病原微生物的感染,在奶牛饲养过程中,要搞好环境卫生,改善饲养环境。应合理地设计牛舍和运动场,保持适当的饲养密度,保证通风良好,向阳性好,防止水气过大。应注意牛舍和运动场的清洁卫生,及时更换垫料,清理粪便,定期消毒。及时擦拭牛体,保持牛体自身的清洁,尤其是乳房的清洁,减少乳房周围病原微生物的感染。同时根据气候条件,做好夏防暑、冬保暖的工作,减少因气候变化引起的应激反应,为奶牛提供一个安静、舒适的生活环境。

2. 规范挤奶 挤奶的方式主要有两种,人工挤奶和机械挤奶。无论采用哪种方式,都要做好消毒工作。因此,挤奶前,要做好挤奶器具和奶牛乳头的消毒工作。挤奶时,注意挤奶卫生,严格执行挤奶操作规程。清洗乳头时,要用温水把乳房擦洗干净,一牛一巾,然后药浴乳头,并认真按摩乳房。应先挤健康牛,再挤假定健康牛,最后挤病牛。人工挤奶时,挤奶员要注意个人卫生,同时也要注意挤奶的手法,以免损伤乳房。机械挤奶时,要注意机器的选用和参数的设置,防止机械损伤。此外,还要按照一定的时间规律进行挤奶,以免因乳池压力过大,引起乳汁倒流,把病原菌带入乳腺组织内部,增加乳腺炎发生的机会。

正确的挤奶程序分以下 4 个步骤。

(1) 温和地对待泌乳牛　在挤奶过程中，由于乳头部分的神经末梢受到刺激，促使脑垂体释放催产素，催产素能促使乳汁排出；如果在挤奶前，粗暴对待泌乳牛，则肾上腺素分泌增加，而肾上腺素抑制催产素的释放，使乳汁排出不完全，影响产奶量。

(2) 清洗乳头与乳头干燥　清洗乳头前先检查和废弃前两把奶（装入专门的容器），以便尽早发现病例和废弃高 SCC 奶，也给乳房一个强烈的放乳刺激。然后用消毒剂清洗或涂抹乳头，同时触摸乳腺有无异常，清洗或涂抹后立即擦干，注意每一头奶牛使用一块毛巾。

(3) 挤奶　乳房清洗后 1 min 内套上挤奶器，挤奶时适当调整奶杯位置，挤完奶先关掉真空，然后再移开挤奶器；手工挤奶则应尽量缩短挤奶时间，挤奶员戴手套。另外应按一定的次序挤奶，一胎牛和健康牛先挤，感染的牛后挤，这可减少乳腺炎在奶牛间相互感染。

(4) 乳头药浴　奶杯移开以后立即进行乳头药浴。挤完奶 15 min 之后，乳头的环状括约肌才能恢复收缩功能，并关闭乳头孔，在这 15 min 之内，张开的乳头极易受到环境性病原菌的侵袭，及时进行药浴，使消毒液附着在乳头上形成一层保护膜，可以大大降低乳腺炎的发病率。乳头药浴杯应每天清洗，药浴液应每天更换。另外要注意挤奶前后的乳头消毒剂的选择，应首选适合本地区牛群的消毒液。

3. 疫苗　国内外对乳腺炎疫苗的研究已经取得了一定的成绩。当前研究的疫苗有多种类型，主要包括灭活疫苗、减毒疫苗、亚单位疫苗，以及遗传重组疫苗、基因工程疫苗等。根据奶牛乳腺炎的发病机制，接种疫苗可以增强机体的免疫防御能力，降低乳腺感染的严重程度。国外有报道，接种大肠杆菌 J5 分子工程苗，机体可产生抗内毒素的抗体，对预防乳腺炎有很好的效果。引起奶牛乳腺炎的病原菌，主要包括细菌、支原体、真菌、病毒等，而目前所研制的疫苗主要针对金黄色葡萄球菌、链球菌和大肠杆菌。因此，接种疫苗预防奶牛乳腺炎还是受到一定的限制（付云贺，2015）。

接种乳腺炎疫苗，能有效地预防乳腺炎，特别是隐性乳腺炎的发生。应用疫苗预防奶牛乳腺炎有很多优点：没有药物残留、有助于降低乳腺感染的严重程度、操作简便、费用低廉等。目前应用较多的主要是奶牛乳腺炎灭活多联苗（主要是由无乳链球菌、停乳链球菌和金黄色葡萄球菌 3 种菌体的抗原及荚膜等与氢氧化铝胶佐剂配合制成的）和亚单位疫苗。具体使用的方法为肩部皮下注射 3 次，每次 5 mL，第 1 次在牛干奶时注射 1 针，30 天后注射第 2 针，并于产后 72 h 内再注射第 3 针，能有效地预防乳腺炎的发生（丁丹丹，2014）。目前关于链球菌疫苗研究得还较少，主要为无乳链球菌和乳房链球菌的灭活苗，但效果均不太理想。李宏胜等（2011）在研究应用疫苗预防奶牛乳腺炎中，先后在兰州地区选取了 3 个规模化奶牛场和 7 个个体奶牛场开展了以乳腺炎疫苗预防为主，结合改善环境卫生和药物防治为辅的综合防治措施试验，结果表明，选取的这 10 个奶牛场经过 1～2 年的试验，隐性乳腺炎乳区发病率比试验前降低了 63.97%（$P<0.01$），临床型乳腺炎发病率比试验前降低 66.36%（$P<0.01$），产奶量与试验前相比提高了 10.18%，达到了良好的防治效果。但目前国内外研制的乳腺炎疫苗也存在一定缺陷，如免疫保护期短等，因此，要取得良好的免疫效果，应注意选择合适的免疫时间。

4. 建立完善的记录系统　　首先，制定奶牛乳腺健康指标，主要是体细胞数和临床乳腺炎的发病率；其次，进行定期的 DHI（包括 SCC）和牛奶质量测定；再次，对临床乳腺炎牛的处理情况做详细的记录，包括发病时间、鉴定情况、感染区、用药及愈后情况等。

5. 干奶期预防　　干奶期用抗生素治疗是控制乳房感染的一项有效措施。在干奶期，奶牛乳房很容易感染环境中的革兰氏阳性菌和革兰氏阴性菌，或者是传染性病原体。处于干奶期的奶牛，乳房退化，乳腺细胞减少，结缔组织增加。在此期间，使用抗生素治疗感染，而且不需要丢弃牛奶。干奶期治疗有助于杀死长期存在于乳腺中的病原菌，控制乳腺炎的发生，而且不会造成乳汁中的药物残留。用于治疗乳腺炎的抗生素有多种。例如，布拉乳腺霉素、日光乳腺霉素、瑞斯托乳腺霉素和列克辛-500 等。采用快速干奶法，在最后一次挤奶后，应立即对奶牛乳房进行药物处理。

6. 补充微量元素和维生素　　微量元素和维生素在机体正常发育中具有重要的作用，补充微量元素和维生素能够增强机体的抵抗力，防止病原微生物的感染。研究发现，适当补充微量元素铜、锌、硒和维生素 A、维生素 E，有助于增强机体对乳腺炎的抗病能力，预防奶牛乳腺炎的发生。微量元素铜与机体的免疫反应之间有重要的联系，补充铜，可以提高机体的免疫力。在饲料中适当添加维生素 E 能够提高奶牛的免疫反应，而且有报道指出，维生素 E 能够增强中性粒细胞的功能。

维生素 E 与硒结合使用可以作为抗氧化剂，增强机体的抗氧化能力。研究发现，在奶牛产前肌内注射维生素 E 和硒，其产后可降低乳腺炎的发病率。

对于高产牛而言，高能量、高蛋白的日粮有助于保护和提高产奶量，但同时也增加了乳房的负荷，使机体的抗病力降低，因此不要喂过多的精料。另有研究表明，日粮中适当补充微量元素硒和维生素 A、维生素 E 有助于抵抗乳腺炎感染。

7. 定期检测　　定期检测是预防乳腺炎的有效措施。利用实验室检测方法，定期检测体细胞、细菌种类和乳成分等，了解奶牛乳房的健康状况，并对可疑患病奶牛及时采取措施，预防乳腺炎的发生。

二、奶牛乳腺炎抗性基因的挑选

对奶牛遗传改良最重要目的是尽量获得最大的经济效益。因此，疾病的抗性选择育种是值得应用的，因为这将降低奶牛生产成本，改善奶牛健康状况和提高牛奶及奶制品的质量。同样应考虑到奶牛乳腺炎的抗性选择对奶牛存活力的影响及由乳腺炎造成个体替换的相关费用，乳腺炎是决定动物个体每年所带来经济价值回报的主要因素之一。

努力提高和衡量整体奶牛经济价值导致遗传评价向选择更健康、更长寿的牛群方向发展。通过直接的自然选择长寿动物，因为有更多后代，通常可以带来更大的经济收益。畜群寿命长导致盈利增加有两种不同的方式：首先，畜群寿命时间的延长降低了奶牛因淘汰或更换所带来的费用；其次，牛群含有较多的成年牛可以始终保持较高的生产水平。畜群寿命长也暗示降低淘汰水平和医疗费用。因此，畜群寿命是在许多育种计划中一个重要的经济性状。畜群寿命主要是由生产者淘汰计划所决定。而淘汰计划一般是由经济利益决定的，期望通过淘汰更换可以产生更高的利润。奶牛淘汰最常见的原因包括：产量低、乳腺炎、繁殖障碍、奶品质差或死亡。许多生产性能如奶产量、乳腺炎的发病率、初产年龄、干奶期等都会影响了畜群的寿命长短和整体经济回报（赵中利，2012）。

直接选择更长寿的畜群是最有效的方法，但畜群寿命的遗传力较低，相对于其他性状，畜群寿命选择指数表达需要更长时间。生产上直接选择一个替代性状——生产寿命（PL），生产寿命的定义是产奶个体总的产奶时间，生产寿命的遗传评估是测量同一时间内同一群体的奶牛不同月份的奶样，同样反映了所有淘汰的原因。因为牛的生产寿命只有一次，所以环境因素的影响在评价中不考虑。此外，体细胞评分可以作为选择低乳腺炎发病率健康长寿奶牛的一个新标准。

体细胞评分与乳腺炎的发病率存在显著相关，可以作为评价乳腺炎的一个标准，可以这样假设，即有低体细胞评分的个体对乳腺炎的抗性较强。由于乳腺炎是作为影响奶牛产奶量和繁殖障碍而造成奶牛淘汰的重要原因之一，因此可以推断乳腺炎抗性与畜群寿命增强存在相关。体细胞评分与生产寿命长度之间存在负相关，选择低体细胞评分的个体具有更长的生产寿命，也能带来更多的经济收益。选择较低体细胞评分的个体不但可减少临床型和隐性乳腺炎的发病率，且可以产生更多的利润（赵中利，2012）。

通过遗传选择提高奶牛乳腺炎抗性（抗乳腺炎育种）被认为是解决乳腺炎问题的一种最好的长期策略。不同动物对许多疾病的抗性都表现出遗传上的差异，奶牛乳腺炎抗性在种间和种内也存在着明显的遗传差异。有报道指出，奶牛乳腺炎的遗传力仅有 0.02～0.04，因此，直接选择奶牛乳腺炎性状的标准性很小。不过，近年来生物信息学和基因组学的迅速发展为奶牛乳腺炎抗病育种提供了新的思路，尤其是不同组织器官转录组及基因芯片的广泛使用，为奶牛乳腺炎的抗病育种带来了新的可能。目前被广泛认可且研究较多的奶牛乳腺炎抗性基因有以下几种：白细胞介素和白细胞介素受体基因、牛锌指蛋白 313（$Znf313$）基因、前脑锌指蛋白基因、Toll 样受体基因（TLR）、乳铁蛋白基因、趋化因子受体基因、热休克蛋白 70（HSP70）、牛主要组织相容性抗原复合体（MHC/BoLA）等（廖想想，2014）。但奶牛乳腺炎抗性相关基因表达的调控机制比较复杂，目前尚无定论，仍需进一步研究证实。

基因芯片是一个高效率、高通量的检测基因突变与表达的方法，临床上可以快速检测。在人医上，基因芯片技术发展已经很成熟。例如，王国建等（2008）将芯片与等位基因特异性引物延伸 PCR 结合开发了一款遗传性耳聋基因诊断芯片。基因芯片可以快速地检测许多基因的表达情况，这也是基因芯片应用最多的一个方面，可以发现大量的新基因，并能够鉴定其功能等（霍金龙等，2007）。基因芯片可以发现致病基因及其相关基因，如 $P53$ 基因的突变会引起肿瘤，Affymetrix 公司通过该基因的序列制成基因芯片，有效地实现了癌症的早期诊断。基因芯片还能够大规模地通过寻找药物作用 IE 位点的方法对药物进行筛选现如今很难分离鉴定药物的有效成分以及西药的开发，运用基因芯片技术能够从基因层面解决这一问题（王小龙，2014）。基因芯片在医学上应用最多的是肿瘤方面，一是诊断治疗方面，二是肿瘤分型。人类基因组计划已完成，但是各种基因的功能还未完全清楚，但基因芯片技术对后基因组计划有重要的意义。基因芯片技术在发现新基因、基因突变检测、后基因组研究、转基因成分和有害微生物等的检测等方面具有深远的意义。2005 年，Pareek 等（2005）利用芯片技术对人肠杆菌刺激乳腺上皮细胞进行了研究，初步确定乳腺上皮细胞在刺激免疫中发挥重要作用。Rainard 等（2006）利用基因芯片技术建立牛、羊的免疫差异基因芯片或开展相应免疫系统研究。Swanson 等（2008）利用表达谱 cDNA 芯片对乳房链球菌引起的奶牛乳腺炎进行了研究。大量研究表明，目前已广泛利用表达谱基因芯片技术研究奶牛乳腺炎。

总之，乳腺炎抗性是一个复杂的性状，受多基因和多因素的控制。因此，在彻底了解奶牛乳腺炎抗性相关基因的基础上，联合应用现代分子育种技术可能是将来发展有效的乳腺炎防治策略的一个关键目标。

附录（Appendix） The research results in dairy mastitis

Section 1 Role of antioxidant vitamins and trace elements in mastitis in dairy cows

Mastitis is associated with release of free radicals, increased total oxidant capacity and decreased total antioxidants capacity in milk. Antioxidant vitamins and minerals protect the body from free radicals either by directly scavenging free radicals or by inhibiting the activity of oxidizing enzymes. The supplementation of mastitis dairy cows with antioxidant vitamins as vitamin A (VA) and β-carotene (BC), vitamin C (VC), vitamin E (VE), and antioxidant minerals as selenium (Se), Zinc (Zn) and copper (Cu) is very important to help the animal recover early. The aim of this section was to discuss the oxidative stress in dairy cows' mastitis, and the roles of VA and BC, VC, VE, Se, Zn, and Cu in mastitis of dairy cows. Before deciding to supplement dairy cow rations with the levels of vitamins and minerals, dairy farmers should have their animal feeds tested and their rations evaluated by a competent dairy cow nutritionist and a trustworthy laboratory to be sure what levels of supplementation may be warranted. While inadequate intake and absorption of certain nutrients may result in a weakened immune system and perhaps more mastitis during the lactation period, unjustified supplementation can be expensive and lead to other animal health problems.

1 Introduction

Inflammatory reaction of mammary gland (i.e., mastitis) usually caused by microorganism, is considered as the most costly disease for dairy cattle. Both clinical mastitis (CM) and subclinical mastitis (SCM) produce great economic losses. CM has elevated body temperature, inappetance, a red, swollen and/or painful udder and/or abnormal milk. However, SCM has no apparent clinical signs but accompanied with elevation of somatic cells count (SCC) in milk (Huijps et al., 2008). The SCC in milk is considered as a well-known indicator that reflects milk quality and health status of mammary gland. Excessive amount of neutrophils, various epithelial cells, macrophages, lymphocytes, eosinophils of mammary tissue in milk is considered as indicative of response of mammary tissue to microorganisms (Atakisi et al., 2010).

Vitamins and minerals have long been recognized as antioxidants in the animal health and production. However, they also have specific roles in mastitis of dairy cows, such as vitamin A (VA) and β-carotene (BC), vitamin C (VC), vitamin E (VE), selenium (Se), Zinc (Zn) and copper (Cu). As described in Table A-1, different antioxidants have crucial role in animals' health. Previous studies revealed that increase in lipid peroxidation in mastitis may cause decrease in levels of some antioxidant molecules leading to an increase in oxidative stress (Weiss et al., 2004). Oxidative stress is generally described as an imbalance between oxidant and antioxidant levels (Lykkesfeldt and Svendsen, 2007). When the production of oxidants exceeds the capacity of antioxidant defense, a condition of oxidative stress is produced resulting in oxidative damage to macromolecules such as lipids, DNA and proteins (Sordillo et al., 2007).

Table A-1　Summary of micronutrient effects on mammary gland immunity (Hayajneh, 2014)

Micronutrient	Observation
Vitamin A	Decreased SCC. Moderated glucocorticoid levels
Beta Carotene	Increased bactericidal function of phagocytes. Increased mitogen-induced proliferation of lymphocytes
Vitamin E	Increased neutrophil bactericidal activity Decreased incidence of clinical mastitis
Se	Decreased efficiency in neutrophils function. Improved bactericidal capabilities of neutrophils. Decreased severity and duration of mastitis
Cu	Deficiency decreased neutrophil killing capability. Deficiency increased susceptibility to bactericidal infection
Zn	Deficiency decreased leukocyte function. Deficiency increased susceptibility to bacterial infection

Milk with higher SCC is positively associated with malondialdehyde (MDA) level in milk (Suriyasathaporn et al., 2006), and the mean level of MDA is significantly higher in SCM milk than in normal milk (Yang et al., 2011b), consequently more free radicals being released and a stat of oxidative stress arise (Su et al., 2002). During the past decade, significant advances have been made in understanding the roles of antioxidants in mastitis. The purpose of this section was to review the current knowledge of the roles of vitamins and minerals in mastitis of dairy cows.

2　Oxidative stress in mastitis of dairy cows

Mastitis could induce the increase of free radicals formation in milk and leading to oxidative stress (Gu et al., 2009), especially during the early lactation period of dairy cows (Sordillo et al., 2007). Both CM and SCM are associated with release of free radicals, increased total oxidant capacity and decreased total antioxidants capacity in milk (Atakisi et al., 2010).

During lactation, mammary epithelial cells exhibit a high metabolic rate and thus produce large amounts of reactive oxygen species (ROS) and lipid peroxides in vivo (Jin et al., 2014). Jhambh et al. (2013) reported that significant ($P<0.05$) decrease in blood superoxide dismutase (SOD) and catalase activities, reduced glutathione (GSH) concentration and an increase in erythrocytic lipid peroxides was observed in cows with clinical mastitis.

Nitric oxide (NO) is one of the most important reactive nitrogen intermediates, which operates in a variety of tissues to regulate a diverse range of physiological processes such as inflammatory response. During inflammation, epithelial cells and macrophage of mammary gland produce a significant amount of NO; this inducible NO mediates inflammation during mastitis. Another source for NO is the mammary epithelial cells and/or mononuclear phagocytes, which contribute to NO production upon stimulation with lipopolysaccharide and cytokines (Boulanger et al., 2001).

NO was reported to increase in CM milk (Atakisi et al., 2010) and in milk and plasma after intramammary infusion of *Escherichia coli* or endotoxin produced by *E. coli* (Komine et al., 2004). Milk SCC is considered as a well-known indicator that reflects mammary health and milk quality; thus, a positive correlation between SCC and NO concentration had been reported

(Atakisi et al., 2010).

NO has an important role in mediating microbistatic and/or microbicidal activity, as the activated macrophages synthesize NO (Jungi, 2000), which is considered as a primary defence system that eliminate intracellular pathogens (O'Flaherty et al., 2003). The antimicrobial property of NO is attributed to peroxynitrite, a reactive nitrogen metabolite, derived from oxidation of NO. In severe mastitis, peroxynitrite is produced in excess, which may result in alterations in the antioxidant balance (Chaiyotwittayakun et al., 2002). This means that excessive release of NO results in oxidative damage to mammary gland secretions (Komine et al., 2004; Atakisi et al., 2010).

The cells contain a variety of antioxidants that play an important role in the protection against excessive release of reactive oxygen species in blood and tissues, body from free radicals either by directly scavenging free radicals or by inhibiting the activity of oxidizing enzymes (Abd Ellah et al., 2009).

3 Mechanisms of defense against oxidative stress

Due to the substantial background exposure to oxidants resulting from a life depending on molecular oxygen, aerobic organisms have adapted to constantly fighting a battle against oxidative stress. Advanced cellular defense strategies have evolved and gradually expanded the possible lifespan for the individual species. The cellular defense mechanisms can be divided into at least three levels according to their function of quenching oxidants, repairing/removing oxidative damage or encapsulating non-repairable damage (Figure A-1).

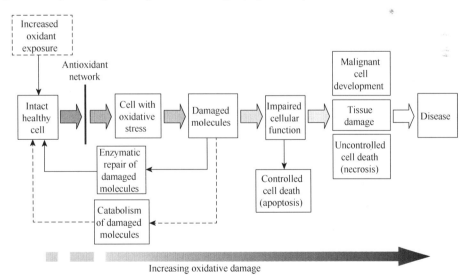

Figure A-1 Schematic outline of cellular defenses against oxidative stress-mediated cellular damage. Increased oxidative stress is initially counteracted by the antioxidant network. Damaged molecules are either repaired or categorized. Controlled cell suicide can be initiated if further oxidative damage leads to impaired cellular function. When these signaling cascades are damaged or the oxidative damage exceeds the capacity of the defense mechanisms, uncontrolled cell death, tissue damage and malignant cell development can progress into disease (Lykkesfeldt and Svendsen, 2007)

As a first level of defense against oxidants, the cell is equipped with a so-called antioxidant network. Antioxidants are capable of donating electrons to oxidants, thus quenching their reactivity under controlled conditions and making them harmless to cellular macromolecules. A second and highly important level of defense is the ability to detect and repair or remove oxidized and damaged molecules. Finally, if the extent of the oxidative damage exceeds the capacity of repair and removal, the organism is equipped with one final weapon, controlled cell suicide or apoptosis. The ability to induce programmed cell death is of major importance in a variety of bodily functions, including control of tissue growth, and is apparently under control by several signaling pathways. However, one of these appears to be that apoptosis is induced by increased oxidative stress and thus constitutes a final resort to encapsulate and isolate the damaged cells (Lykkesfeldt and Svendsen, 2007).

4 The roles of vitamin A and β-carotene in mastitis of dairy cows

VA and its precursor-BC are important in maintaining epithelial tissue health and play a vital role in mucosal surface integrity and stability. These functions may affect cow resistance to pathogen entry into the mammary gland as well as resistance post entry. VA is an important factor in improving immune function and attenuating oxidative stress (Jin et al., 2014). In addition, BC appears to function as an antioxidant, reducing superoxide formation within the phagocyte, and it play an important role in protecting udder tissue and milk from the harmful effect of free radicals. This is potentially important due to phagocytosis and internal killing of bacteria once engulfed are primary mechanisms used by phagocytic cells to eliminate bacteria. Both VA and BC also appear to have stimulatory effects on immune cells (Heinrichs et al., 2009). Cows with higher California Mastitis Test (CMT) scores had significantly lower plasma VA and BC concentrations than cows with CMT scores indicating no mastitis (Yang et al., 2011a).

BC functions independently of VA in mastitis and reproduction. It was reported that, low concentrations of plasma VA (<0.8 μg/mL) and BC (<2 μg/mL) were linked with severity of mastitis. Jukola et al. reported that VA and BC supplementation have an effect on udder health only when the plasma level of VA is lower than 0.4 mg/L and BC is lower than 3.0 mg/L. LeBlanc et al. (2004) showed a 60% relative reduction in the risk of clinical mastitis for every 100 ng/mL increase in serum retinol. Data showing the association of VA and BC with mastitis have been variable at best, and supplementation is likely not needed in excess of normal recommendations. However, several studies have shown no impact of supplementing BC on intramammary infection level (Heinrichs et al., 2009).

5 The roles of vitamin C in mastitis of dairy cows

Ascorbic acid is the most abundant and important water-soluble antioxidant for mammals. Even though it can be synthesized in the body of most mammals except primates and guinea pigs which have a dietary vitamin C requirement. Thus, ascorbic acid is not a required nutrient for dairy cows; some data are accumulating that suggest vitamin C is related to mastitis. The cows suffering from mastitis have lower concentrations of vitamin C in their milk and plasma (Weiss et al., 2004; Kleczkowski et al., 2005). The severity of clinical signs is related with magnitude of the decrease in concentrations (Weiss et al., 2004). Ascorbic acid scavenges aqueous reactive oxygen species by rapid electron transfer, thus inhibiting lipid peroxidation, and represents one of the important antioxidant defences against oxidative damage. In bovine mastitis, it has been identified as oxidative stress biomarkers. The number of the leukocytes per milliliter blood was correlated positively with VC content of plasma. Decreased concentration of ascorbic acid has been recorded from mastitis milk of cows (Naresh et al., 2002). Many studies had been reported that its milk concentration significantly decreased in acute mastitis and SCM especially when the condition is accompanied by an increase in the levels of lipid hydroperoxide in erythrocytes (Weiss et al., 2004; Kleczkowski et al., 2005; Ranjan et al., 2005).

Ascorbic acid along with cupric ions was found successful to prevent and treat the mastitis of dairy cows as teat dip or intramammary infusion (Naresh et al., 2002). Vitamin C administered to cows by subcutaneous injection may have therapeutic value in mastitis (Ranjan et al., 2005). However, its therapeutic effect decreased in the presence of lipid peroxidation with moderate improvement in clinical signs of mastitis (Ranjan et al., 2005).

6 The roles of vitamin E in mastitis of dairy cows

The VE is the most important lipid soluble membrane antioxidant, and the biologically active form is known as α-tocopherol. The VE is an integral component of lipid membrane, and has an important role in protecting lipid membranes from attack of reactive oxygen. VE enhances the functional efficiency of neutrophils by protecting them from oxidative damage following intracellular killing of ingested bacteria. Fresh green forage is an excellent source of VE; however, concentrates and stored forages (hays, haylages, and silages) are generally low in VE (NRC, 2001). The best understood the role of VE on mastitis is that it acts as a lipid soluble cellular antioxidant, free radical scavenger, and protects against lipid peroxidation (Yang et al., 2011a).

Many experiments have shown that plasma concentrations of α-tocopherol in dairy cows are low at parturition. Cows are immunosuppressed during the time when plasma concentrations of VE are low. Supplementation of antioxidant vitamins may reduce inflammatory response and oxidative stress during mastitis. Smith et al. were the first to report that supplemental VE and Se reduced CM. That study was conducted in Ohio where the soil concentration of Se is very low, and used dry cows fed hay-based diets (should be very low in VE). Cows were either injected with a placebo or 50 mg of Se at 21 days before calving and were fed either 0 or 1000 IU/day of supplemental VE (dl-α-tocopheryl acetate). Se without supplemental VE reduced the incidence and duration of CM, but the largest response was caused by VE with or without Se. However, Chandra et al. (2013) indicated that supplementation of vitamin E and Zn in food of Sahiwal cows during peripartum period enhanced milk production by reducing negative energy balance.

The VE deficiencies are frequently observed in peripartum dairy cows. Most cases of CM occur during the first month of lactation and originate in the dry period (Green MJ et al., 2002), and coincide with the lowest VE blood concentration. It was suggested that maintaining an optimal VE level, together with low levels of oxidative stress is an important factor in dry cow management and improvement of udder health.

7　The roles of selenium in mastitis of dairy cows

Se is recognized as an essential trace element for domestic animals, the majority of Se in body tissues and fluids is present as either selenocysteine, which functions as an active center for selenoproteins, or selenomethionine, which is incorporated into general proteins and acts as a biological pool for Se (Juniper, 2006). It functions in the antioxidant system as an essential component of the glutathione peroxidase (GSH-Px), which is responsible for reduction of H_2O_2 and free O_2 to H_2O (NRC, 2001). It also plays a vital role in protecting both the intra- and extra-cellular lipid membranes against oxidative damage. The activity of GSH-Px in milk varies with the species and diet (Fox and Kelly, 2006). GSH-Px catalyzes the reduction of various peroxides, protecting the cell against oxidative damage (Abd Ellah et al., 2007) and protects milk lipids from oxidation. Vitamin E and selenium supplements in diets appear to have a preventive effects against acute infections in which high polymorphonuclear response occur in the infected gland (Ata and Zaki, 2014).

Dairy cattle have several known endogenous antioxidant defense mechanisms that can counteract the harmful effects of ROS accumulation. The Se-dependent selenoproteins had been studied extensively with respect to mammary gland health. Se supplementation to periparturient cows reduces the incidence and severity of mastitis. The effects of feeding organic Se in the form of selenized yeast and sodium selenite were compared in a feeding experiment on 100 dairy cows. Feed was supplemented at a rate of 0.2 ppm for 8 weeks. The results showed that the yeast Se yielded the greater blood level of Se. The GSH-Px level went from 0.22 to 3.0 and to 2.3 microKat/g of hemoglobin from Se yeast and selenite, respectively. Blood GSH-Px continued to increase up to 10 weeks after the supplementation had stopped. The bioavailability of yeast Se was found to be superior to selenite. The percentage of quarters harboring mastitis dropped along with milk SCC. It is important to note that the test cows began the trial with a significantly low blood Se status. Moeini et al. (2009) reported that milk SCC of heifers at 8 weeks lactation was significantly decreased (193, 000/mL *vs.* 179, 000/mL, $P<0.05$) by 20 mg Se and 2000 IU dl-α-tocopheryl acetate supplemental at 4 and 2 weeks before expected calving. The beneficial effects of Se supplementation are thought to be due to the actions of certain antioxidant Se-dependent enzymes (Papp et al., 2007). NRC (2001) recommended the level of Se in dairy diets is 0.3 mg/kg dry matter (DM) and should be closely monitored to ensure that over supplementation does not occur.

8 The roles of zinc in mastitis of dairy cows

Zn is an essential trace mineral found to be an integral component of over 300 enzymes in metabolism (Yang et al., 2011a), the functions of Zn include tissue or cell growth, cell replication, bone formation, skin integrity, cell-mediated immunity, and generalized host defense (Gropper et al., 2005). The mammary gland is an organ that is derived from the skin, thus making Zn necessary to maintain the integrity of the keratin that lines the streak canal. Zn has a significant effect on gene expression and cellular growth.

Cell mediated immunity has also been found to be altered by Zn deficiency. Zn deficiency has been associated with reduced formation of both T and B lymphocytes and phagocytes. T and B cells are the major cellular components of the adaptive immune response. Once they have recognized an invader, the cells generate specific responses that work to eliminate pathogens or pathogen infected cells. As previously discussed, Zn is also involved in the removal of free radicals by superoxide dismutase (SOD). Extracellular and cytosolic SOD requires both Zn and Cu. Another important factor to note in regard to the cow's immune response is that immunoactive substances such as VA have been found to react with Zn in several ways. Zn is necessary for the hepatic synthesis of retinol-binding protein, which transports VA in the blood (Gropper et al., 2005).

Zn is required for the formation of Mn-Zn SOD, deficiency of Zn affect the activity of SOD in blood and tissues, which results in increased superoxide radicals. It is known that mastitis associated with increased SCC counts in milk, which act as a source for free radicals and hence oxidative stress. Low Zn status leads to low quality milk with high SCC and increased incidence of mastitis (Gaafar et al., 2010). An experiment was done that included 12 lactation trials addressing Zn supplementation. The result indicates that SCC was reduced from 294,000/mL to 196,000/mL (as a 33.3% reduction, $P<0.01$) in cows receiving Zn methionine complex (Kellogg et al., 2004). In an experiment, Popovic (2004) replaced 33% of the supplemental inorganic Zn-sulphate with organic Zn for 45 days of pre-calving until 100 days of post-calving. The cows receiving organic Zn had significantly ($P<0.05$) lower SCC at day-10 of lactation (158,840/mL vs. 193,530/mL), and at the end of the trial (62,670/mL vs. 116,440/mL). In addition, Kinal et al. (2005) reported that replacing 30% of the inorganic Cu, manganese (Mn), and Zn for 6 weeks pre-calving until 305 days of lactation in dairy cows could result 34% reduction in SCC (270,000/mL vs. 409,000/mL, $P<0.01$). NRC (2001) recommended the level of Zn supplementation for lactating dairy cows at 40~60 mg/kg DM.

9 The roles of copper in mastitis of dairy cows

Cu is a component of ceruloplasmin, which facilitates iron absorption and transport. In addition, Cu is considered as an important part of SOD, an enzyme that protects cells from the toxic effects of oxygen metabolites produced during phagocytosis (Yang et al., 2011a).

Therefore, lactating cows were recommended for Cu supplementation at 11 mg/kg DM (NRC, 2001). Cu deficiency in cattle is generally due to the presence of dietary antagonists, such as sulfur, molybdenum and iron that reduce Cu bioavailability (Spears, 2003).

As a modulator of the inflammatory process, ceruloplasmin serves as an acute-phase protein. Acute phase proteins rise in the blood with infection and other inflammatory events (Gropper et al., 2005). The enzyme SOD, which is found both in the cytosol of cells and extracellularly, is Cu and Zn dependent. Without the presence of SOD, superoxide radicals can form more destructive hydroxyl radicals that damage both unsaturated double bonds in cell membranes, fatty acids, and other molecules in cells. Therefore, SOD assumes a very important protective function (Gropper et al., 2005). A study was conducted on first lactation Holstein heifers to assess potential role of dietary Cu in enhancing resistance to *E. coli* mastitis, and conclusions were made that Cu supplementation reduced the severity of clinical signs during experimental *E. coli* mastitis but the duration of mastitis was unaffected (O'Rourke, 2009). Experimental studies approved that Cu supplementation reduced the severity of clinical signs of *E. coli* mastitis (Scaletti et al., 2003).

10 Conclusion

Mastitis is associated with release of free radicals, increased total oxidant capacity and decreased total antioxidants capacity in milk. Antioxidant vitamins and minerals protect the body from free radicals either by directly scavenging free radicals or by inhibiting the activity of oxidizing enzymes. The supplementation of mastitis dairy cows with antioxidant vitamins as vitamin A, C, E and β-carotene, and antioxidant minerals as selenium, Zinc and copper is very important to help the animal recover early. Before deciding to supplement dairy cow rations with the levels of vitamins and minerals indicated above, dairy farmers should have their animal feeds tested and their rations evaluated by a competent dairy cow nutritionist and a trustworthy laboratory to be sure what levels of supplementation may be warranted. While inadequate intake and absorption of certain nutrients may result in a weakened immune system and perhaps more mastitis during the lactation period, unjustified supplementation can be expensive and lead to other animal health problems.

Section 2 Effects of vitamins and trace-elements supplementation on milk production in dairy cows

During the past decade, significant advances were made in understanding the effects of vitamins and trace-elements supplement on milk production of dairy cows. This work discussed the effects of vitamins and trace-elements supplementation on milk production of dairy cows. Studies have indicated that vitamin A (VA) and β-carotene (BC) supplementation have some effects on udder health and milk yield in dairy cows whose intake is below 110 IU/kg BW/day. If low quality forage is fed, supplementation of VA should be considered. Supplementation of

B-vitamin has important effects on milk production and could increase milk yield and milk component production. The effect of vitamin E (VE) and selenium (Se) supplementation on the milk yield and milk components are not unified due to the optimum dose, route and timing of VE administration in lactational dairy cows. Zinc (Zn) supplementation increases lactation performance and reduces milk somatic cell count (SCC) in most studies. Limited research has indicated that copper (Cu) supplementation could reduce milk SCC. Before deciding to supplement vitamins and trace-minerals indicated earlier for the improvement of milk production of lactational cows, farmers should have their animals fed with tested and evaluated rations to be sure of the levels of supplementation which may be warranted.

1 Introduction

Vitamins and trace-elements have long been recognized as a requirement for reproduction due to their cellular roles in metabolism and growth. However, they also have specific roles and requirements in lactational cows, they are: vitamin A (VA), β-carotene (BC), B-vitamin, vitamin E (VE), selenium (Se), zinc (Zn) and copper (Cu). Many of these micronutrients have antioxidant activities that are beneficial to animal health and milk yield.

Their concentrations often decrease around calving and extra supplementation is recommended in some dairy herds. However, their requirement varies, for example, depending on the quantity and quality of feedstuffs in the diet. During the past decade, significant advances have been made in understanding the effects of vitamins and trace-element supplements on the milk production of dairy cows (Girard and Matte, 2005; Griffiths et al., 2007; Bourne et al., 2008). Recent reviews have focused on the roles of vitamins and trace-minerals in immune function and disease resistance in cattle (Spears, 2000; Weiss and Spears, 2006), and the roles of antioxidants and trace minerals on immunity and health in transition cows (Spears and Weiss, 2008). However, there is very limited literature on the roles of micronutrients on the milk production of dairy cattle. This section reviews the current knowledge of the effects of vitamins and trace-element supplements on the milk production of dairy cows.

2 Vitamin A and β-carotene

VA and its precursor-BC are important in maintaining epithelial tissue health and play a vital role in mucosal surface integrity and stability. These functions may affect cow resistance to pathogen entry into the mammary gland as well as resistance to postentry. In addition, BC appears to function as an antioxidant, reducing superoxide formation within the phagocyte, and it is an important free radical scavenger. Cows with higher California mastitis test (CMT) scores had significantly lower plasma VA and BC concentrations than cows with CMT scores, indicating no mastitis. Jukola et al. reported that VA and BC supplementation have an effect on udder health only when the plasma level of VA is lower than 0.4 mg/L and BC is lower than 3.0 mg/L. However, several studies have shown no impact of supplementing BC on intramammary infection level (Heinrichs et al., 2009).

An increase in milk yield to the extent of 6% to 11% on BC supplementation at 400 mg/day per cow has been reported. However, supplementation with VA (550, 000 IU/day) exceeded daily requirements about 8-fold for up to 2 months to dairy cows during the dry period, which have slightly decreased energy-corrected 100-d milk yield and milk fat yield (Puvogel et al., 2005), possibly because of enhanced apoptic rates of mammary cells.

Diets based on good quality silage or fresh forage probably provide adequate BC, and supplementation would not be economical. If low quality forage is fed, supplementation should be considered. Supplementation of VA is likely not needed in excess of normal recommendations. The level currently recommended (NRC, 2001) is 110 IU/kg BW/day for milking cows.

3 B-vitamin

In the past several centuries, dairy cows have greatly increased their average milk and milk component yields. It is likely that their B-vitamin requirements increased accordingly and that ruminal synthesis alone may not be sufficient to meet the new needs, even though mature ruminant animal's rumen microflora could synthesize a certain amount of B-vitamins. In recent years, several studies have well documented that B-vitamin supplementation has important effects on the milk production of dairy cows.

Majee et al. (2003) used supplemental biotin (20 mg/day) and a B-vitamin blend [thiamin (150 mg/day), riboflavin (150 mg/day), pyridoxine (120 mg/day), B_{12} (0.5 mg/day), niacin (3000 mg/day), pantothenic acid (475 mg/day) and folic acid (100 mg/day)] in early lactation multiparous cows with a 28 day period and the results showed that milk yield was increased (1.7 kg/day) for supplemental biotin at 20 mg/day alone, while yields of milk protein and lactose but not fat were higher for supplemental biotin and the B-vitamin blend. In a study by Sacadura et al. (2008), supplying early lactation cows with a ruminally protected B-vitamin blend (3 g/cow/day), which contained biotin (3.2 mg/g), folic acid (4 mg/g), pantothenic acid (40 mg/g) and pyridoxine (25 mg/g) for a 35 day period, resulted in milk and milk component yields increasing with B-vitamin feeding, especially milk protein yield. It can be suggested that the mechanism leading to the positive overall production response with B-vitamin supplementation was due to improvements in metabolic efficiency of intermediary metabolism, rather than increased metabolic activity. However, in a study by Rosendo et al. (2004), multiparous Holstein cows received 0 or 20 mg of biotin/day starting at an average of 16 days prepartum and then switched to 0 or 30 mg of biotin/day from calving through 70 days postpartum. The results showed that milk production (35.8 *vs.* 34.8 kg/day) and milk fat concentrations (3.59 *vs.* 3.69%) were similar, and indicated that lactation performance was not improved by supplemental biotin.

Shaver and Bal (2000) evaluated the effects of dietary thiamin supplementation on milk production of dairy cows. 88 Holstein cows were blocked by parity and assigned randomly to either placebo or thiamin top-dress for the 8-week experiment to provide a supplemental thiamin intake of 0 or 150 mg/day per cow. Within each of these groups, cows were further assigned randomly to two total mixed rations (TMR) for 4 weeks, with the TMR treatments which then reversed a second 4-week experimental period. Milk yield was 2.7 kg/day higher for thiamin-supplemented cows. Yields of milk fat and protein were increased (0.13 and 0.10 kg/day, respectively) by dietary thiamin supplementation.

Graulet et al. (2007) demonstrated that supplementary folic acid (2.6 g/day) from 3 weeks before to 8 weeks after calving increased milk production by 3.4 kg/day ($P = 0.01$) and milk crude protein yield by 0.08 kg/day, and between 45 days of gestation and drying off, supplementary folic acid tended to increase milk production by 1.5 kg/day ($P = 0.09$). Girard and Matte (2005) found that milk production was increased linearly with the quantity of folic acid ingested (0, 2 or 4 mg/kg BW/day) in multiparous cows from 4 weeks before the expected time of calving until 305 days of lactation.

4 Vitamin E

VE is an important lipid soluble membrane antioxidant that enhances the functional efficiency of neutrophils by protecting them from oxidative damage following intracellular killing of ingested bacteria. Fresh green forage is an excellent source of VE; however, concentrates and stored forages (hays, haylages, and silages) are generally low in VE (NRC, 2001). The best understanding of the role of VE on mastitis and milk production is that it acts as a lipid soluble cellular antioxidant, free radical scavenger and protects against lipid peroxidation.

The beneficial effect of supplementation of VE and BC on milk production has been documented by Chawla and Kaur (2004). Cows were supplemented with 1000 IU α-tocopheryl acetate and 1000 IU α-tocopheryl acetate + 300 mg BC from 30 days prepartum to 2 weeks postpartum, which resulted in increased milk production of cows supplemented VE (14.1 vs. 11 kg/day, $P<0.01$) and VE + BC (14.6 vs. 11 kg/day, $P<0.01$). The reason was fewer udder infections during the period in supplemented cows.

However, in a recent study by Brozos et al. (2009), daily administration of a blend containing 60 g ammonium chloride, 1000 IU VE and 0.05 ppm Se throughout the dry period seemed to be safe, but without any effect on milk yield at 30 and 60 days postpartum. In addition, cows were given intramuscular injections of 2100 mg of VE (and 7 g of sodium selenite) 2 weeks before calving and on the day of calving, but there was no effect on milk yield (Bourne et al., 2008).

Questions still remain on the benefit, optimum dose, route and timing of VE administration in lactational dairy cows. Based on these and numerous other studies, NRC (2001), recommended the VE requirement for milking cows to be 15 to 20 IU/kg dry matter (DM).

5 Selenium

Se is recognized as an essential trace element for domestic animals, and it functions in the antioxidant system as an essential component of the glutathione peroxidase, which is responsible for the reduction of H_2O_2 and free O_2 to H_2O (NRC, 2001). It also plays a vital role in protecting both the intra- and extra-cellular lipid membranes against oxidative damage. The majority of Se in body tissues and fluids is present as either selenocysteine, which functions as an active center for selenoproteins, or selenomethionine, which is incorporated into general proteins and acts as a biological pool for Se (Juniper et al., 2006).

Moeini et al. (2009) reported that daily milk production of heifers at 8 weeks lactation was significantly increased (26.1 vs. 29.4 L/day, $P<0.05$) and milk SCC decreased (193, 000/mL vs. 179, 000/mL, $P<0.05$) by 20 mg Se and 2000 IU D, L-α-tocopheryl acetate supplemented at 4 and 2 weeks before expected calving. Wang et al. (2009) evaluated the effects of selenium-yeast supplementation on lactation performance in dairy cows. Treatments were: control, LSY, MSY and HSY with 0, 150, 300 and 450 mg selenium yeast per kg of diet dry matter, respectively. Experimental periods were 45 days with 30 days of adaptation and 15 days of sampling. The results indicate that supplementation of diet with selenium-yeast improved the milk yields. Milk yields were higher ($P<0.05$) for LSY and MSY than for HSY and the control but proportions and yields of milk fat, protein and lactose were not affected by selenium-yeast supplementation ($P>0.05$). The optimum selenium-yeast dose was about 300 mg per kg diet dry matter.

However, Juniper et al. (2006) found no significant effects for different levels of selenized yeast and sodium selenite on milk yield, milk composition (fat, protein, lactose, urea nitrogen and SCC), and yield of milk constituents. Results of selenized yeast on milk yields and milk components are inconclusive. It could be due to differences in the composition of the diet and/or the dose of selenized yeast and lactation period of dairy cows.

The differences presented earlier were due to the Se levels of blood before the experiment and the form, optimum dose, route and timing of administration of Se in lactational dairy cows. NRC (2001) recommended that the level of Se in dairy diets is 0.3 mg/kg DM and should be closely monitored to ensure that over supplementation does not occur.

6 Zinc

Zn is an essential trace mineral which is found to be an integral component of over 300 enzymes in metabolism, and NRC (2001) recommended the level of Zn supplementation for lactating dairy cows to be 40 to 60 mg/kg DM. Kellogg et al. (2004) indicated that Zn methionine increased lactation performance [produced more ($P<0.01$) milk, energy-corrected milk and fat-corrected milk] and improved udder health (as a 33.3% reduction in SCC), but milk composition did not change ($P>0.15$). Popovic (2004) evaluated replacing 33% of the supplemental inorganic Zn sulphate with organic Zn for 45 days pre-calving until 100 days post-calving. Cows that received the organic Zn had significantly ($P<0.05$) lower SCC by day 10 of lactation (158, 840/mL vs. 193, 530/mL) and at the end of the trial (62, 670/mL vs. 116, 440/mL), the average milk yield was numerically greater (27.75 kg/day vs. 26.22 kg/day). In addition, Kinal et al. (2005) reported that replacing 30% of the inorganic Cu, Zn and manganese

(Mn) for 6 weeks pre-calving until 305 days of lactation in dairy cows resulted in a 6.5% increase in milk yield (22.35 vs. 21.20 kg/day, $P<0.05$) and a 34% reduction in SCC (270, 00/mL vs. 409, 000/mL, $P<0.01$).

In a study by Griffiths et al. (2007), supplementing cows with CTM (providing daily 360 mg Zn, 200 mg Mn, 125 mg Cu as amino acid complexes and 12 mg cobalt (Co) from Co glucoheptonate) resulted in ($P\leq0.05$) a 6.3% increase in milk production, 5.6% increase in milk energy, 6.4% improvement in fat yield, 6.5% improvement in crude protein yield, 5.8% improvement in production of milk solids, and a trend for a reduction in mastitis cases ($P\leq0.10$). Milk composition was not affected by treatment, although fat, crude protein and the solids content of milk produced by CTM supplemented cows were numerically ($P\geq0.10$) higher than that produced by the control cows. Cope et al. (2009) evaluated the effects of the level and form of dietary Zn on milk performance, and found that cows supplemented with organically chelated Zn at the recommended level had a higher milk yield (37.6 kg/day) than those fed inorganic Zn at the recommended level (35.2 kg/day), or organically chelated Zn at low level (35.2 kg/day), but there was no difference from those fed inorganic Zn at the low level (36.0 kg/day). Milk composition was unaffected by dietary treatment. Animals that received the low level of Zn had higher SCC.

Nocek et al. (2006) reported that first lactation cows had no differences in SCC when fed Zn, Mn, Cu and Co in complex or inorganic form at 75 or 100% of NRC above the basal diets, but a small significant milk production response was noted between the organic and the inorganic minerals, even when the inorganic minerals were fed at 75% of NRC. However, Uchida et al. (2001) reported that feeding a combination of Zn amino acid (AA), Mn AA and Cu AA complexes, and Co glucoheptonate to early lactation Holstein cows had no effect on milk production, milk fat and protein content, and linear SCC.

7 Copper

Cu is a component of ceruloplasmin, which facilitates iron absorption and transport. In addition, Cu is considered as an important part of superoxide dismutase, an enzyme that protects cells from the toxic effects of oxygen metabolites produced during phagocytosis. Therefore, lactating cows were recommended for Cu supplementation at 11 mg/kg DM (NRC, 2001). Copper deficiency in cattle is generally due to the presence of dietary antagonists, such as sulfur, molybdenum and iron that reduce Cu bioavailability (Spears, 2003). Dietary requirements for Cu are greatly increased by high concentrations of molybdenum and sulfur.

Scaletti et al. (2003) evaluated the effect of dietary Cu on the responses of heifers to an intramammary *Escherichia coli* challenge at 34 days of lactation. Primigravid Holstein heifers were supplemented with Cu sulfate (20 ppm) beginning from 60 days prepartum through 42 days of lactation. The results suggested that Cu supplementation reduced the clinical response; a reduction in bacterial and SCC lowered the clinical mammary scores and lowered the peak rectal temperatures during experimental E. coli mastitis, but the duration was unchanged. In a study designed to evaluate the effects of dietary organic sources of Zn, Cu and Se for dairy cows on SCC and the occurrence of subclinical mastitis, it was found that the number of new and total cases of subclinical mastitis was lower for the group of cows fed with organic sources of Zn, Cu and Se when compared with animals that received inorganic sources. Average SCC during the first 80 days of lactation tended to be lower for the group fed with organic Zn, Cu and Se (Cortinhas et al., 2010).

Assuming normal bioavailability and typical ingredients, an average lactating Holstein cow producing 100 pounds (1 pound = 0.4536 kg) of milk needs to consume about 300 mg Cu/day. Excessive intake of Cu can be toxic (only four to five times more than the requirement) for animals, and Cu supplementation should be avoided unless feed analysis data is indicated.

8 Conclusion

In this review, recent researches on vitamins and traceelement supplements for improving the milk performance of dairy cows was considered. Some vitamins and traceminerals are clearly documented with their influence, while the impacts of supplementation of some micronutrients on milk performance are less clear. Continued research using field data and controlled studies are needed to further define the role of nutrition on the milk performance of dairy cows.

Before deciding to supplement dairy cow rations with the levels of vitamins and traceminerals indicated earlier, dairy farmers should have their animal feeds tested and their rations evaluated by a competent dairy cow nutriationist and a trustworthy laboratory to be sure of the level of supplementation that may be warranted. While inadequate intake and absorption of certain nutrients may result in a weakened immune system and perhaps more mastitis and less milk production during the lactation period, unjustified supplementation can be expensive and lead to other animal health problems.

Section 3　Bovine mastitis in subtropical dairy farms

The purpose of this section was to investigate the incidence of clinical mastitis (CM) and subclinical mastitis (SCM), isolation and identification of the major pathogens and test the antimicrobial resistance of milk bacterial isolates in subtropical dairy farms in Guangxi region (south of China) between 2005 and 2009. The average percentages of blind quarter(s) at cow and quarter level were 11.5% and 3.7%, respectively. The incidence of CM at cow and quarter level were 8.7% and 3.7%, respectively, while that of SCM at cow and quarter level were 48.8% and 19%, respectively. A total of 105 and 56 microorganisms were isolated from the 109 CM and 67 SCM samples, respectively. The most common bacterial isolates from CM cases were *S. aureus* (29.5%), *E. coli* (25.7%), and *C. neoformans* (16.2%), however, in SCM they were *S. aureus* (32.1%), CNS (19.7%), and *St. agalactiae* (17.9%). The antimicrobial sensitivity test indicated that all the antimicrobial agents (except for ampicillin) showed lower proportion of resistant isolates of all the isolated bacteria (except for *C. neoformans*), among the employed antimicrobials, Ruyanxiao showed the lowest proportion of resistant isolates.

1 Introduction

Mastitis, an inflammatory reaction of the mammary gland that is usually caused by a microbial infection, is recognized as the most costly disease in dairy cattle despite of the progress in improving general udder health in recent years. When a modern dairy farm in the tropics was first adopted, mastitis was predicted to be an important disease in dairy cattle (Mungube et al. 2004). In the efforts to avoid mastitis, some vaccines that can reduce the severity of this illness were generated. However, these vaccines still do not control efficiently the development of mastitis (Leitner et al., 2003). On the other hand, it has been verified that the indiscriminate treatment with antibiotics, without either a technical prescription or identification tests of the pathogen, can contribute to an increased resistance of these micro-organisms, making the cure of mastitis still more difficult (Gruet et al., 2001).

It is very limited that published articles on the incidence of dairy clinical mastitis (CM) and subclinical mastitis (SCM), major pathogens and their antimicrobial resistance in subtropics of China. To fill the vacancy of current research on the incidence of bovine mastitis and antibiotic resistance patterns in subtropical dairy farms of China, this section was designed to investigate the incidence of dairy CM and SCM, isolate the major microorganisms responsible for CM and SCM, and test their antimicrobial sensitivity between 2005 and 2009.

2 Materials and methods

2.1 Study area and animals The study was conducted at two Holstein dairy farms from 2005 to 2009 in Guangxi region, where is locates in the area of subtropical monsoon climate. The annual average milk production is 4813 kg per cow in 2008. Milking was done three times per day, and all lactation cows except CM cows were machine-milking. And they are treated with antibiotics at the time of drying off. The animals are allowed feeding for 7 h indoors every day and outdoors for the rest of time.

2.2 Clinical examination and California Mastitis Test (CMT) All the lactating cows of the two dairy farms were clinically examined for the blind quarters and diagnosis of CM and SCM in the spring every year during the study period. CM was diagnosed on the basis of visible signs of inflammation, if a quarter or the milk was shown abnormal and inflammation response will be considered the cow had CM.

Somatic cell count (SCC) has been accepted as the best index in both evaluating milk quality and predicting udder infection in cow (Pyörälä, 2003). Under field conditions, determination of SCC in cow's milk is usually performed by CMT. If a cow was considered for CM, the cow was didn't test SCC. The test result was interpreted based on the thickness of the gel formed by CMT reagent and milk mixture and scored as 0 (negative), T (trace), 1 (weak positive), 2 (distinct positive), and 3 (strong positive). Quarters with CMT score of 1 or above were judged as positive for SCM; otherwise negative. When more than one quarter turned out to be positive for CMT, cows were considered positive for CMT.

2.3 Collection of milk samples Milk was sampled from CM and SCM quarters of the cows in accordance with standard milk sampling techniques (Getahun et al., 2008).

2.4 Bacteriological examination of milk samples Milk samples were examined following the standard procedures (Getahun et al., 2008). Isolation of two or more types of colonies from a sample was considered as mixed growth.

2.5 Antibacterial sensitivity test Kirby-Bauer disk diffusion method was employed to test *in-vitro* antibiotic sensitivity and the antibacterial activity of Ruyanxiao (Chinese medicine) (Song et al., 2009). The following seven antimicrobial drugs were used: carbenicillin (CAR) (100 μg), cephazolin (CEP) (30 μg), kanamycin (KAN) (30 μg), chloramphenical (CHL) (30 μg), erythromycin (ERY) (15 μg), ciprofloxacin (CIP) (5 μg) and Ruyanxiao (RYX). The cut off values for the evaluation of the susceptibility of isolates were according to Getahun et al.(2008) for antibiotics and Song et al.(2009) for Ruyanxiao.

2.6 Statistical analysis CM incidence was calculated based on clinical examination, and SCM prevalence rate was calculated as defined by the CMT score. In addition, on the analysis of pathogen, a ratio between contagious and environmental pathogens was calculated to summarize the contribution of contagious and environmental pathogens to the infection pattern. Contagious pathogens in this calculation were *S. aureus*, *St. agalactiae* and coagulase- negative staphylococci (CNS). Environmental pathogens were *E. coli*, *St. uberis*, *St. dysgalctiae*, and *C. neoformans*.

3 Results

The average of lactating cows which had blind quarter(s) from 2005 to 2009 is presented in Table A-2. The details of CM incidences at cow and quarter levels of the lactating cows in the period between 2005 and 2009 are given in Table A-3. The details of SCM incidences at cow and quarter levels of the lactating cows in the period between 2005 and 2009 are shown in Table A-4. Distributions of different CMT scores of the milk from 2005 to 2009 are given in Table A-5.

Table A-2 The average lactating cows had blind quarter (s) of the animals from 2005 to 2009

	2005	2006	2007	2008	2009	Mean
NC	973	1032	1105	1151	1184	1089
NCBQ (%)	111 (11.4)	113 (10.9)	135 (12.2)	128 (11.1)	139 (11.7)	125 (11.5)
NBQ (%)	143 (3.7)	152 (3.7)	170 (3.8)	169 (3.7)	183 (3.9)	163 (3.7)
NCBQ 1 (%)	79 (8.1)	74 (7.2)	100 (9)	87 (7.6)	95 (8)	87 (8)
NCBQ 2 (%)	32 (3.3)	39 (3.8)	35 (3.2)	41 (3.6)	44 (3.7)	38 (3.5)

Note: NC: Number of lactating cows; NBQ: Number of blind quarters; NCBQ: Number of cows had blind quarter(s); NCBQ 1: Number of cows had one blind quarter; NCBQ 2: Number of cows had two blind quarters

Table A-3 The incidence of CM at cow and quarter levels of the animals from 2005 to 2009

Observation level	2005		2006		2007		2008		2009		Mean	
	TN	PN (%)	TN	PN (%)	TN	PN (%)	TN	PN (%)	TN	PN (%)	TN	PN (%)
Cow level	973	75 (7.7)	1032	84 (8.1)	1105	106 (9.6)	1151	103 (8.9)	1184	107 (9)	1089	95 (8.7)
Quarter level	3749	127 (3.4)	3976	141 (3.5)	4250	165 (3.9)	4435	167 (3.8)	4553	173 (3.8)	4193	155 (3.7)

Note: TN: Total number; PN: Positive number

Table A-4 The incidence of SCM at cow and quarter levels of the animals from 2005 to 2009

Observation level	2005		2006		2007		2008		2009		Mean	
	TN	PN (%)	TN	PN (%)	TN	PN (%)	TN	PN (%)	TN	PN (%)	TN	PN (%)
Cow level	898	405 (45.1)	948	439 (46.3)	999	547 (54.8)	1048	506 (48.3)	1077	528 (49)	994	485 (48.8)
Quarter level	3466	621 (17.9)	3659	664 (18.1)	3859	827 (21.4)	4046	739 (18.3)	4159	806 (19.4)	3838	731 (19)

Note: TN: Total number; PN: positive number

Table A-5 The distributions of CMT scores of the milk from 2005 to 2009

CMT score	2005	2006	2007	2008	2009	Mean
0 + T	2845 (82.1)	2995 (81.8)	3032 (78.6)	3307 (81.8)	3353 (80.6)	3106 (81)
1	320 (9.2)	339 (9.3)	386 (10)	356 (8.8)	386 (9.3)	357 (9.3)
2	167 (4.8)	171 (4.7)	235 (6.1)	205 (5)	227 (5.5)	201 (5.2)
3	134 (3.9)	154 (4.2)	206 (5.3)	178 (4.4)	193 (4.6)	173 (4.5)
Total	3466 (100)	3659 (100)	3859 (100)	4046 (100)	4159 (100)	3838 (100)

A total of 109 clinical mastitis milk samples and 67 subclinical mastitis milk samples were examined in the period between 2008 and 2009. The number and proportions of microorganism isolated from clinical and subclinical mastitis milk samples are presented in Table A-6. From clinical and subclinical mastitis milk samples, 4 (3.7%) and 11 (16.4%) showed negative, 96 (88.1%) and 55 (82.1%) had single growth, and 9 (8.2%) and 1 (1.5%) had mixed growths, respectively. The common isolated microorganisms from the clinical mastitis milk samples were *S. aureus* (29.5%, *n*=31), *E. coli* (25.7%, *n*=27), and *C. neoformans* (16.2%, *n*=17), while from subclinical mastitis milk samples were *S. aureus* (32.1%, *n*=18), CNS (19.7%, *n*=11), and *St. agalactiae* (17.9%, *n*=10). Other isolates include *St. uberis* and *St. dysgalctiae*.

Table A-6 Number and proportion of microorganism isolates from clinical and subclinical mastitis milk samples

Type of bacterial isolates	CM No. (%)	SCM No. (%)	Total No. (%)
S. aureus	31 (29.5)	18 (32.1)	49 (30.4)
E. coli	27 (25.7)	4 (7.1)	31 (19.3)
C. neoformans	17 (16.2)	5 (8.9)	22 (13.7)
St. agalactiae	8 (7.6)	10 (17.9)	18 (11.2)
St. uberis	5 (4.7)	5 (8.9)	10 (6.2)
St. dysgalctiae	3 (2.9)	2 (3.6)	5 (3.1)

续表

Type of bacterial isolates	CM No. (%)	SCM No. (%)	Total No. (%)
CNS	5 (4.7)	11 (19.7)	16 (9.9)
Mixed growths			
S. aureus + C. neoformans	3 (2.9)	1 (1.8)	4 (2.5)
E. coli + C. neoformans	3 (2.9)	-	3 (1.9)
S. aureus + E. coli	2 (1.9)	-	2 (1.2)
S. aureus + St. agalactiae	1 (1)	-	1 (0.6)
Total isolates	105 (100)	56 (100)	161 (100)

Note: No.: number of milk samples positive for the specific bacterial isolate; %: proportion from the total of the same column

The results of antimicrobial sensitivity test for the different isolates are given in Table A-7. This study was employed 7 antimicrobial agents, all the antimicrobial agents (except for ampicillin) showed lower proportion of resistant isolates of all the isolated bacteria (except for *C. neoformans*).

Table A-7 *In-vitro* antimicrobial susceptibility test results of bacterial isolates recovered from mastitis milk samples

Isolates	No.	AMP (%)		CEP (%)		KAN (%)		CHL (%)		ERY (%)		CIP (%)		RYX (%)	
		Sus	Int	Sus	Int	Sus	Int	Sus	Int	Sus	Int	Sus	Int	Sus	Int
S. aureus	49	8.2	22.4	55.1	38.8	22.5	36.7	24.5	46.9	18.4	49	26.5	49	55.1	38.8
E. coli	29	6.9	13.8	55.2	44.8	34.5	37.9	37.9	41.4	27.6	20.7	48.3	51.7	37.9	55.2
C. neoformans	22	0	0	0	9.1	0	4.5	0	4.5	0	22.7	0	27.3	36.4	50
St. agalactiae	18	5.6	22.2	38.9	55.5	22.2	50	22.2	55.6	22.2	38.9	27.8	50	44.4	50
St. uberis	9	0	44.4	44.4	55.6	11.1	33.3	33.3	44.5	11.1	44.4	22.2	44.5	22.2	77.8
St. dysgalactiae	4	0	25	75	25	0	50	50	50	25	75	0	50	25	75
CNS	15	0	40	73.3	20	13.3	53.4	26.7	40	6.7	40	40	46.7	40	60

Note: No.: number of observations; AMP: ampicillin; CEP: cephazolin; KAN: kanamycin; CHL: chloramphenicol; ERY: erythromycin; CIP: ciprofloxacin; RYX: Ruyanxiao

4 Discussion

In this study, the average of lactating cows which had blind quarter(s) at quarter level is 3.7%, which is in agreement with the reports by Mungube et al. (2004) (3.7%) and Almaw et al. (2008) (3.8%), it is higher than the reports by Kivaria et al. (2007) (2.1%) and Getahun et al. (2008) (2.3%), while at cow level (11.5%), it is higher than the reports by Kivaria et al. (2004) (5%). The blind quarters observed in this study might be an indication of a serious mastitis problem on the farms, such as in 2007.

The average incidence of CM at cow and quarter levels in the study period (8.7% and 3.7%) are higher than the reports by Mungube et al. (2004) (6.6% and 2.8%), Kivaria et al. (2004) (3.8% and 2.1%) and Getahun et al. (2008) (1.8% and 0.51%). The average prevalence rate of SCM (based on CMT) at cow level (48.8%) is in agreement with the reports by Mungube et al. (2004)

(46.6%) and Gianneechini et al. (2002) (52.4%). It is higher than the report by Getahun et al. (2008) (22.3%). However, our study is by far lower than the reports of Kerro Dego and Tareke (2003) (62.9%) and Kivaria et al. (2004) (90.3%). The average incidence of SCM at quarter level in this study (19%) is higher than that of Getahun et al. (2008) (10.1%), and is lower than that of Mungube et al. (2004) (27.8%) and Gianneechini et al. (2002) (27.6%). The climatic differences of the farms may also have been a factor for CM and SCM. As in the area of subtropical monsoon climate, the higher humidity and temperature conditions at the two farms may enhance the survival of bacteria in the environment, where they remain as sources of infection.

Although calving is not strictly seasonal in the investigated area, a relatively higher calving rate in the wet season than in the dry season may contribute to the high prevalence CM and SCM during the wet season, because of birth-related problems (Kerro Dego and Tareke, 2003). High serum selenium concentrations were associated with reduced rates of mastitis and lower bulk-tank somatic cell counts in Ohio dairy herds (Jerry and William, 2008), while it is significance higher selenium level in the soil of Guangxi than the average of China and USA, so it is relatively high in serum selenium concentration of cows in this area, and this is might be a factor that the incidence of CM and SCM in this study is lower than some areas.

The average distributions of 1, 2 and 3 CMT score of SCM quarters is 9.3%, 5.2% and 4.5% (Table A-5), respectively. The result is lower than reported by Contreras et al. (27.6%, 20.9% and 4.6%, respectively) and Sung et al. for goat. The animals had the best score in 2005 and the worst in 2007.

In this study, the most frequently bacterial isolates from clinical and subclinical mastitic samples was *S. aureus*, the high prevalence of *S. aureus* mainly attributed to the wide distribution of microorganism inside the mammary gland and on the skin of teat and udder. They usually establish chronic, subclinical infections, which serves as a source of infection for other healthy cows during the milking process (Workineh et al., 2002). In addition, It is difficult to eliminate the bacteria from the mammary gland due to the very low rate of self cure and a number of factors affect the rate of cure after treatment which is in general low (Getahun et al., 2008).

However, mixed growths in CM and SCM are 8.7% and 1.8%, respectively. As far as SCM, it is lower than the report by Almaw et al. (2008) (9.9%). Miscellaneous bovine mastitis pathogens, such as *S. aureus* + *C. neoformans*, are associated with poor and unhygienic housing and milking, unsanitary intramammary infusion practices, indiscriminate use of antibiotics, and non implementation of mastitis control programme. The most important findings were 16.2% and 8.9% microorganisms isolates from CM and SCM was *C. neoformans*, showing *C. neoformans* constitute a real problem once the therapy against bacterial mastitis is unsuccessful in eliminating the *C. neoformans* present in the quarter. Conditions like the chronic form of mastitis, frequently subjected to prolonged, excessive and repeated antibacterial therapy, help the establishment of *C. neoformans* or mixed infections (bacteria and *C. neoformans*).

The antimicrobial sensitivity test indicated that all the antimicrobial agents were employed (except for ampicillin) showed lower proportion of resistant isolates of all the isolated bacteria (except for *C. neoformans*), Interestingly, among the antimicrobial agents tested, cephazolin and Ruyanxiao showed the least proportion of resistant isolates of all the bacteria tested.

The finding is attributed to the rare use of cephazolin the treatment of bovine CM, and Ruyanxiao, whose main components are Tokyo violet herb, honeysuckle flower, Angelica sinesis, Angelica Dahurica, and white wax, was Chinese medicine. Ruyanxiao had significant antimicrobial effects and was used about 4 years at this farm.

In addition, for the CM cases, the veterinaries usually do intramammary injection of Ruyanxiao accompanying with external used Xiaoyangao, which is a compound traditional Chinese herb medicine ointment, and its main components are snakegourd root, phellodendri cortex, rhubarb, curcuma, and dahuria angelica root. By this method of treatment, the disease was cured in 2 days on average.

5 Conclusion

The important findings in this section were investigated the lactating cows which had blind quarter(s), the incidence of CM and SCM, and the distributions of different CMT scores of SCM quarters for the first time in subtropical dairy farms in Guangxi region, and the most important findings were 16.2% and 8.9% microorganisms isolates from CM and SCM were *C. neoformans*.

Section 4   The prevalence of heifer mastitis and its associated risk factors in Huanggang

The purposes of this section were to determine the prevalence of heifer clinical mastitis (HCM) and heifer sub-clinical mastitis (HSCM), isolate HCM-causing bacteria, and assess the association of some risk factors in Huanggang, Central China. A total of 1374 lactating heifers from three dairy farms were examined in the present study; 22.64% of heifers were positive for mastitis, out of which, 3.86 and 18.78% were with HCM and HSCM, respectively. Of the 67 HCM samples, 91.05% were single growth, 7.46% mixed growth, and 1.49% no bacterial growth. Coagulase-negative staphylococci (CNS) accounted for 30.98% of the isolates followed by *Escherichia coli* (29.58%), *Staphylococcus aureus* (16.9%), and *Streptococcus uberis* (11.27%). Logistic regression analyses showed that heifer milk yield lower than 3 L had the highest prevalence of HCM compared to heifermilk yield of more than 3 L ($P<0.01$), and prevalence of HCM was significantly high in heifers with the presence of teat lesions ($P<0.01$). Moreover, the heifer milk yield lower than 3 L had the highest prevalence of HSCM compared to heifermilk yield of more than 3 L ($P<0.01$). These two factors were significantly associated with the occurrence of heifer mastitis, which needs to be considered in the prevention and control strategies of the disease.

1 Introduction

Mastitis in heifer is a permanent concern of dairy farmers worldwide because it potentially threatens production and udder health in the first and subsequent lactations. Nearly 11% of heifers were culled within 28 days after mastitis treatment; udder damage caused by mastitis was the only or main reason for culling in 96% of those heifers (Waage et al. 2000). The incidence of clinical mastitis (CM) in heifers is higher than in cows (De Vliegher et al. 2012). There are a number of papers that have reported on the pathogens associated with CM in dairy heifers and primiparous cows (Kalmus et al. 2006; Compton et al. 2007; Tenhagen et al. 2009).

Heifers have not been challenged with milking vacuums that have been associated with a deleterious effect on the structure of the teat end. For most of the heifer's life, the mammary gland has been immature and it would seem less likely to be in close physical contact to the environment, as contrasted with multiparous cattle. There are a few risk factors identified by several researchers that influence the prevalence of heifer clinical mastitis (HCM) and heifer subclinical mastitis (HSCM) such as herd size, floor type, juvenile intersucking, removing supernumerary teats, age at first calving, location of close-to-calving heifers, number of calves, udder and/or teat wounds, unhygienic udder postcalving, milk production, blocked milk secretion during the first milking, and oxytocin administration necessary for the first milking (Parker et al. 2007a; Kromker et al. 2012; Santman-Berends et al. 2012). However, mastitis is a multifactorial disease, requiring exposure to a combination of environmental and pathogenic factors and with variable responses between animals. Identification of causal factors for mastitis is important for development of control and prevention strategies in heifers. In China, there are some reports on the prevalence and major pathogens causes of the disease (Yang et al. 2011), but information relating to its risk factors is scant. The purposes of this study were to identify the prevalence of HCM and HSCM in heifers and discuss the possible risk factors associated with heifer mastitis in Huanggang.

2 Material and methods

2.1 Farms and animals This study was conducted from February 2013 to January 2014 in three commercial dairy farms in Huanggang, Hubei Province. All the studied dairy farms practiced similar feeding and management. All cows in the study area were machinemilked and milked two times a day. Prepartum diets consisted of hay, straw, and silage. Heifers in the first 14 days of lactation were excluded from this study. The number of study subjects was 1374 lactating heifers: 868 heifers fromXiandai, 349 from Yili, and 157 from Yangtze.

2.2 Collection of milk samples Each heifer was sampled once, just before the milking; milk samples were collected from these functional quarters. Strict foremilk (first stream) was discharged, and then 15 mL to 20 mL aseptic milk samples were collected separately from all the four quarters, viz., the left fore (LF), the left hind (LH), the right fore (RF), and the right hind (RH). All collected CM milk samples were subjected to bacteriological examination. And, the other milk samples were detected for subclinical mastitis (SCM).

2.3 Examination of clinical mastitis The udder and milk of the heifers were thoroughly examined for the presence of CM. CM was diagnosed at the quarter level based on visible and palpable signs, and there could be alteration in the color and consistency of milk. In this case, pathogens were detected by bacteriological culture. Additionally, blind teats were also considered as clinical cases. A heifer was considered to have CM if it had at least one quarter with CM.

2.4 Detection of subclinical mastitis All milk samples, except for CM milk samples, were detected for SCM by performing the California Mastitis Test (CMT, Shanghai Dairy Science Institute). In the present study, samples with negative or trace CMT scores (0, t) were assigned to healthy quarters, while those with positive (1+ to 3+) scores were assigned to SCM quarters. A heifer was considered to have SCM if it had at least one quarter with positive CMT reaction.

2.5 Bacteriological examination of clinical mastitis milk samples For culturing, a wire loop (about 10 μL) of each CM milk sample was spread on 5% defibrinated sheep blood agar plates. The inoculated plate then was incubated aerobically at 37℃ for up to 72 h. The plates were examined for growth, morphology, hemolytic features, and the numbers of each colony type at 24, 48, and 72 h after inoculation. Representative colonies were then subcultured on sheep blood agar plates and incubated aerobically at 37℃ for 24 h to obtain pure cultures. Catalase and coagulase production was tested for Gram-positive cocci. Specific identification of staphylococci was done using the coagulase test. Gram-positive, catalase-negative isolates were tested by CAMP, aesculin reaction, growth at 45℃. Gram-negative rods were subcultured on violet red bile agar, tested afterward for oxidase and indole reaction and additionally cultured in triple sugar iron agar and simmons citrate agar. This method allowed identification of bacteria at genus or species level in most cases; otherwise, unidentified organisms were recorded as other bacteria.

2.6 Data analysis The response variable analyzed was mastitis status of the heifers, and potential risk factors considered were lactation stage, body condition score (BCS), the milk yield on the test day, age at first calving, presence of teat lesions, and pregnant status. Data were collected by abstracting the farmrecords and examining the heifers. Stage of lactation was categorized as 15~90 days, 91~180 days, and＞180 days in milk (DIM); BCS was categorized as＜2, 2~3, and＞3 on a scale of 1 to 5 (1=thin, 5=fat). The milk yield on the test day was categorized as＜3 L, 3~5 L, and＞5 L. Age at first calving was categorized as＜750 days, 751~900 days, and＞900 days. The presence of teat lesions and pregnant status were categorized as yes and no. Collected data were recorded and analyzed using Microsoft Excel and SPSS version 19 statistical software, respectively. The prevalence of HCM and HSCM was calculated as percentage value, and univariable logistic regression analyses were used to assess the association between mastitis and individual risk factors considered in the study. Odds ratio (OR) and 95% confidence interval (95% CI) around the OR were calculated.

3 Results

3.1 Prevalence A total of 1374 lactating heifers were examined during the study period clinically as well as subclinically by using CMT; 311 (22.64%) heifers were positive for mastitis. From which, 3.86% and 18.78% showed clinical and subclinical mastitis, respectively. The quarter-level prevalence of mastitis was 7.99% (439/5496). Out of these, 1.22% and 6.77% were found to be of clinical and subclinical mastitis, respectively (Table A-8). The detail of heifers with the number of mastitis quarters is shown in Table A-9. The individual quarter-level prevalence of clinical and subclinical mastitis is shown in Table A-10.

Table A-8 The prevalence of HCM and HSCM at cow ($n = 1374$) and quarter ($n = 5496$) level

	Cow level		Quarter level	
	No. of affected	Prevalence (%)	No. of affected	Prevalence (%)
HCM	53	3.86%	67	1.22%
HSCM	258	18.78%	372	6.77%
Total	311	22.64%	439	7.99%

Table A-9 Frequency and percentage of heifers by the number of mastitis affected quarters ($n = 1374$)

The No. of mastitis quarter	No. (%) of clinical mastitis heifers	No. (%) of subclinical mastitis heifers	Total No. (%) of mastitis heifers
Single	40 (2.91)	175 (12.74)	215 (15.65)
Two	12 (0.87)	61 (4.44)	73 (5.31)
Three	1 (0.07)	13 (0.95)	14 (1.02)
Four	0 (0)	9 (0.65)	9 (0.65)
Total	53 (3.85)	258 (18.78)	311 (22.63)

Table A-10 Quaterwise distribution of the CM and SCM prevalence

| Quarter | Clinical mastitis No. (%) | Subclinical mastitis | | | Total No. (%) of subclinical mastitis | No. (%) of health quarters (CMT negative or trace) |
| | | No. (%) of subclinical mastitis quarters in each category | | | | |
		1+	2+	3+		
LF	15 (1.09)	55 (4)	26 (1.89)	8 (0.58)	89 (6.48)	1270 (92.43)
LH	17 (1.24)	60 (4.37)	22 (1.6)	11 (0.8)	93 (6.77)	1264 (91.99)
RF	16 (1.16)	71 (5.17)	18 (1.31)	3 (0.22)	92 (6.7)	1266 (92.14)
RH	19 (1.38)	57 (4.15)	27 (1.97)	14 (1.02)	98 (7.13)	1257 (91.49)
Total	67 (1.22)	243 (4.42)	93 (1.69)	36 (0.66)	372 (6.77)	5057 (92.01)

3.2 Bacteriological examination result Sixty-seven CM milk samples were subjected to bacteriological examination. Out of these samples, 61 (91.05%) were single growth, and 5

(7.46%) were mixed growth, the remaining 1 sample was no bacterial growth (Table A-11). Based on the culture growth, the most common isolates were recorded for coagulase-negative staphylococci (CNS), which accounted for 30.98% of the total isolates, followed by *Escherichia coli* (29.58%), *Staphylococcus aureus* (16.9%), *Streptococcus uberis* (11.27%), *Streptococcus agalactiae* (2.82%), *Corynebacterium bovis* (2.82%), *Streptococcus dysgalactiae* (1.41%), and Other (4.22%).

Table A-11　Frequency of bacterial isolates from heifers clinical mastitis milk ($n=67$)

Bacterial species	No. of milk samples (%)	Total No. of bacteria isolates (%)
Single growth	61 (91.05)	—
E. coli	19 (28.36)	21 (29.58)
CNS	17 (25.37)	22 (30.98)
S. aureus	11 (16.42)	12 (16.9)
S. uberis	7 (10.44)	8 (11.27)
S. agalactiae	2 (2.99)	2 (2.82)
C. bovis	2 (2.99)	2 (2.82)
S. dysgalactiae	1 (1.49)	1 (1.41)
Other	2 (2.99)	3 (4.22)
Mixed growth	5 (7.46)	—
E. coli + CNS	2 (2.99)	—
S. aureus + CNS	1 (1.49)	—
S. uberis + CNS	1 (1.49)	—
Other + CNS	1 (1.49)	—
No growth	1 (1.49)	—
Total	67 (100)	71 (100)

3.3　Analysis of risk factors　Association of the investigated risk factors is shown in Table A-12. Risk factor logistic regression analyses showed that the milk yield on the test day and the presence of teat lesions had significant effect ($P \leqslant 0.05$) on the HCM prevalence, and the milk yield on the test day had significant effect ($P \leqslant 0.05$) on the HSCM prevalence. As expected, the heifer milk yield on the test day lower than 3 L had the highest prevalence of HCM ($P \leqslant 0.05$); likewise, the prevalence of HCM was significantly high in heifers with the presence of teat lesions ($P \leqslant 0.05$). Moreover, the milk yield on the test day lower than 3 L had the highest prevalence of HSCM ($P \leqslant 0.05$). However, age at first calving, lactation stage, BCS, and pregnant status had no statistically significant ($P > 0.05$) association with the HCM prevalence. And, age at first calving, lactation stage, BCS, presence of teat lesions, and pregnant status had no statistically significant ($P > 0.05$) association with the HSCM prevalence.

Table A-12 Bivariable analysis of potential risk factors for mastitis in lactating heifers

Risk factors	Categorize	No. of examined	Heifers clinical mastitis			Heifers subclinical mastitis		
			No. (%) of positive	Odds ratio (95% CI)	P value	No. (%) of positive	Odds ratio (95% CI)	P value
Age at first calving	<750 d	348	16 (4.6)	Reference		71 (20.4)	Reference	
	751~900 d	910	32 (3.52)	0.935 (0.335~2.61)	0.897	169 (18.57)	0.89 (0.653~1.213)	0.46
	>900 d	116	5 (4.31)	0.756 (0.41~1.396)	0.372	18 (15.52)	0.717 (0.407~1.262)	0.249
Lactation stage	15~90 DIM	374	18 (4.81)	Reference		69 (18.45)	Reference	
	90~180 DIM	504	17 (3.37)	0.69 (0.351~1.358)	0.283	88 (17.46)	0.935 (0.66~1.324)	0.705
	>180 DIM	496	18 (3.63)	0.745 (0.382~1.452)	0.387	101 (20.36)	1.13 (0.804~1.589)	0.481
Body condition score	<2	105	6 (5.71)	Reference		22 (20.95)	Reference	
	2~3	908	35 (3.85)	0.662 (0.271~1.612)	0.363	180 (19.82)	0.933 (0.567~1.534)	0.19
	>3	361	12 (3.32)	0.567 (0.208~1.55)	0.269	56 (15.51)	0.693 (0.4~1.2)	0.784
Milk yield	<3 L	294	40 (13.61)	Reference		74 (25.17)	Reference	
	3~5 L	553	12 (2.17)	0.14 (0.072~0.272)	<0.01	107 (19.35)	0.713 (0.509~1.000)	0.050
	>5 L	527	1 (0.19)	0.012 (0.002~0.088)	<0.01	77 (14.61)	0.509 (0.356~0.727)	<0.01
Presence of teat lesions	Yes	247	34 (13.77)	Reference		44 (17.81)	Reference	
	No	1127	19 (1.69)	0.107 (0.06~0.192)	<0.01	214 (18.99)	1.081 (0.756~1.547)	0.669
Pregnant status	Yes	616	26 (4.22)	Reference		127 (20.62)	Reference	
	No	758	27 (3.56)	0.838 (0.484~1.452)	0.529	131 (17.28)	0.804 (0.613~1.055)	0.116

4 Discussion

4.1 Prevalence The overall prevalence of mastitis in heifers was 22.64% at cow level and 7.99% at quarter level in the current study. The prevalence of HCM recorded in this study was 3.86%, while it is lower than several earlier findings of Parker et al. (2007a, 2007b) and Clyne (2013). And, the prevalence of HSCM at cow level based on CMT in the present study (18.78%) was in general agreement with the finding of Santman-Berends et al. (2012).

Heifer mastitis is a complex disease, and the variation in the prevalence of mastitis between this study and the others might be due to differences in climate, season, and region, and environmental and management practices. Besides, differences in diagnostic techniques used and in cut points used to define "infection" likely contribute to the reported variation in prevalence and incidence of heifermastitis (Sampimon et al. 2010).

The fact that the prevalence of mastitis was higher in single quarter (15.65%) and reduced as more quarters are affected (5.31, 1.02, and 0.65% in two, three, and four quarters, respectively) is an indication that, possibly, one quarter is usually first infected and the others become affected through contamination and other means especially during the milking procedures (Shittu et al. 2012).

In the current study, the prevalence of 1.22% of HCM and 6.77% of HSCM was observed at quarter level. The LH and RH quarters were more affected in comparison to LF and RF quarters (Table A-10), and among the hind quarters, there is more susceptibility of the RH quarters than the LH. It is in agreement with Khan and Muhammad (2005). Although an immediate explanation cannot be established for this observation, it might simply suggest that the relationship between quarter position and mastitis prevalence is not straightforward but influenced by other factors like breed and management.

4.2 Risk factors The potential risk factors of lactation stage, BSC, the milk yield on the test day, age at first calving, presence of teat lesions, and pregnant status on the prevalence of HCM and HSCM were studied. The result indicates that there is a statistical association between two risk factors and HCM and a statistical association between one risk factor and HSCM in this study.

The association between the prevalence of mastitis and age at first calving was in disagreement with Kromker et al. (2012), who reported that older heifers at first calving were more likely than younger heifers to have mastitis.

The result of this study indicated that there was no significant difference in the prevalence of heifer mastitis between the three categorizes of lactation stage. While, it was not in agreement with previous studies (Abera et al. 2012a; Haftu et al. 2012; Clyne 2013; Khanal and Pandit 2013; Yohannis and Molla 2013).

BSC is a subjective measure of the amount of metabolizable energy stored on a live animal and could be an indicator trait for health and fertility. The finding of this study was also assessed for BCS predisposition to HCM and HSCM, but no significant difference in the prevalence was detected between the three groups. This was consistent with the results by Berry et al. (2007) and Breenetal. (2009). However, Sarker et al. (2013) reported that SCM was significantly associated with BCS of 2~2.5 compared to other scores in dairy cows. Other studies indicated that cows with poor BCS have a significantly higher risk for CM and SCM (Kivaria et al. 2007; Uddin et al. 2009).

Prevalence of HCM and HSCM was significantly higher among the heifers producing less than 3 L compared to heifers producing 3 or more than 3 L which was similar to the view of Khanal and Pandit (2013). While, it was in no agreement with the view of Piepers et al. (2010) and Sarker et al. (2013).

The presence of teat lesions was observed to be one of the important risk factors for the occurrence of HCM. Of the 247 heifers having teat lesions in this study, 13.77% of heifers were examined for HCM and 17.81% for HSCM. It is one of many factors that predispose heifers to HCM as the lesions facilitate bacterial entry and leaves behind permanent tissue damage. Similar observation has been recorded by Getahun et al. (2008), Lakew et al. (2009), and Mekibib et al. (2010). It suggests the prevailing inadvertence to udder management in heifers.

The pregnant status was not significantly affecting the prevalence of HCM and HSCM in our study. This is agreement with Sarker et al. (2013) and Gunawardana et al. (2014).

4.3 Bacteriological examination   In the current study, CNS were the most prevalent bacteria isolated from HCM, accounting for 30.98% of positive samples. After CNS, *E. coli* was most frequent, followed by *Staphylococcus aureus* and *Streptococcus uberis*. *Staphylococcus species* and coliforms accounted for about 77% of the total isolates. The high prevalence of staphylococci species and coliforms in this study is in accordance with other workers (Compton et al. 2007; Tenhagen et al. 2009; Abera et al. 2012b; Sharma et al. 2013). *E. coli* was the most frequently isolated single pathogen in our study which is consistent with the previous studies (Kalmus et al. 2006). The high prevalence of staphylococci species may contribute to the presence of these agents on the skin and mucus membranes of various parts of the animal body and their contagious nature, especially Staphylococcus aureus and Streptococcus agalactiae (Abera et al. 2012b). Streptococcus species other than Streptococcus agalactiae accounted for 12.68% of the total isolates, and this result is in agreement with Abera et al. (2012a). A few of HCM milk samples were mixed growth in this study; miscellaneous heifer mastitis pathogens are associated with poor and unhygienic housing and milking, unsanitary intramammary infusion practices, indiscriminate use of antibiotics, and non-implementation of mastitis control program.

Section 5   Somatic cell counts positive effects on the DNA yield extracted directly from Murrah buffalo milk

The aim of this section was to verify the relationship between somatic cell counts (SCC) and the DNA yield extracted directly from buffalo milk. Twenty-five Murrah buffalo milk samples were measured SCC and extracted DNA. Milk SCC was between 16,000 and 1,934,000 cells/mL. DNA yield was within the range from 3747.2 ng/μL to 7080.7 ng/μL. The ratio of A_{260}/A_{280} and A_{260}/A_{230} of genomic DNA was ranged from 1.66 to 2.17 and 1.71 to 2.13, respectively. The results indicated that DNA can be extracted directly from Murrah buffalo milk, but may be presence of some contaminants. The Pearson's correlation analysis showed a very strong positive correlation between the DNA yield extracted directly from buffalo milk and the milk SCC ($r = 0.927$, $P<0.001$). Milk sample was preferred to blood as a source of DNA, due to milk collection was routinely performed, as well as less expensive and more easily accomplished than blood collection. Moreover, milk collection was also less stressful to animals, given that capture, handling, and venipuncture were not required for milk sampling.

1   Introduction

Milk somatic cells are mainly milk-secreting epithelial cells that have been shed from the lining of the gland and white blood cells (leukocytes) that have entered the mammary gland in response to injury or infection (Sharma et al., 2011). The milk somatic cells include 75% leukocytes (i.e. neutrophils, macrophages, lymphocytes, erythrocytes, which are derived from blood circulation) and 25% epithelial cells. While, the proportions of neutrophils), macrophages, and lymphocytes for healthy milk are approximately 12, 60, and 28%, respectively (Kelly et al., 2000). The epithelial cells of the glands are normally shed and get renewed, however, during infection the numbers increase.

The normal composition of milk somatic cells varies with the type of secretion or lactation cycle. Normally, in cows' milk from a healthy mammary gland, the SCC is lower than 100, 000 cells/mL, while bacterial infection can make it increase to above 1, 000, 000 cells/ml (Bytyqi et al., 2010), the stage of lactation, season, milk yield and number of lactations are all also known to influence milk SCC (Kelly et al., 2000). Whole blood can contain up to 350 times more nucleated cells than milk, when comparing lower cell counts for the same volumes of whole blood and milk (Murphy et al., 2002).

Genetic research has recently shown a marked increase as interest in understanding the genetic basis of diseases and drug regimens increases. Almost of all studies require DNA isolations. Blood leukocytes are generally used as an excellent source of genomic DNA. Nevertheless, collection of blood samples causes technical difficulties and leads to stress for animals. Other tissues, such as skin biopsies or hair follicles, can be also used as a source of genomic DNA (Healy et al., 2002). However, milk, which is widely available and obtained noninvasively, is the source of choice in lactating animals, and dose not require trained personnel or the milker.

Several studies have developed methods to extract genomic DNA from somatic cells of bovine milk (Murphy et al., 2002), caprine milk (Lopez-Calleja et al., 2004; D'Angelo et al., 2007), and human milk (Abdalla et al., 2009; Haas et al., 2011). However, there is scarcity of literature on the relationship between milk SCC and DNA extracted directly from milk samples. Therefore, the purpose of the section was to evaluate the effects of SCC on the DNA yield extracted directly from Murrah buffalo milk.

2 Material and methods

2.1 Samples   Twenty-five Murrah Buffaloes at 5 to 8 weeks postpartum, in their second to fourth lactation feeding in Guangxi Buffalo Research Institute. All the buffaloes did not show clinical signs of mastitis or other illnesses. After a quarter had been washed with tap water and dried, the teat end was disinfected by wiping with cotton wool soaked in 70% ethyl alcohol. Milk samples were collected prior to the morning milking into sterile test tubes (10 mL) hand-stripping after discarding the first three squirts of milk, and then placed in an icebox and transported to the laboratory for examination within one hour after collection. Milk sampling was done only once from each animal.

2.2 Somatic cell count   Milk SCC was measured quantitatively using the Fossomatic Minor analyzer (Foss Electric, Hillerød, Denmark) according to the International Dairy Federation standard.

2.3 DNA extraction   The total DNA was extracted from milk samples according to the previously described method with some modifications (De et al., 2011; Lopez-Calleja et al., 2004). Briefly, 0.5 mL of normal saline solution was added to 1 mL of the milk sample. Samples were mixed by inverting them 10 times before they were centrifuged at 13,000 r/min for 10 min at 4℃. The result was a cream pad on the top of a clear supernatant. Both the pad and the supernatant were carefully removed, and the pellet left at the bottom of the tube was resuspended in 1 mL of normal saline solution, then the mixture was centrifuged again and the pad on the top and the supernatant were carefully discarded. The resulting pellet, which contained cells and caseins, was resolved in 860 μL of TES buffer (pH 8.0; 50 mmol/L Tris-HCl, 10 mmol/L EDTA, and 1% SDS), 100 μL of 5 mol/L guanidine hydrochloride, and 40 μL of 20 mg/mL proteinase K. The mixtures were incubated in a water bath at 50℃ overnight, and they were left to cool at room temperature. The cellular debris and proteins were removed by adding 500 μL Tris saturated phenol, and the inverted them 10 times before centrifugation at 13,000 r/min for 5 min at 4℃. The clear aqueous supernatant obtained after the centrifugation was carefully transferred to a new Eppendorf tube. The DNA was precipitated by adding 0.8 volume of isopropanol in the presence of 0.1 volume of 3 M sodium acetate (pH 5.3). The DNA pellet, obtained after centrifugation for 10 min at 13,000 r/min, was air dried and subsequently dissolved in 100 μL of sterile double distilled water.

2.4 Spectrophotometer measurements Quality of the DNA extracted was determined by a UV-VIS spectrophotometer (Nanodrop, Thermo Scientific, USA). The concentration of DNA was calculated based on the approximation that an absorbance reading of 1 of the purified DNA at 260 nm was taken to correspond to 50 ng/μL. The ratio of absorbance at 260 nm and 280 nm was used to assess protein or RNA contamination while the ratio of absorbance at 260 nm and 230 nm was calculated to assess organic solvent contamination (Turashvili et al., 2012). Both spectrophotometer measurements constituted criteria for DNA quality assessment with higher values associated with better DNA quantity and purity. An A_{260}/A_{280} ratio of 1.8~2.0 is indicative of high purity (Pirondini et al., 2010).

2.5 Statistical analysis Statistical analyses were carried out using SPSS software, version 19.0 for windows (SPSS Inc., Chicago, IL, USA). Pearson's correlation coefficients and linear regression were used to investigate the relationship between milk SCC and DNA concentration extracted. Differences were considered significant when $P<0.01$.

3 Results

The SCC, concentrations and purities of the DNA extracted from the 25 buffalo milk samples, as determined using absorbance values at 230 nm (A_{230}), 260 nm (A_{260}), and 280 nm (A_{280}), are presented in Table A-13. Milk SCC of the samples was between 16,000 and 1,934,000 cells/mL. DNA yield obtained was within the range of 3747.2 ng/μL to 7080.7 ng/μL. The ratio of A_{260}/A_{280} and A_{260}/A_{230} of genomic DNA was ranged from 1.66 to 2.17 and 1.71 to 2.13, respectively.

The Pearson's correlation analysis showed a very strong positive correlation between the

DNA yield extracted directly from buffalo milk and the milk SCC ($r = 0.927$, $P < 0.001$). The linear regression equation was $y = 1.429 x + 4208.97$, where x and y represent milk SCC and DNA yield, respectively (Figure A-2).

Table A-13   The milk SCC, DNA yield and UV parameters

No.	SCC (10³)	DNA yield (ng/μL)	A_{260}	A_{280}	A_{260}/A_{280}	A_{260}/A_{230}
1	56	4052.4	81.05	44.03	1.84	1.98
2	160	4492.6	117.85	62.61	1.88	1.84
3	1859	6753.1	135.06	79.21	1.71	1.95
4	343	4613.8	88.28	46.54	1.9	2.03
5	16	4394.7	95.9	52.23	1.84	1.78
6	71	4279.1	57.58	26.48	2.17	2.13
7	38	4293.5	95.87	51.15	1.87	1.85
8	66	4428.9	96.58	50.86	1.9	1.92
9	219	4430.1	84.6	46.67	1.81	1.82
10	49	4245.4	84.91	45.88	1.85	1.87
11	24	4603.1	92.06	49.26	1.87	1.82
12	251	4720.5	102.41	53.37	1.92	1.86
13	55	4869.3	109.39	57.06	1.92	1.94
14	30	3747.2	69.94	39.6	1.77	1.71
15	44	4100.5	82.01	43.91	1.87	1.87
16	128	3877.8	67.56	37.83	1.79	1.94
17	123	4680.8	95.62	50.36	1.9	1.88
18	91	4299.2	73.98	40.58	1.82	1.94
19	273	4863.8	97.28	51.88	1.88	1.84
20	153	4553.5	93.07	50	1.86	1.91
21	121	4239.4	84.79	44.8	1.89	2.08
22	182	4987.5	109.75	57.02	1.92	1.91
23	133	3916.1	78.32	41.97	1.87	2.07
24	242	4220.9	60.42	29.76	2.03	2.1
25	1934	7080.7	145.61	87.68	1.66	1.72

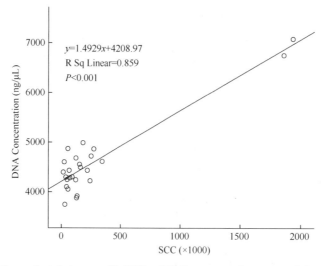

Figure A-2   Correlation analysis between milk SCC and DNA concentration extracted directly from buffalo milk

4 Discussion

The normal composition of milk somatic cells varies with the type of secretion or lactation cycle. The milk somatic cells include 75% leucocytes, i.e. neutrophils, macrophages, lymphocytes, erythrocytes, and 25% epithelial cells (Sharma et al., 2011). Milk sample is preferred to blood as a source of DNA due to milk collection is routinely performed and is less expensive and more easily accomplished than blood collection. Moreover, milk collection is also less stressful to animals, given that capture, handling, and venipuncture are not required for milk sampling. In recent years, several studies have developed methods to extract DNA directly from milk somatic cells instead of blood and use organic extraction, overnight incubation, or expensive commercial kits (Lopez-Calleja et al., 2004). A rapid simple salting-out method for DNA extraction from caprine milk was proposed but only 75% of tested samples were suitable as a substrate for PCR-RFLP genotyping (D'Angelo et al., 2007). In another study, a solid phase absorption commercial kit (Wizard DNA cleanup kit, Promega) was tested in ruminant's milk with reliable results (Lopez-Calleja et al., 2004). In most literatures, the methods extracted milk DNA were used low milk volumes and required overnight incubation of samples with proteinase K.

D'Angelo et al. (2007) extracted the DNA from goat milk sample by a simple salting-out method, the total DNA yield ranged from 2.12 to 610.12 μg per goat milk sample (40 mL of raw milk), and SCC for these samples ranged from 79,000 to 4,000,000 cells/mL. While, in this study, the method extracted DNA from milk samples was according to De et al. (2011) and Lopez-Calleja et al. (2004) with some modifications, the volume of each sample extracted for DNA is 1 mL Murrah buffalo milk, the final DNA concentrations range from 374.72 to 708.07 μg per sample, and SCC between 16,000 and 1,934,000 cells/mL.

The ratio of A_{260} and A_{280} is used to assess the purity of DNA. If the ratio is appreciably lower than 1.8, it may indicate the presence of protein, phenol or other contaminants that absorb strongly at or near 280 nm. DNA extractions with a A_{260}/A_{280} ratio of above 1.80 are deemed to be of high quality (Anonymous, 2008). A_{260}/A_{280} ratio values ranged from 1.66 to 2.17 in this study, with 84% of samples tested achieving a value of 1.8 or greater (Table A-12). The ratio of absorbance at 260 nm and 230 nm is used as a secondary measure of nucleic acid purity. The A_{260}/A_{230} values for "pure" nucleic acid are often higher than the respective A_{260}/A_{280} values (Anonymous, 2008). Values of 2 or more would be considered high quality samples, less than this indicates the presence of contaminants that absorb light at 230 nm, such as carbohydrates, guanidine thiocyanate, phenols and humic acids (Eland et al., 2012). In the samples, A_{260}/A_{230} ratios values ranged from 1.71 to 2.13, with 20% of samples tested achieving a value of 2 or greater (Table A-13). Therefore, the DNA extracted directly from Murrah Buffalo milk samples may be presence of contaminants, which need confirmed in the future.

A significant positive correlation ($P<0.001$) between milk SCC and DNA extracted from buffalo milk was demonstrated in the present study, which is consistent with the research by

D'Angelo et al. (2007) who reported that a positive correlation ($P<0.001$) was found between DNA yield and SCC in goats' milk. The result indicated that most of the DNA extracted from milk originated from SCC. However, further studies are required to confirm the composition of DNA extracted.

5 Conclusions

The result of the study indicated that DNA can be extracted directly from Murrah Buffalo milk, and a very strong positive correlation between the DNA yield extracted directly from buffalo milk and the milk SCC ($r = 0.927$, $P<0.001$). Milk sample was preferred to blood as a source of DNA, due to milk collection was routinely performed, as well as less expensive and more easily accomplished than blood collection.

Section 6 Detection of virulence-associated genes in *Staphylococcus aureus* isolated from bovine clinical mastitis milk samples in Guangxi

Staphylococcus aureus is recognized worldwide as a pathogen causing many serious diseases in humans and animals and is one of the most common etiological agents of clinical and subclinical bovine mastitis. The purpose of this section was to determine the presence of genes encoding clfA, fnbA, fnbB, cap5, cap8, hla, hlb, nuc, sea and tst of *S. aureus* strains (n=39) isolated from bovine clinical mastitis in Guangxi by polymerase chain reaction amplification. The results of the present study indicated that all isolates were found to contain one or more virulence-associated genes. The most frequently encountered genes were *fnbA* (97%) and *nuc* (90%), followed by *hla* (85%) and *hlb* (82%), respectively. None of the investigated *S. aureus* strains harbored *fnbB* and *sea* genes. The data in the present study showed a relatively wide distribution of the genes *fnbA* and *nuc* among the investigated isolates, indicating that they play an important role on bovine mastitis pathogenesis. The study provides a valuable insight into the virulence-associated genes of this important pathogen.

1 Introduction

Staphylococcus aureus is recognized worldwide as a pathogen causing many serious diseases in humans and animals and is the most common etiological agent of clinical and subclinical bovine mastitis, which is a considerably costly disease for dairy cows industry all over the world (Barkema et al., 2006; Ote et al., 2011; Rich, 2005).

S. aureus has a capacity to produce a large number of putative virulence factors (Fitzgerald et al., 2000; Kalorey et al., 2007), including surface-associated adhesins, a capsular polysaccharide, exoenzymes, and exotoxins. Some of these factors may be of more importance than others in different diseases or at different stages of the pathogenesis of particular infections, as not all factors are produced by each strain.

Presence of the *clfA* gene is considered as the *Staphylococcus* spp. virulence genes in the development and severity of mastitis (Akineden et al., 2001; Sharma et al., 2000). Fibronectin-binding protein (FnBP) A (FnBPA) and B (FnBPB), encoded by the *fnbA* and *fnbB* genes, respectively, play prominent roles in *S. aureus* attachment and colonization of host tissues or implanted biomaterials (Renzoni et al., 2004). FnBPs also promote endocytic uptake of *S. aureus* by epithelial and endothelial cell lines and fibroblasts (Fowler et al., 2000). The enhanced virulence of encapsulated bacteria is generally attributed to their resistance to phagocytosis and polymorphonuclear phagocytes, and capsular polysaccharide (CP) may also be involved in the adhesive capacity of some bacterial. Although 11 capsular polysaccharide types are described, only two types, types 5 and 8, are clinically relevant in that they are predominant among clinical infection isolates of varied geographic origin (Roghmann et al., 2005; Verdier et al., 2007). *S. aureus* produces an extracellular nuclease protein which is heat stable with activity retained after incubation at 97℃ for 60 min. The activity of the enzyme has been shown to be preserved for at least 10 months in various foods at temperatures of (37 ± 5)℃. Thermonuclease activity has been used for the rapid and direct detection of *S. aureus* in blood cultures and foods. Staphylococcal hemolysins are identified as important virulence factors that contribute to bacterial invasion and escape from the host immune response (Da Silva et al., 2005). Alpha-hemolysin is the most studied and characterized *S. aureus* cytotoxin and is considered a main pathogenicity factor because of its hemolytic, dermonecrotic, and neurotoxic effects (Dinges et al., 2000). Additionally, beta-hemolysin is a sphingomyelinase that is highly active against bovine erythrocytes (Larsen et al., 2002). Enterotoxigenic S. aureus is one of the major pathogens causing food poisoning cases world-wide (Dinges et al., 2000); in addition, the presence of the thermonuclease gene (*nuc*) showed strong correlation with enterotoxin production and it is a marker of food contamination with enterotoxigenic *S. aureus* (Tamarapu et al., 2001). Enterotoxins and toxic shock syndrome toxin 1 are responsible for foodborne illnesses and toxic shock syndrome in humans, respectively. Both toxins are included in the pyrogenic toxin superantigen family (Gunaydin et al., 2011).

The importance of evaluating the combination of *S. aureus* virulence-associated factors has been emphasized both in human and veterinary medicine (Peacock et al., 2002; von Eiff et al., 2004; Zecconi et al., 2006), and knowledge about the genetic variability within different *S. aureus* populations would help in the design of efficient treatments. *S. aureus* (47%) was the major pathogen responsible for infected milk samples in Guangxi (Yang et al., 2011a). Thus, the present study was to determine virulence-associated genes in *S. aureus* strains isolated from clinical bovine mastitis cases in Guangxi by specific polymerase chain reaction (PCR) amplification.

2 Materials and methods

2.1 Bacterial isolates A total of 39 *S. aureus* isolates were used in this study. All the pathogens were isolated from 110 bovine milk samples for clinical mastitis. These milk samples were collected from different dairy cows (2009 to 2010) in Guangxi, China. Milk samples were examined following the standard procedures (Getahun et al., 2008; Yang et al., 2011b). Briefly, these isolates were cultured on Columbia sheep blood agar medium (Beijing Land Bridge Technology Co., Ltd., Beijing, China), containing 5% of sheep blood and 0.1% of esculin, and incubated at 37℃ for 24 h. *S. aureus* strains were identified according to conventional methods, including Gram staining, colony morphology, hemolysis, and tests for catalase, clumping factor, tube coagulase, DNAse, acetoin, and anaerobic fermentation of mannitol (Winn et al., 2006). The isolates were additionally investigated by PCR amplification of species-specific parts of the gene encoding the 23S rRNA with the oligonucleotide primers shown in Table A-14. All the isolates were stored at −20℃ in trypticase soy broth (TSB; tryptone, 15 g; Soytone, 5 g; NaCl, 5 g; ddH$_2$O, 1000 mL; pH 7.2) (Zhang et al. 2009) containing 10% (*V/V*) of glycerol until further use.

Table A-14 Target genes, primer sequences, expected sizes of the products, and PCR conditions

Gene	Primer sequence (5'-3')	Product size (bp)	Temperature (℃)	Reference
23S rRNA	ACG GAG TTA CAA AGG ACG AC AGC TCA GCC TTA ACG AGT AC	1250	57	Yang et al. (2012)
clfA	GGC AAC GAA TCA AGC TAA TAC AC TTG TAC TAC CTA TGC CAG TTG TC	719	57	Yang et al. (2009)
fnbA	GCG GAG ATC AAA GAC AA CCA TCT ATA GCT GTG TGG	1279	51	Yang et al. (2012)
fnbB	GGA GAA GGA ATT AAG GCG GCC GTC GCC TTG AGC GT	812	54	Yang et al. (2012)
cap5	ATG ACG ATG AGG ATA GCG CTC GGA TAA CAC CTG TTG C	881	53	Moore and Lindsay (2001)
cap8	ATG ACG ATG AGG ATA GCG CAC CTA ACA TAA GGC AAG	1148	53	Moore and Lindsay (2001)
hla	GGT TTA GCC TGG CCT TC CAT CAC GAA CTC GTT CG	534	55	Salasia et al. (2004)
hlb	GCC AAA GCC GAA TCT AAG GCG ATA TAC ATC CCA TGG C	833	53	Salasia et al. (2004)
nuc	GCG ATT GAT GGT GAT ACG GTT ACG CAA GCC TTG ACG AAC TAA AGC	279	56	Yang et al. (2012)
sea	AAA GTC CCG ATC AAT TTA TGG CTA GTA ATT AAC CGA AGG TTC TGT AGA	216	55	Yang et al. (2012)
tst	ATG GCA GCA TCA GCT TGA TA TTT CCA ATA ACC ACC CGT TT	350	54	Yang et al. (2012)

2.2 DNA extraction DNA extraction was carried out using the bacteria genomic DNA purification kit (Tiangen BioTech, Beijing) according to the manufacturer's instructions for staphylococci. DNA from 2 mL of a single colony overnight Luria-Bertani broth culture was purified and stored at −20℃ until further use.

2.3 PCR amplification The target genes, primer sequence, expected sizes of PCR products (bp), and specific temperature (℃) are reported in Table A-14. All the oligonucleotide primers were synthesized by Beijing Sunbiotech Co., Ltd. The PCR reactions were performed in a final volume of 20 μL reaction mixture consisted of 20 ng genomic DNA, 20 pmol of each primer, 10 μL 2×Taq PCR MasterMix (Tiangen Biotech; 0.1 U Taq polymerase/μL, 0.5 mmol/L dNTP each, 20 mmol/L Tris-HCl (pH 8.3), 100 mmol/L KCl, 3 mmol/L $MgCl_2$). The cycling conditions were the following: an initial denaturation at 94℃ for 5 min; 30 cycles of 94℃ for 30 s, specific temperature for 30 s (Table A-14), and 72℃ for 30 to 90 s depending on the PCR product length; and a final extension at 72℃ for 10 min. All PCR products (5 μL) were analyzed by electrophoresis in a Tris-acetate-EDTA, 2% (m/V) agarose gel. Gels were then stained with ethidium bromide (0.5 μg/mL) and photographed under UV light. A 100-bp DNA ladder (Tiangen BioTech, Beijing) was used as a marker.

3 Results

According to the result of culture properties and amplification of the gene encoding the 23S rRNA (1250 bp), all 39 isolates used in the present study were identified as *S. aureus*. Among the 39 *S. aureus* cultures investigated, 3 (7.7%) isolates had one single toxin gene, 3 (7.7%) isolates had two toxin genes, 2 (5.1%) isolates had four toxin genes, 8 (20.5%) isolates had five toxin genes, and remaining 23 (59%) isolates had six different toxin genes. Interestingly, all of the isolates were found to be positive for one or more virulence-associated genes (Table A-15). The relative frequency of the 10 genes whose presence was investigated by PCR among the 39 isolates of the collection is reported in Table A-16 as percentage of positive isolates. The *fnbA* gene was the most widespread gene among the target genes (97%), followed by *nuc* (90%), *hla* (85%) and *hlb* (82%). None of the investigated isolates were carried any of *fnbB* and *sea* genes.

Table A-15 Distribution of target genes among *S. aureus* strains isolated from bovine clinic mastitis

Genotypes	Number of isolates (%)
clfA, nuc, fnbA, hla, hlb, cap5	17 (43.6)
clfA, nuc, fnbA, hla, hlb, cap8	6 (15.4)
clfA, nuc, fnbA, hla, hlb	1 (2.6)
fnbA	2 (5.1)
fnbA, hla, hlb, cap5, cap8	2 (5.1)
fnbB	—
nuc	1 (2.6)
nuc, fnbA	3 (7.7)
nuc, fnbA, hla, hlb	2 (5.1)
nuc, fnbA, hla, hlb, cap5	4 (10.3)
sea	—
tst, nuc, fnbA, hla, cap8	1 (2.6)

Table A-16　Relative frequency (%) of the target genes

Target gene	No. of isolates (%)
clfA	24 (62)
fnbA	38 (97)
fnbB	0 (0)
cap5	23 (59)
cap8	9 (23)
hla	33 (85)
hlb	32 (82)
nuc	35 (90)
sea	0 (0)
tst	1 (3)

4　Discussion

S. aureus is the bacteria most frequently isolated from bovine mastitis and thus considered one of the most important etiological agents for this infection. The identification of the 39 S. aureus isolates of the present study could be performed by conventional methods and by PCR technology. The latter uses oligonucleotide primers targeted to species-specific parts of the gene encoding the 23S rRNA.

Comparable PCR-based systems for the identification of S. aureus isolates from various origins have been used by numerous authors (Akineden et al., 2001; Momtaz et al., 2010; Nashev et al., 2004). The target gene allowed a rapid identification of this species with high sensitivity and specificity. As was found by Straub et al., the amplification of the gene encoding an S. aureus-specific part of the 23S rRNA revealed an amplicon with a size of 1250 bp for all S. aureus isolates investigated.

The findings of this study indicated that all of the S. aureus isolates harbored one or more virulence-associated genes in dairy herds of cows suffering from clinical mastitis, and it was higher than that reported from other areas in China. Recently, Wang et al. (2009) investigated the distribution of superantigenic toxin genes in S. aureus isolates from bovine subclinical mastitis in InnerMongolia and Shanghai regions of China, and they found that 65% of the isolates possessed at least one toxin gene. In the present study, none of 39 strains contained fnbB and sea genes. The fnbB gene had a certain positive rate for S. aureus isolated from stable nasal carriers (36.7%) (Nashev et al., 2004). The sea gene was the most frequently encountered from food poisoning cases in humans and subclinical mastitis cases in bovine (Balaban and Rasooly, 2000; Wang et al., 2009). Presence of fnbA, hla, and hlb genes were frequently reported in S. aureus strains from bovine mastitis (Momtaz et al., 2010; Ote et al., 2011).

In agreement with previous studies, we observed that vast majority of the isolates harbored the clfA, fnbA, hla, hlb, and nuc genes and nearly no sea and tst genes (Gunaydin et al., 2011;

Kalorey et al., 2007; Momtaz et al., 2010; Ote et al., 2011); however, the *fnbB* gene was not present, comparing with other studies (36.7%), in which the pathogenic *S. aureus* strains were isolated from stable nasal carriers (Nashev et al., 2004).

The ability of *S. aureus* to adhere to extracellular matrix proteins was thought to be essential for the colonization and the establishment of infections (El-Sayed et al., 2005). *S. aureus* possesses various adhesion genes, including *clfA* and *fnbA*. Presence of the *clfA* gene was considered as the *Staphylococcus* spp. virulence genes in the development and severity of mastitis (Akineden et al., 2001). Momtaz et al. (2010) indicated their existence of a statistically significant relationship between *clfA* gene and the developed mastitis. This study showed that *fnbA* (97%) is the most frequent gene isolated from the pathogenic *S. aureus* strains, which is consistent with the reports by Nashev et al. (2004) and Ote et al. (2011).

The predominant capsule polysaccharide types were reported to be the 5 and 8, which are expressed by about 80% bovine *S. aureus* clinical isolates (Ote et al. 2011), although the serotype prevalence can vary according to geographical regions of the world (Salasia et al. 2004). Similarly, we observed that 23 of the isolates of this collection carried the *cap8* gene, while the majority carried the *cap5* gene (59%). Nine isolates (23%) were non-*cap5* or non-*cap8*. However, Ote et al. (2011) observed that one third of *S. aureus* harbored the *cap5* gene, and the majority carried the *cap8* gene. Our results were in agreement with the findings of Guidry et al. (1998) and Reinoso et al. (2008), suggesting that capsular types 5 and 8 are the most prevalent capsular types in bovine *S. aureus*. Tuchscherr et al. (2005) considered that lack of capsular expression was suggested to be associated with persistent infection.

About 85% of the total strains harbored hla and/or hlb gene, and this finding was in agreement with Da Silva et al. (2005). According to Cifrian et al., interaction between α- and β-hemolysin increased both the adherence to bovine mammary epithelial cells and the proliferation of *S. aureus*. In combination with the earlier finding of the β-hemolysin gene in 96% of bovine *S. aureus* isolates in contrast to 56% of the examined human isolates, this fact suggested that these hemolysins may be an active factor in the pathogenesis of bovine mastitis. The remaining *S. aureus* isolates that were negative for *hla* and *hlb* may either have lost *hla* and *hlb* genes or the ability to express it or carried a variant gene and failed to express hemolytic activity *in vitro* as demonstrated by Aarestrup et al.

Some *S. aureus* strains would produce an extracellular thermostable nuclease (thermonuclease), which is an endonuclease, degrading both DNA and RNA, and the enzymatic activity can resist 100°C for at least 1 h. The results of this study showed that 90% of the *S. aureus* strains contained the *nuc* gene. It was consistent with other studies (Cremonesi et al., 2005; Kalorey et al., 2007).

In agreement with other studies, we observed that the *tst* gene was only found in one isolate (3%) overall (Gunaydin et al., 2011; Momtaz et al., 2010; Nashev et al., 2004). However, these results are in disagreement with previous studies (Ote et al., 2011; Peacock et al., 2002); Peacock et al. (2002) presented data that 44% of investigated carriage isolates harbored the gene *tst*. Chiang et al. (2008) reported higher tst gene positivity among *S. aureus* isolates from food poisoning outbreaks in Taiwan.

In conclusion, a PCR assay was used to investigate the presence of virulence-associated genes in *S. aureus* isolates from clinical bovine mastitis cases for the first time in Guangxi. Our results demonstrated that all of the tested *S. aureus* isolates harbored one or more virulence-associated genes of dairy cows suffering from clinical mastitis in Guangxi. The *fnbA*, *hla*, *hlb* and *nuc* genes were the most frequent genes isolated from the pathogenic *S. aureus* strains, and none of 39 strains contained *fnbB* and *sea* genes. The study provided a valuable insight into the virulence-associated genes of this important pathogen.

Section 7 Malonaldehyde level and some enzymatic activities in subclinical mastitis milk

The purpose of this section was to evaluate the changes occurring in milk malondialdehyde (MDA) level and some enzymatic activities as a result of subclinical mastitis (SCM) in dairy cows. A total of 124 milk samples were collected from 124 lactating cows from the same herd in the period between the 2nd week after calving and the 10th week postpartum. They were classified by bacterial culture and the California mastitis test (CMT) as positive were deemed to have glands with SCM, and the periodic incidence rate of SCM was 26.6%. The most common bacterial isolates from SCM cases were Staphylococcus aureus (47%) and coagulase negative Staphylococci (CNS) (27%). The mean level of MDA and activities of lactate dehydrogenase (LDH) and alkaline phosphatase (ALP) were significantly higher in SCM milk than in normal milk, while the mean activity of glutathione peroxidase (GPx) was significantly lower in SCM milk than in normal milk. There were no differences in the activities of superoxide dismutase (SOD) and aspartate aminotransferase (AST) between normal milk and SCM milk. Therefore, the measurement of milk MDA level and GPx, LDH and ALP activities, appears to be a suitable diagnostic method for identifying SCM in dairy cows.

1 Introduction

Mastitis, an inflammatory reaction of the mammary gland is the most dreaded disease of dairy farmers because of reduced milk production, increased treatment costs, labour, milk discarding following treatment, death and premature culling. It is very important to determine efficient techniques that are able to identify the presence of mastitis early in the disease syndrome.

A positive diagnosis of mastitis should fulfill two criteria: a positive bacteriological test and an inflammatory change. Cell counts have generally been used for the latter purpose (Babaei et al., 2007). The quantification of cells in milk or somatic cell count (SCC), is estimated using direct microscopic analysis or by an indirect method of estimating SCC using the California mastitis test (CMT) (Babaei et al., 2007). In milk-testing laboratories, the most commonly used method of enumerating somatic cell count is the fluoro-optical electronic or fossomatic counter method

(Sierra et al., 2006). However, fossomatic counting is considered as a costly and sophisticated equipment which is not available everywhere. The CMT is already used on-farm to diagnose indirectly subclinical mastitis (SCM) for lactating cows. The sensitivity and specificity of the CMT reported in the literature is variable (Pyörlä, 2003).

Malondialdehyde (MDA), the lipid peroxidation end product is one of the most reliable and widely used indexes of oxidative stress. Milk with higher somatic cell count has been shown to have more infiltrated polymorphonuclear cells, and this caused an increase of oxidative reactions (Su et al., 2002). In cow's milk, MDA levels were measured to evaluate the peroxidation status when milk was kept under different circumstances (Cesa, 2004; Miranda et al., 2004), and milk SCC is positively associated with malondialdehyde level in milk (Suriyasathaporn et al., 2006). Glutathione peroxidase (GPx) has no known enzymatic function in milk, in which it binds 30% of the total selenium (Se), an important trace element in the diet. The level of GPx in milk varies with the species (human>caprine>bovine) and diet (Fox and Kelly, 2006). In cows with mastitis, serum lipid peroxidation levels were increased, and the level of blood glutathione peroxidase was decreased when compared to the levels in healthy cows. Several workers have observed superoxide dismutase (SOD) in bovine milk and suggested that the enzyme may play an important role in the oxidative stability of milk. However, no study has determined the malondialdehyde level and glutathione peroxidase and superoxide dismutase activities in milk in relation to subclinical mastitis. The activities of milk lactate dehydrogenase (LDH) and alkaline phosphatase (ALP) changes have been used as an indicator of SCM in dairy cows (Babaei et al., 2007).

The present research seeks: (i) To investigate the incidence and etiology of SCM in a commercial subtropical dairy farm and (ii) to study the changes occurring in the level of MDA and activities of GPx, SOD, LDH, ALP and aspartate aminotransferase (AST) in cow's milk as a result of subclinical mastitis.

2 Materials and methods

2.1 Animals   A total of 124 milk samples were collected from 124 multiparous lactating dairy (Holstein) cows at the same time, and they were all in the same herd of a subtropical commercial farm in Nanning, China. Every milk sample was collected from four quarters of each cow randomly. All milk samples were collected in the period between the 2nd week after calving and the 10th week postpartum. Cows about 4 to 5 years old selected for this study did not show clinical signs of mastitis or other illnesses.

2.2 Collection of samples   After a quarter had been washed with tap water and dried, the teat end was swabbed with cotton wool soaked in 70% ethyl alcohol. Milk samples were collected prior to the morning milking in sterile test tubes (10 mL) after discarding the first three squirts of milk, and then placed in an icebox and transported to the laboratory for examination within two hours after collection.

2.3 California mastitis test (CMT)　　The California mastitis test was used on all milk samples, using the method by Schalm et al. The CMT score is based on the number of leukocytes in milk. The reaction involved in the CMT is the disintegration of leukocytes when milk is mixed with the reagent (Babaei et al., 2007). According to the visible reactions, the results were classified in four scores: 0 = negative or trace, 1 = weak positive, 2 = distinct positive and 3 = strong positive.

2.4 Bacteriological examination of milk samples　　Milk samples were examined following standard procedures (Batavani et al., 2003; Yang et al., 2011).

2.5 Definition of subclinical mastitis　　Mammary glands without clinical abnormalities and with apparently normal milk that were bacteriologically negative and negative on the California mastitis test were considered to be normal milk, while those that were bacteriologically positive and with the CMT positive were considered to have subclinical mastitis.

2.6 Measurement of milk MDA level and enzyme activity　　Normal and subclinical mastitic milk samples were skimmed by centrifugation at 10,000 g for 20 min at 4℃. Defatted milk was used for MDA level and enzyme activity estimations. Milk MDA level and enzyme activity were measured by spectrophotometric techniques. Malondialdehyde level was determined by the thiobarbituric acid (TBA) method, which was modified from methods of Satoh and Yagi. Total (Cu-Zn and Mn) superoxide dismutase (EC 1.15.1.1) activity was measured according to the method described by Sun et al. and glutathione peroxidase (EC 1.11.1.9) using the method described by Paglia and Valentine and Goldberg and Spooner. Lactate dehydrogenase (LDH) and alkaline phosphatase (ALP) activity were determined by the method of Bergmeyer and Goldberg and Spooner, respectively, while aspartate aminotransferase (AST) was carried out by the method of Reitman and Frankel.

2.7 Statistical analysis　　The sensitivity and specificity of the CMT results were calculated by using standard 2-by-2 contingency tables. A 95% confidence interval was calculated for the sensitivity and specificity of the CMT. Data for milk MDA level and enzymatic activities were expressed as mean±standard error of mean (S.E.M) and Student's t-test was used to evaluate differences between subclinical mastitis and healthy milk. Differences with $P<0.01$ were considered to be significant.

3　Results

A total of 124 milk samples were collected from lactating glands in the period between the second week after calving and the tenth week postpartum. Positive CMT were recorded from 65 (52.4%) glands. Bacteria were isolated from 45 (36.3%) milk samples. According to the definition of subclinical mastitis described earlier, the incidence of SCM was 26.6% (n = 33). The specificity, sensitivity, positive predictive value (PPV) and negative predictive value (NPV) of California mastitis test in detecting subclinical mastitis were 59.49, 73.33, 50.8 and 79.7%, respectively (Table A-17). The Kappa value (κ = 0.3) demonstrated weak (poor) agreement between the CMT results and culture test.

Table A-17　Two-way contingency table to investigate agreement between bacteriological and the CMT results for 124 milk samples

	Culture +	Culture −	Total
CMT +	33	32	65
CMT −	12	47	59
Total	45	79	124

Note: κ Value: 0.3; specificity: 59.49%; sensitivity: 73.33%; proportion positive by the CMT: 52.4%; proportion positive by culture: 36.3%; positive predictive value (PPV) : 50.8%; negative predictive value (NPV) : 79.7%

Distributions of microbial isolates responsible for infected milk samples were: *S. aureus* (47%), coagulasenegative staphylococci (CNS) (27%), *Escherichia coli* (9%), *Streptococcus agalactiae* (9%), *Streptococcus uberis* (4%) and *Cryptococcus neoformans* (4%) (Table A-18).

Table A-18　Frequency distribution of microorganisms isolates from milk samples positive by culture

Microorganism	Frequency (%)
S. aureus	21 (47)
CNS	12 (27)
E. coli	4 (9)
St. agalactiae	4 (9)
St. uberis	2 (4)
C. neoformans	2 (4)
Total	45 (100)

The level of malondialdehyde (MDA) and activities of glutathione peroxidase (GPx), superoxide dismutase (SOD), lactate dehydrogenase (LDH), alkaline phosphatase (ALP) and aspartate aminotransferase (AST) in normal and subclinical mastitis milk samples are presented in Table A-19. The mean level of MDA and activities of LDH and ALP were significantly higher in SCM milk than in normal milk ($P<0.01$), while, the mean activity of GPx was significantly lower in SCM milk than in normal milk ($P<0.01$). There were no differences in the activities of SOD and AST between normal milk and SCM milk ($P>0.05$).

Table A-19　Mean level of MDA and activities of GPx, SOD, LDH, ALP and AST in normal and subclinical mastitis milk samples of lactation cows

	Normal milk (47)	Subclinical mastitis milk (33)
MDA (nmol/mL)	24.37 ± 0.9^A	28.45 ± 0.96^B
GPx (IU/mL)	32.81 ± 1.41^A	26.41 ± 2.03^B
SOD (IU/mL)	1.82 ± 0.11	1.6 ± 0.12
LDH (IU/L)	177.94 ± 12.55^A	724.49 ± 34.91^B
ALP (IU/L)	71.85 ± 3.71^A	116.58 ± 6.18^B
AST (IU/L)	151.99 ± 11.56	140.98 ± 11.76

Note: Different superscript capital letters within the same row means $P<0.01$

4 Discussion

According to the results of this study, the incidence of subclinical mastitis (SCM) in dairy cows in subtropical China is in agreement with the report by Getahun et al. (2008), but it seems to be higher than that reported by Mungube et al. (2004) and Gianneechini et al. (2002). A direct comparison between these earlier results is difficult, because many factors could affect the incidence of SCM in dairy cows, especially management, case definition and the diagnostic criteria used.

The California mastitis test (CMT) has been standardized for cow's milk and only reacts with liberated nuclear DNA (Batavani et al., 2003), while bacteriological culture of milk is a definition of intramammary infections (IMI) in dairy cows. The CMT showed a higher incidence of SCM than bacteriological culture (52.4% versus 36.3%). The sensitivity and specificity of the CMT reported in the literature is variable (Pyörlä, 2003). Sargeant et al. (2001) found that sensitivity and specificity to identify any IMI at a quarter level at 3 days in milk were 57% and 56%, respectively. On the other hand, Vijaya Reddy et al. reported a sensitivity of 71% and a specificity of 75%. According to our result, the specificity (59.49%) and sensitivity (73.33%) of the CMT in detecting subclinical mastitis was quite high. However, the California mastitis test was not a very good tool to correctly diagnose subclinical mastitis quarters by any pathogen in subtropical dairy farms. Low specificity and a low-to-moderate prevalence yielded low positive predictive values (PPV). False positives, associated with a low specificity, would occur when somatic cells are present in the milk with bacteria not being isolated. False negative reactions, associated with a low sensitivity, would occur when bacteria are indeed present in the gland but somatic cells are not.

The common isolate from subclinical cases were *S. aureus* (47%) and coagulase negative staphylococci (27%). These organisms have been considered to be the major cause of non-clinical IMI in a number of previous investigations (Getahun et al., 2008). The high prevalence of S. aureus is mainly attributed to the wide distribution of microorganism inside the mammary gland and on the skin of teat and udder (Workineh et al., 2002). Classically, coagulase negative staphylococci (CNS) were classified as minor pathogens and their importance as an independent cause of subclinical or clinical mastitis was judged to be limited. The impact of CNS IMI on cow level SCC was intermediate when compared to culture negative animals and cows infected with major pathogens (Schukken et al., 2009). In view of this, the sensitivity of the CMT was lower by the high prevalence of coagulase negative staphylococci IMI. False negative reactions occur when bacteria are indeed present in the gland but somatic cells are not.

Several studies have evaluated milk LDH, ALP and AST activities changes to diagnose udder infections in dairy cows (Babaei et al., 2007), but little information was available in relation to changes in milk malondialdehyde level and GPx and SOD activities for cows SCM. IMI increases the permeability of microcirculatory vessels by secretion of various chemical mediators such as histamine, prostaglandin, kinins and free oxygen radicals from inflammatory cells.

Malondialdehyde (MDA) is the final product of lipid peroxidation and therefore is used as index of this process. In our study, the mean level of MDA was significantly higher in SCM milk at 28.45 ± 0.96 nmol/mL when compared to 24.37 ± 0.9 nmol/mL in normal milk ($P<0.01$), and MDA would be considered as an indicator of subclinical mastitis udders. The higher malondialdehyde levels in subclinical mastitis milk reported in this study demonstrated that the auto-oxidative activity of SCM milk is higher than normal milk. Similar results have been reported by Suriyasathaporn et al. (2006). Milk from mastitic udders is of low quality and less suitable for consumption. In addition, the increased malondialdehyde due to high somatic cell count further reduces milk quality (Suriyasathaporn et al., 2006). In fact, malondialdehyde is known to be a mutagen and a suspected carcinogen; it can react with DNA to generate mutations.

Glutathione peroxidase (GPx) has been established as a selenium-containing enzyme catalyzing the reduction of various peroxides and protecting the cell against oxidative damage, and sufficient GPx could protect milk lipids from oxidation. Glutathione peroxidase activity has been detected in bovine milk at levels between 12 and 32 IU/m, and its activity correlated significantly with selenium concentration (Przybylska et al., 2007). In cows with mastitis, serum lipid peroxidation levels were increased, and the level of blood glutathione peroxidase was decreased when compared to the levels in healthy cows. In this study, the mean activity of GPx was significantly higher in normal milk at 32.8 ± 1.41 IU/mL when compared to 26.41 ± 2.03 IU/mL in subclinical mastitis milk ($P<0.01$). Thus, normal milk with higher glutathione peroxidase activity may be better as a source of milk products than subclinical mastitic milk with lower GPx activity, because glutathione peroxidase exhibits a cellular defense function through decomposition of hydroperoxides. Atroshi et al. found that the decline in GPx in mastitis cows may be related to the changes in lipid peroxidation and prostaglandin formation. Human milk glutathione peroxidase activity have been shown to decrease within the time of lactation, but there is no published literatures pertaining to bovine milk.

Superoxide dismutase (SOD) was reported first in milk by Hicks et al., and they suggested that the enzyme may play an important role in the oxidative stability of milk. The average activity of milk SOD was 1.82 ± 0.11 IU/mL with a range of 0.67 to 3.23 IU/mL in normal udders, while in subclinical mastitis udders, it was 1.6 ± 0.12 IU/mL with a range of 0.39 to 3.02 IU/mL. The activity of milk SOD varies between cows and breeds (Lindmark-Mansson and Akesson, 2000). In cow's milk, it is not affected by stage of lactation or age of cow, and high SCC (Przybylska et al., 2007). In present study, we also demonstrated that the pattern of distribution of superoxide dismutase in the milk of normal and subclinical mastitis udder showed no significant difference.

Changes in enzyme activities in blood or other biological fluids such as milk can be a consequence of cell structural damage. Our findings showed that the mean activities of lactate dehydrogenase (LDH) and alkaline phhosphatase (ALP) were significantly higher in subclinical mastitis milk than in normal milk ($P<0.01$), but there was no significant difference in the activity of aspartate aminotransferase ($P>0.05$). Similar results have been reported previously (Batavani et al., 2003; Babaei et al., 2007). LDH activity in milk has been considered as a sensitive indicator of changing mammary gland function due to disease and the ALP activities test was reliable in the early diagnosis of subclinical mastitis (Babaei et al., 2007).

In conclusion, the present study indicated that: (1) CMT has a certain value for diagnosing subclinical mastitis in subtropical dairy cows; (2) S. aureus and coagulase negative staphylococci are predominant causes of SCM in the commercial dairy farm and; (3) the measurement of milk MDA level and GPx, LDH and ALP activities appear to be suitable diagnostic methods for identifying subclinical mastitis in dairy cows.

Section 8 Clinical mastitis from calving to next conception negatively affected reproductive performance of dairy cows in Nanning, China

The purpose of this section was to evaluate the effect of clinical mastitis between calving and next conception on reproductive performance in Chinese Holstein cows. Six hundred and three multiparous Holstein cows from a commercial dairy farm were divided into three groups retrospectively: cows with first clinical mastitis before first artificial insemination (AI) (MG1; $n=113$), cows with first clinical mastitis between first AI and pregnancy diagnosis (MG2; $n=36$), and cows with out any clinical diseases (CG; $n=454$). Clinical cases of mastitis were identified at every milking by the trained milkers or the herd manager based on abnormal milk or signs of inflammation of the mammary gland. Number of days from calving to first AI and days from calving to conception, number of AI per conception, and the conception rate at first AI were evaluated in each group. The number of days to first AI was significantly greater for cows in MG1 than MG2 and CG ($P<0.01$). The number of days to conception was similar for cows in MG1 and MG2 ($P>0.05$), but they were all greater than cows in CG ($P<0.01$). The number of services per conception was significantly greater for cows in MG1 and MG2 than CG ($P<0.01$), and cows in MG1 had fewer number of services per conception compared to MG2 ($P<0.05$). Conception rate at first service was similar for cows in MG1 and MG2 ($P>0.05$), however, conception rate for those groups were both lower than for CG ($P<0.01$). In conclusion, clinical mastitis during early lactation markedly negatively influenced reproductive performance of dairy cows. Therefore, reduction of clinical mastitis in early lactation should also improve reproductive performance of dairy cows. Further study is needed to better understand the mechanisms of how clinical mastitis affects reproductive performance in dairy cows should lead to better strategies to avoid such negative effects.

1 Introduction

Mastitis, an inflammatory reaction of the mammary gland is the most dreaded disease of dairy farmers because of reduced milk production, increased treatment costs, labour, milk discarding following treatment, death and premature culling (Yang et al., 2011). Besides, mastitis has been recently reported to have a detrimental effect on reproductive performance in lactating dairy cows (Hertl et al., 2010; Nava-Trujillo et al., 2010; Santos et al., 2004; Schrick et al., 2001).

The first published report suggested that clinical mastitis were caused by Gram negative bacteria, Escherichia coli, could alter the interestrous interval. Cows with clinical mastitis due to Gram negative bacteria were 1.6 times more likely to have an altered interestrous interval compared to herdmates without clinical mastitis. Barker et al. indicated that cows that developed clinical mastitis during early lactation could have a markedly negative impact on the reproductive performance of dairy cows, such as the number of days to first artificial insemination (AI), artificial inseminations per conception and the number of days to conception. Moreover, both Gram negative and Gram positive pathogens may act through similar mechanisms to increase inflammatory mediators, leading to reproductive failure during early lactation. Schrick et al. (2001) reported that subclinical mastitis reduced reproductive performance of lactating cows similar to clinical mastitis. Huszenicza et al. (2005) indicated that clinical mastitis could affect the resumption of ovarian activity in postpartum dairy cows. And Nava-Trujillo et al. (2010) suggested that clinical mastitis before first service would increase the days to first service and the days to conception in dual-purpose cows. However, Peake et al. (2011) found that cows with neither subclinical nor clinical mastitis in the first two months of lactation had a negative effect on fertility parameters.

Reproductive losses in lactating dairy cows have increased in recent years (Lucy, 2001), and those losses seem to be multifactorial. Efficient reproductive performance is essential for the maintenance of consistently high level of milk production. However, the negative effect of clinical mastitis from calving to next conception on reproductive performance in Chinese Holstein cows is in general unknown. Therefore, the purpose of this study is to determine the effects of clinical mastitis, occurring from calving to next conception, on reproductive performance in Chinese Holstein cows, under subtropical conditions.

2 Materials and methods

2.1 Animals and management　　Six hundred and three multiparous Holstein dairy cows from a commercial dairy farm in Nanning were used in this study. All lactating cows in the farm were milked twice daily in a double-14 fishbone milking parlor equipped with automatic milking machine take-offs (Westfalia-Surge, Naperville, IL), with herd average 305-d milk yield between 4000 kg/cow and 7200 kg/cow. Cows stood on an elevated platform in a fishbone fashion facing away from the operator area, which exposed enough of the back half of the cow to allow access to milk her from the side. Teat dipping was routinely performed before and after each milking. Milking machines were backflushed (Surge Backflush II, Westfalia- Surge) after removal from cows. Milking equipment was evaluated routinely and maintained per the recommendation of the manufacturer. Cows were housed in free-stall barns and fed via a feed alley, which were bedded in stalls with straw on rubber mats, and scrapers automatically cleaned the concrete floor 3 times daily. Within each site, all lactating cows were fed with the same diet as a total mixed ration that was formulated to meet or exceed the nutrient requirements for a lactating Holstein cow weighing 650 kg and producing 45 kg of 3.5% fat-corrected milk (NRC, 2001).

All cows were dried off approximately 7 weeks to 9 weeks before expected calving, and all quarters of cows were infused with an antibiotic preparation approved for use in non-lactating cows following the last milking of lactation. And subsequently, non-lactating cows were moved into a separate pen.

Data were collected from cows that calved between September 2009 and January 2010. Retrospectively, cows were divided into three groups according to time in lactation when the first clinical mastitis case was identified or absence of clinical mastitis: first clinical mastitis prior to first postpartum artificial insemination (AI) (MG1; $n = 113$); first clinical mastitis between first postpartum AI and pregnancy diagnosis (MG2; $n = 36$); cows that were either not diagnosed with mastitis or were diagnosed after confirmed pregnancy (CG; $n = 454$).

2.2 Diagnosis of clinical mastitis　　At each milking, all cows were examined for symptoms of clinical mastitis by trained milkers or by the herd manager immediately before milking. Clinical mastitis cases were characterized by the presence of abnormal milk or by signs of inflammation in one or more quarters, and were treated by intramammary infusion of antibiotics according to treatment protocols established by the herd veterinarian. Periodically milk samples from the identified quarter were obtained by the herd health veterinarian and cultured for microbiological status. Treated cows were moved to the hospital pen and the incidence, infected quarter(s), and treatment type were recorded by using computer software records (Microsoft Excel 2003).

2.3 Reproductive management　　Cows were observed for estrus for 30 min at least three times daily. In addition, milkers observed cows at milking time, and all farm personnel regularly participated in estrus detection throughout the day. Cows in estrus were confined to tie stalls to minimize the chance of injury. As calving following, cows were generally subjected to a voluntary waiting period of 40 days before first AI. Pregnancy diagnosis was performed by herd veterinarians via palpating per rectum the uterus and its contents approximately 50 to 65 days after insemination. The farm maintained computerized and paper records of the reproductive performance of each cow, including the number of days to first AI and days from calving to conception, the number of AI per conception, and the conception rate at first AI.

2.4 Statistical analyses　　The statistical analyses were performed using SAS/STAT software [Statistical Analyses Systems SAS system for Windows, Version 9.1.3 (SAS Institute Inc. SAS Institute Inc.)]. Binomially distributed data such as the conception rate at first AI were analyzed by logistic regression by the LOGISTIC procedure of SAS [SAS system for Windows, Version 9.1.3 (SAS Institute Inc. SAS Institute Inc.)]. The number of days to first AI, the number of AI per conception, and the number of days from calving to conception were analyzed according to General Linear Models (GLM) procedure of SAS [SAS system for Windows, Version 9.1.3 (SAS Institute Inc. SAS Institute Inc.)]. Data are presented as least squares mean (L.S.M.) ± standard error of mean (S.E.M). Treatment differences with $P \leq 0.05$ were considered significant and $0.05 < P \leq 0.01$ were considered a tendency.

3 Results

The incidence of clinical mastitis at cow level was 24.7% (149/603) in the study period. The reproductive parameters of cows in MG1, MG2, MG1 and MG2, and CG were presented in Table A-20. And the percentages of cows in each group with the times of AI per conception were shown in Table A-21. In detail, the number of days to first AI was significantly greater for cows in MG1 than MG2 and CG (73.84±1.23 vs. 54.98±0.34; $P<0.01$; Figure A-3). For pregnant cows, the number of days from calving to conception was similar for cows in MG1 and MG2 (121.82±5.03 vs. 133.31±11.36; $P>0.05$; Figure A-4), but means for both those groups were greater than for cows in CG (121.82±5.03 vs. 89.74±2.17 and 133.31±11.36 vs. 89.74±2.17; $P<0.01$; Figure A-4). Furthermore, the number of services per conception was significantly greater for cows in MG1 and MG2 than cows in CG (1.88±0.08 vs. 1.53±0.03 and 2.19±0.16 vs. 1.53±0.03; $P<0.01$; Figure A-5), and cows in MG1 had fewer number of services per conception compared to cows in MG2 (1.88±0.08 vs. 2.19±0.16; $P<0.05$; Figure A-5). Conception rate at first service was similar for cows in MG1 and MG2 (38% vs. 28%; $P>0.05$; Figure A-5), however, all of them were lower than cows in CG (38% vs. 55% and 28% vs. 55%; $P<0.01$; Figure A-5).

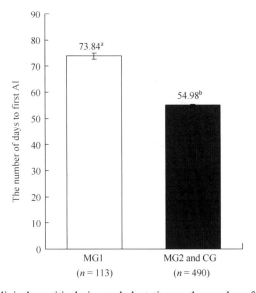

Figure A-3 Influence of clinical mastitis during early lactation on the number of days to first AI of lactating Holstein dairy cows. a, b Different superscripts denotes differences ($P<0.01$) between groups

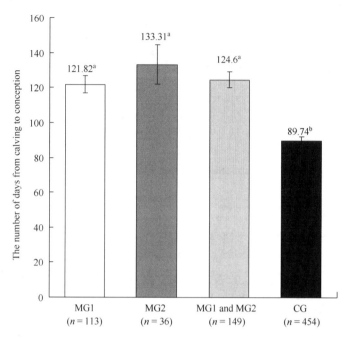

Figure A-4　Influence of clinical mastitis during early lactation on the number of days from calving to conception of lactating Holstein dairy cows. a, b Different superscripts denotes differences ($P<0.01$) between groups

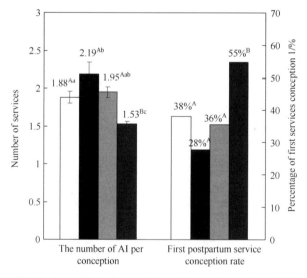

Figure A-5　Influence of clinical mastitis during early lactation on the number of AI per conception and first service conception rate of lactating Holstein dairy cows. White bars represent MG1, hatched bars represent MG2, gray bars represent MG1 and MG2, and black bars represent CG. Means or percentages with different superscripts lower case letters (a, b) within bar groups differ at $P<0.05$, and different superscripts capital letters (A, B) differ at $P<0.01$

Table A-20 Effect of clinical mastitis on reproductive parameters of lactating Holstein dairy cows
(L.S.M.±S.E.M.)

	Mastitis Group 1 (n = 113)	Mastitis Group 2 (n = 36)	Mastitis Group 1 and 2 (n = 149)	Control Group (n = 454)
The number of days to first AI	$73.84^{Aa} \pm 1.23$	$58.19^{Bb} \pm 1.69$	$70.06^{Cc} \pm 1.16$	$54.73^{Bd} \pm 0.34$
The number of days from calving to conception	$121.82^{A} \pm 5.03$	$133.31^{A} \pm 11.36$	$124.6^{A} \pm 4.69$	$89.74^{B} \pm 2.17$
The number of AI per conception	$1.88^{Aa} \pm 0.08$	$2.19^{Ab} \pm 0.16$	$1.95^{Aab} \pm 0.07$	$1.53^{Bc} \pm 0.03$
Conception rate at first AI	$38.1\%^{A}$	$27.8\%^{A}$	$35.6\%^{A}$	$54.9\%^{B}$

Note: a, b, c, d means in a row with different superscripts lower case letters differ ($P<0.05$). A, B, C, D means in a row with different superscripts capital letters differ ($P<0.01$)

Table A-21 Percentages of pregnant cows with once, twice, three times and four times services per conception in MG1, MG2, and CG

	Mastitis Group 1 (n = 113)	Mastitis Group 2 (n = 36)	Control Group (n = 454)
Percentage of cows with once AI per conception	$38.1\%^{A}$	$27.8\%^{A}$	$54.0\%^{B}$
Percentage of cows with twice AI per conception	40.7%	36.1%	40.1%
Percentage of cows with three times AI per conception	$16.8\%^{A}$	$25.0\%^{A}$	$5.0\%^{B}$
Percentage of cows with four times AI per conception	$4.4\%^{A}$	$11.1\%^{A}$	$0.9\%^{B}$
Percentage of cows conceived at the end of the research	100%	100%	100%

Note: A, B, C, D means in a row with different superscripts capital letters differ ($P<0.01$)

4 Discussion

Mastitis is one of the most costly and common diseases affecting dairy cows throughout the world. More recently, compromised reproductive performance has also been recognized as a detrimental effect of clinical mastitis (Ahmadzadeh et al., 2009; Gunay and Gunay, 2008; Nava-Trujillo et al., 2010; Santos et al., 2004). However, it is essentially unknown that the effect of clinical mastitis from calving to next conception on reproductive performance in Chinese dairy cows under subtropical conditions.

Past research on the influence of mastitis on reproductive performance focused on experimental coliform mastitis and infusion of endotoxin from Gram-negative pathogens. However, there were no differences in reproductive performance between clinical mastitis caused by Gram-negative or Gram-positive pathogens (Santos et al., 2004). In our study, we did not isolate the pathogens responsible for mastitis. The occurrence of clinical mastitis cases early in lactation, either before breeding or between first postpartum AI and pregnancy diagnosis resulted in detrimental effects on reproduction in Holstein cows.

In the present study, the number of days to first AI was significantly greater for cows with clinical mastitis before first AI (73.84 d) than all other groups (54.98 d). This finding was similar to that in the previous study (Santos et al., 2004; Schrick et al., 2001). It could have been due to insufficient follicular development; anovulation, resulting from blockage of the LH surge; or decreased estrogen synthesis, resulting in the loss of behavioral estrus (Schrick et al.,

2001). Those factors could explain the delayed first postpartum AI in cows experiencing mastitis before first AI in the current and previous studies.

We observed a much lower conception rate at first AI when cows had their clinical mastitis cases before first AI, even the effect of mastitis on first AI conception rate was exacerbated when the clinical case occurred between first AI and pregnancy diagnosis (38.1% and 27.8% vs. 54.9%), which is in accordance with the results of Santos et al. (2004) (22%, 10% and 29%, respectively). The bovine mammary gland synthesizes prostaglandin F2α (PGF2α), and increasing PGF2α was found in milk from cows with clinical mastitis. This increase in plasma concentrations of PGF2α could induce luteolysis. Therefore, decreased luteal function or complete luteal regression caused by premature release of PGF2α induced by mastitis could potentially explain the conception rate at first AI in lactating dairy cows in MG1 and MG2. Whether the possible impacts of bacterial infections on luteal function are the exact cause for reduced conception in cows affected by clinical mastitis still remain to be proved in controlled studies.

The number of AI per conception was greater for cows experiencing clinical mastitis after the first service than cows with clinical mastitis before first service and cows with no clinical mastitis (2.19 vs. 1.88 and 1.53, respectively). Cows with clinical mastitis after first service typically had an increased number of services per conception compared to the control group. Our findings were in agreement with the results reported by Barker et al., Santos et al. (2004), and Gunay and Gunay (2008). Elevated concentrations of serum PGF2α associated with mastitis may cause a decrease in embryonic development and ultimately result in an increased number of services per conception and number of days to conception.

The number of days from calving to conception for cows in MG1 (121.82 d) and MG2 (133.31 d) were significantly greater than that for control group cows (89.74 d), which is in agreement with the reports by Barker et al., Schrick et al. (2001) and Gunay and Gunay (2008). While the days to conception for cows in MG1, MG2 and CG were all longer than Santos et al. (2004) reported in the literature (165, 189 and 139 days, respectively). The longer days to conception for cows in MG1 and MG2 than CG was mainly due to lower conception rates at first AI and additional AI per conception for cows in MG1 and MG2 than CG. These results could have been due to luteolysis, subsequent loss in progesterone, and early embryonic death.

Barker et al. reported that both Gram-negative and Gram-positive pathogens may act through similar mechanisms to increase inflammatory mediators, leading to reproductive failure during early lactation. Clinical mastitis leaded to the production of bioactive molecules in the reproductive tract tissues. For example, *E. coli* endotoxin didn't usually penetrate from the udder in to the blood, but could induce massive release of cytokines. Those cytokine-mediated neural and endocrine changes played a key role in the inflammatory process (Hansen et al., 2004; Huszenicza et al., 2005). Cullor suggested that endotoxin may induce premature luteolysis and influence conception by release of inflammatory mediators. Specifically, endotoxin stimulates synthesis of PGF2α, glucocorticoids, and interleukin-1. In addition, infusion of endotoxin resulted in a pyretic response in dairy cows. Elevated body temperature was a symptom often associated with clinical mastitis. Experimentally induced S. uberis clinical mastitis resulted in elevated body temperatures (Hockett et al., 2000). Edwards and Hansen reported that exposure of bovine oocytes to elevated temperatures decreased blastocyst formation.

McCann et al. indicated that cytokines, released following endotoxin challenge, blocked the pulsatile secretion of luteinizing hormone (LH) but not follicle stimulating hormone (FSH) through alterations in nitric oxide production to inhibit gonadotropin releasing hormone (GnRH). Battaglia et al. reported significant inhibition of GnRH pulse amplitude and total GnRH following intravenous endotoxin infusion. Consequently, insufficient follicular development and (or) oocyte maturation could lead to insufficient estrogen production and subsequent anovulation, eventually leading to lack of behavioral estrus and failure of artificial insemination.

Results of this study demonstrated clearly that clinical mastitis either before first AI or between first AI and pregnancy diagnosis can have a markedly negative impact on the reproductive performance of dairy cows. More attention and effort needs to be assigned to prevention of this disease that decrease reproductive efficiency. Proper dry cow treatment, adequate nutrition, housing comfort and cleanliness, and adequate milking procedures would reduce the incidence of clinical mastitis, which is expected to result in decrease days to first AI, days from calving to conception, and umber of AI per conception and increase the conception rate in lactating dairy cows.

The negative impact of mastitis on reproduction underlines the importance of avoiding mastitis early in lactation, highlighting the significance of mastitis preventive measures as the rule but not the exception. Further study is needed to better understand the mechanisms by which clinical mastitis affects reproductive performance in dairy cows and to design better strategies to avoid those negative effects.

5 Conclusions

In conclusion, to our knowledge this study is the first report of the negative impact of clinical mastitis occurring from calving to pregnancy on the reproductive performance in Chinese dairy cows. Mainly delays the calving to first service interval, increases the calving to conception interval, elevates the number of AI per conception, and reduces the conception rate at first AI. Reproductive efficiency is of a great concern to dairy producers. Therefore, it is important to reduce the incidence of clinical mastitis in early lactation for improving the reproductive performance of dairy cows.

参 考 文 献

白峰，王文魁，王广彬，等. 22 味中草药对 3 种奶牛乳腺炎病原菌的体外抑菌试验[J]. 中国奶牛，2010，(2)：39-41.

白瑞景，孙少华，马亚宾. 乳中体细胞数或体细胞评分变化规律及其与乳成分的关系研究[J]. 畜牧与兽医，2010，42（12）：79-81.

鲍士兵. 天津市北辰区奶牛隐性乳房炎的调查及治疗[D]. 长春：吉林大学，2014.

曹铎耀. γδ T 细胞及相关细胞因子在 C57BL/6J 小鼠金黄色葡萄球菌乳房炎模型中的变化[D]. 西安：西北农林科技大学，2012.

曹立亭，陈星，胡松华. 奶牛真菌性乳房炎研究进展[J]. 中国奶牛，2009，(05)：38-39.

曹随忠，杜立新，赵兴绪. 16S-23S rDNA 间隔区在奶牛乳房炎诊断中的应用[J]. 中国畜牧兽医，2005，(02)：26-27.

车车. 无乳链球菌、停乳链球菌、乳房链球菌 GapC 蛋白的表达与免疫原性研究[D]. 大庆：黑龙江八一农垦大学，2009.

陈福广. 分选酶 A 在金黄色葡萄球菌引起的乳腺炎、菌血症和肺炎致病过程中的作用[D]. 长春：吉林大学，2014.

陈建坡，陈树兴，李丽丽，等. 不同体细胞数的原料乳品质变化的比较研究[J]. 食品科学，2013，34（7）：34-37.

陈廷祚. 金黄色葡萄球菌肠毒素、超级抗原及其抗肿瘤作用的研究进展[J]. 微生物学免疫学进展，2005，33（3）：66-73.

陈伟. 奶牛乳房炎葡萄球菌生物被膜形成机制研究进展[J]. 中国预防兽医学报，2010，31（12）：996-1000.

程安春，汪铭书，方鹏飞，等. 间接酶联免疫吸附试验（ELISA）检测血清 4 型鸭疫里默氏杆菌抗体的研究[J]. 中国家禽，2004，26（16）：11-14.

程广龙，江喜春，赵辉玲，等. 牛奶中高体细胞数的危害及调控措施[J]. 畜牧与饲料科学，2009，30（4）：65-66.

崔焕忠. 奶牛乳房炎基因工程亚单位疫苗实验免疫研究[D]. 长春：吉林农业大学，2004.

戴鼎震，赵永前，王晓丽，等. 南京地区奶牛乳房炎的现状、成因与防制[J]. 江苏农业科学，2002，(04)：59-61.

邓海平，蒲万霞，倪春霞，等. 甘肃、贵州地区奶牛乳房炎病原菌分离鉴定及耐药性调查[J]. 畜牧与兽医，2010，42（7）：76-79.

丁丹丹. 河南省奶牛隐性乳房炎主要病原菌的分离鉴定与药敏试验[D]. 郑州：河南农业大学，2014.

丁月云. 21 种中药对奶牛乳房炎 3 种病原菌的体外试验[J]. 畜牧与兽医，2004，(12)：5-7.

董全，吕景松，姜小平，等. 奶牛隐性乳房炎检测与防治[J]. 北京农业，2010，(24)：12-16.

杜红芳，窦爱丽，杨威，等. 热应激奶牛的生理变化及缓解措施[J]. 中国畜牧杂志，2007，43（12）：59-62.

杜健，张庆茹，史书军，等. 增乳康散对奶牛泌乳内分泌的影响[J]. 中国兽医杂志，2007，(03)：37-38.

樊杰. 奶牛乳房炎病原菌耐药性及无乳链球菌抗原基因研究[D]. 兰州：甘肃农业大学，2014.

方光远，张志成，晏文梅. 几种不同条件下细菌菌种保存试验[J]. 金陵科技学院学报，2004，(03)：30-31，50.

方磊，郭玉江. 奶牛乳房炎及其综合治疗措施[J]. 中国畜牧兽医文摘，2012，28（8）：100-101.

方维焕，张晓峰，Stephen P. Oliver. 乳铁素参与乳房链球菌对乳腺上皮细胞的黏附[J]. 中国兽医学报，2003，23（1）：35-37.

方振华, 高腾云, 丁利. 奶牛乳腺炎的研究进展[J]. 上海畜牧兽医通讯, 2007, (2): 13-15.
房强. 耐甲氧西林金黄色葡萄球菌临床检测及耐药性分析[D]. 长春: 吉林大学, 2009.
冯超. 呼和浩特地区某奶牛场奶牛乳腺炎病原菌的分离鉴定和药敏试验[D]. 呼和浩特: 内蒙古农业大学, 2013.
冯士彬, 李志明, 孟庆娟, 等. 中药复方透皮贴剂对奶牛临床型乳房炎的疗效及其机理[J]. 江苏农业学报, 2012, 27 (4): 790-794.
冯万新. 奶牛乳房炎主要病原菌分离鉴定及其多联苗抗体检测研究[D]. 兰州: 甘肃农业大学, 2008.
付秀花, 王恬. 中草药防制奶牛乳房炎的研究进展[J]. 动物科学与动物医学, 2001, (02): 33-36.
付云贺. 核受体 LXRα 抗奶牛乳腺炎作用及其激动剂筛选[D]. 长春: 吉林大学, 2015.
高海慧. 陕西某奶牛场隐性乳房炎的调查与病原菌分离鉴定及药敏试验[D]. 西安: 西北农林科技大学, 2014.
高瑞峰. 绿原酸抗乳腺炎作用及机制研究[D]. 长春: 吉林大学, 2014.
高树新. BoLA 基因多态性及其与奶牛乳房炎的关联研究[D]. 呼和浩特: 内蒙古农业大学, 2005.
戈胜强, 柴同杰, 吕静. 泰安地区部分奶牛场乳腺炎病原菌的分离鉴定[J]. 畜牧与兽医, 2008, 40 (7): 84-85.
耿梅英, 薛爱红, 贾亚池. 常用中药防治奶牛乳房炎病原菌效果研究[J]. 今日畜牧兽医, 2006, (04): 3-4.
龚平阳, 肖兵南, 任慧波, 等. 中草药复方搽剂防治奶牛乳房炎的效果观察[J]. 湖南畜牧兽医, 2009, (1): 5-6.
顾有方, 陈继红, 陈会良, 等. 奶牛隐性乳房炎与被毛中 Zn、Cu、Mn 含量的关系[J]. 动物医学进展, 2005, 26 (3): 82-83.
关红, 李成应, 李培锋, 等. 正交法筛选抗奶牛乳房炎主要致病菌的中药组方[J]. 黑龙江畜牧兽医, 2010, (9): 137-139.
郭昌明, 张乃生, 周昌芳, 等. 髓过氧化物酶检测奶牛隐性乳房炎研究进展[J]. 动物医学进展, 2006, (06): 54-57.
郭梦尧. 黄芩苷对小鼠金黄色葡萄球菌性乳腺炎的作用及机制研究[D]. 长春: 吉林大学, 2014.
郭庆, 王鲁, 朱伟, 等. 牛乳电导率值与体细胞数相关性的研究[J]. 贵州畜牧兽医, 2009, (05): 1-3.
郭庆, 朱伟, 易琼, 等. 8 味中药对奶牛乳房炎病原菌的体外抑菌试验[J]. 中国兽医杂志, 2012, 48 (05): 63-65.
郭时金, 张志美, 付石军, 等. 奶牛热应激的危害及控制[J]. 家畜生态学报, 2013, 34 (3): 65-67, 71.
郭旭东, 刁其玉, 汪新建. 中草药防治奶牛乳房炎的研究[J]. 中国饲料, 2012 (08): 24-25.
韩玉婷, 崔彦召, 徐国忠, 等. 热应激对上海地区奶牛生理指数和产奶量的影响[J]. 上海畜牧兽医通讯, 2011, 2: 12-14.
郝建国. 日本学者对体细胞数与乳房炎关系的论述[J]. 中国奶牛, 2002, (1): 52-54.
何钦. 热应激对不同泌乳阶段奶牛生产性能及其营养代谢的影响[D]. 重庆: 西南大学, 2012.
贺宽军. 金黄色葡萄球菌 fnbA 基因的克隆、表达载体的构建及原核表达[D]. 西安: 西北农林科技大学, 2006.
洪帮兴, 江丽芳, 胡玉山, 等. 23S rRNA 基因序列分析及其在细菌鉴别诊断中的应用[J]. 中华微生物学和免疫学杂志, 2004, 24 (03): 241-244.
侯伟杰. 奶山羊乳房炎灭活疫苗的制备及其免疫效果评估[D]. 西安: 西北农林科技大学, 2014.
侯引绪, 张凡建, 魏朝利, 等. 奶牛抗热应激物理性措施应用效果研究[J]. 中国奶牛, 2012, 1: 013.
黄光红. 奶牛肢蹄病发病原因及其对生产性能影响的调查分析[J]. 上海畜牧兽医通讯, 2011, (4): 28-29.

黄权利, 施明华, 蔡荣湘. 隐性乳房炎中体细胞和病原菌及酶间的相关性[J]. 浙江农业大学学报, 1995, 21 (3): 243-246.
黄远全, 吴建英, 陈橙, 等. 自制中草药制剂治疗奶牛乳房炎[J]. 中国畜牧兽医文摘, 2012, 28 (2): 147-148.
霍建丽. 奶牛隐性乳房炎的调查与分析[J]. 畜牧与兽医, 2013, (09): 115-116.
霍金龙, 苗永旺, 曾养志. 基因芯片技术及其应用[J]. 生物技术通讯, 2007, 18 (02): 329-332.
贾敏, 吴聪明, 沈建忠, 等. 重组溶葡萄球菌酶对奶牛乳房炎和子宫内膜炎常见病原菌的体外抑杀菌试验[J]. 中国奶牛, 2007, (06): 38-42.
金凤, 王若君, 杨英. 中药治疗奶牛隐性乳房炎临床试验[J]. 中国兽医杂志, 2005, (08): 33-34.
金兰梅, 伍清林, 马高民, 等. 奶牛隐性乳房炎发生规律的研究[J]. 金陵科技学院学报, 2010, 26 (1): 85-89.
金耀忠, 万世平, 姜法铭, 等. 奶牛乳房炎流行情况调查及主要病原分离与药敏试验[J]. 畜牧与兽医, 2011, 43 (5): 64-67.
敬晓棋. 乳腺炎细胞因子及T淋巴细胞亚群研究[D]. 西安: 西北农林科技大学, 2013.
阚威. 奶牛乳房炎链球菌和肠球菌的分离鉴定及毒力基因的检测[D]. 兰州: 甘肃农业大学, 2014.
孔祥峰. 中药新促孕液防治母畜子宫内膜炎的机理研究[D]. 南京: 南京农业大学, 2006.
孔雪旺, 陈功义. 奶牛乳房炎病原菌的分离鉴定及药敏试验[J]. 湖北畜牧兽医, 2005, (06): 27-29.
李成应. 治疗奶牛乳房炎的复方中药制剂的开发研究[D]. 呼和浩特: 内蒙古农业大学, 2008.
李德鑫. "乳炎消"注射液治疗奶牛急性乳房炎[J]. 青海畜牧兽医杂志, 2003, (03): 53.
李宏胜, 罗金印, 李新圃, 等. 以疫苗预防为主的奶牛乳房炎综合防治措施应用效果观察[J]. 中国草食动物, 2011, 30 (6): 69-71.
李宏胜, 罗金印, 王旭荣, 等. 我国奶牛乳房炎无乳链球菌抗生素耐药性研究[J]. 中兽医医药杂志, 2012, 31 (6): 5-7.
李宏胜, 郁杰, 李新圃, 等. 乳牛乳腺炎多联疫苗的研制及其临床应用效果[J]. 中国兽医科学, 2007, (04): 363-368.
李会, 孙志文, 汤树生, 等. 复方阿莫西林乳房注入剂治疗临床型奶牛乳房炎的疗效观察[J]. 中国兽医杂志, 2012, 48 (03): 71-74.
李慧. 金黄色葡萄球菌A型肠毒素单克隆抗体的制备及ELISA检测方法的建立[D]. 武汉: 华中农业大学, 2007.
李建军, 杨学云, 王治才, 等. 新疆集约化奶牛场隐性乳房炎流行病学和病原学研究[J]. 新疆农业科学, 2008, (02): 369-373.
李金平, 陈杰, 庄青叶. 麻鸡源性粪肠球菌的分离及初步鉴定[J]. 中国动物检疫, 2011, 28 (07): 61.
李进. 太原地区奶牛乳房炎流行病学调查及复方中草药对奶牛乳房炎病原菌体外抑菌效果的研究[D]. 晋中: 山西农业大学, 2014.
李庆国. 奶牛乳房炎发病规律的调研[J]. 中国畜牧兽医文摘, 2006, 42 (19): 60-61.
李诗莹. 甘草总黄酮混悬乳房注射剂对奶牛乳房炎治疗效果研究[D]. 大庆: 黑龙江八一农垦大学, 2015.
李书文, 易琼, 王鲁. 贵阳市奶牛隐性乳房炎调查分析[J]. 贵州畜牧兽医, 2012, 36 (04): 29-31.
李晓香, 杨改青, 王林枫, 等. 不同饲养模式对奶牛健康状况和牛奶品质的影响[J]. 中国奶牛, 2013, (17): 34-38.
李兴禄, 张莉萍, 黄长武, 等. 耐甲氧西林金黄色葡萄球菌MecA基因快速检测方法的评价[J]. 中华医院感染学杂志, 2003, 13 (02): 105-106.
李英. 新疆南疆地区奶牛乳房炎性表皮葡萄球菌aap基因的原核表达及功能研究[D]. 阿拉尔: 塔里木大

学，2015.

李玉文，秦建华. 中西药对奶牛临床型乳房炎的效果观察[J]. 河北北方学院学报：自然科学版，2010，26（01）：42-45.

廖想想. 奶牛金黄色葡萄球菌人工诱导型乳腺炎基因表达谱分析[D]. 扬州：扬州大学，2014.

林锋强，潘杭君，胡松华. 奶牛乳房炎疫苗研究进展[J]. 中国奶牛，2002，（01）：40-42.

刘博. 小鼠巨噬细胞 TLR2、TLR4 及 RP105 在金黄色葡萄球菌感染中的天然免疫应答机制[D]. 长春：吉林大学，2013.

刘朝. 奶牛隐性乳腺炎病原菌的分离鉴定以及生物被膜的检测[D]. 武汉：华中农业大学，2007.

刘东梅，毕建成，郄会卿，等. 黄芩、黄连、乌梅、金银花、败酱草对产 AmpC β-内酰胺酶细菌的体外抑菌作用[J]. 河北中医，2008，（06）：654-655.

刘峰，迟玉杰. 乳房炎乳的检测方法[J]. 现代食品科技，2005，（01）：129-131.

刘海林，高帅，段洪峰，等. 喷淋降温对荷斯坦奶牛抗热应激的研究[J]. 家畜生态学报，2013，（1）：44-46.

刘华芬，余康霞，余康健，等. 奶牛肢蹄病的发病原因及防治措施[J]. 中国畜牧兽医，2009，（5）：134-135.

刘纪成，张敏，李建柱，等. 猪链球菌毒力因子的研究进展[J]. 黑龙江畜牧兽医，2012，（1）：23-25.

刘建文. 药理实验方法学[M]. 北京：化学工业出版社，2003.

刘箭. 生物化学实验教程[M]. 北京：科学出版社，2004.

刘俊杰. 自制中药灌注液对奶牛临床型乳房炎的疗效观察[J]. 中国兽医杂志，2010，（12）：45-46.

刘澜，周德刚，莫云，等. 26 种中草药对奶牛隐性乳房炎病原菌体外抗菌作用的研究[J]. 中国农业大学学报，2009，（03）：83-88.

刘纹芳，李成，李磊，等. 复合溶菌酶制剂对奶牛乳房炎病原菌的抑菌试验[J]. 中国乳业，2006（6）：33-35.

刘营. 奶牛乳房炎病原菌分离鉴定及复方中草药治疗的研究[D]. 济南：山东师范大学，2008.

刘珍. 呼和浩特地区临床型奶牛乳腺炎病原分离鉴定及其对小白鼠致病性的研究[D]. 呼和浩特：内蒙古农业大学，2005.

吕平，韦丽君，黄强，等. 中草药提取物与酶复合物对牛乳房炎病原菌的抑菌效果[J]. 西北农业学报，2012，21（03）：32-37.

马保臣，秦卓明，蔡玉梅，等. 多重 PCR 检测奶牛乳腺炎金黄色葡萄球菌、无乳链球菌、停乳链球菌和酵母菌方法的建立与应用[J]. 畜牧兽医学报，2006，（11）：1202-1208.

马定坤，伍杰，刘娟，等. γ-干扰素对金黄色葡萄球菌性临床型乳房炎的疗效研究[J]. 中国畜牧兽医，2011，（07）：206-209.

马吉锋，潘忠学，黎玉琼，等. 奶牛隐性乳房炎研究进展[J]. 畜牧与饲料科学，2010，30（9）：79-81.

马径军. 奶牛胎衣不下的初步研究[D]. 南京：广西大学，2008.

马立艳，许淑珍，马纪平. 肠球菌致病机制的研究进展[J]. 中华医院感染学杂志，2005，15（3）：356-360.

马晓艳. 抗菌肽对奶牛乳房炎病原菌的抑菌作用及该病与酮病的相关性研究[D]. 乌鲁木齐：新疆农业大学，2007.

马玉臣，马俊飞，李长海，等. 链球菌致病机理及防治[J]. 畜牧与饲料科学，2010，（2）：180-181.

买尔旦·马合木提，古丽仙·胡加，王晓雯. 细菌耐药机制的研究进展[J]. 新疆医科大学学报，2003，（06）：609-610.

毛翔光，毛华明，朱新培. 规模化奶牛场乳腺炎发病规律的调查分析[J]. 中国牛业科学，2009，（05）：66-69.

毛永江，杨章平. 南方地区中国荷斯坦牛乳中体细胞数变化规律的研究[J]. 中国牛业科学，2007，33（5）：1-3.

孟庆娟. 中药复方灌注剂对奶牛乳房炎的防治效果及作用机理的研究[D]. 合肥：安徽农业大学, 2011.

苗晋锋, 马海田, 邹思湘, 等. 内毒素对山羊乳腺组织中与乳腺炎相关的酶和细胞因子的影响[J]. 福建农林大学学报（自然科学版）, 2007, 36（6）：608-613.

莫放. 养牛生产学[M]. 北京：中国农业大学出版社, 2003.

牟珊. 关中奶山羊临床型乳房炎模型的建立及疫苗研制[D]. 西安：西北农林科技大学, 2010.

齐亚银, 剡根强, 王静梅, 等. 致羔羊脑炎型粪肠球菌的分离及鉴定[J]. 石河子大学学报：自然科学版, 2005, 23（02）：200-202.

秦春香. 禽呼肠孤病毒非结构蛋白的克隆表达及重组蛋白ELISA方法的建立[D]. 南宁：广西大学, 2007.

秦璐璐. 合肥某场奶牛乳房炎流行病学调查及病原菌的分离鉴定[D]. 合肥：安徽农业大学, 2009.

邱家章. 黄芩苷抗金黄色葡萄球菌α-溶血素作用靶位的确证[D]. 长春：吉林大学, 2012.

伞治豪. 蒲公英甾醇对LPS诱导的乳腺炎抗炎作用及调节机制研究[D]. 长春：吉林大学, 2014.

沈诚, 林源, 叶承荣, 等. 重组溶菌酶质粒pcDNAKLYZ治疗泌乳期奶牛乳房炎[J]. 中国奶牛, 2011,（04）：4-6.

石祖星. 改进荷斯坦奶牛乳房线性性状降低隐性乳房炎的可行性分析[J]. 安徽农学通报, 2010, 16（7）：179-180.

史文军. 微生态制剂治疗奶牛乳房炎的疗效观察[J]. 河南畜牧兽医：综合版, 2006,（3）：51.

史彦斌, 薛明, 罗永江, 等. 消炎醍的抗炎实验研究[J]. 中兽医医药杂志, 2000,（1）：12-13.

双金, 嘎尔迪, 包鹏云, 等. 奶牛隐性乳房炎的发生规律及其致病菌的分离鉴别与药物敏感性试验[J]. 内蒙古农业大学学报：自然科学版, 2001,（01）：18-23.

宋波, 刘颋, 吕丽艳, 等. 苦参、黄芩、乌梅对金黄色葡萄球菌、大肠埃希菌和白假丝酵母菌抗菌活性的研究[J]. 中国微生态学杂志, 2010,（06）：507-508.

宋华容, 李正国, 关辉, 等. 复方中药对乳房炎奶牛细胞免疫功能的影响[J]. 中国动物检疫, 2009,（04）：65-67.

宋丽华, 杨帆, 韩吉雨, 等. 抗菌肽制剂对奶牛隐性乳房炎防治效果的研究[J]. 中国奶牛, 2011,（4）：58-61.

宋战辉. 奶牛乳房炎金黄色葡萄球菌亚单位疫苗的构建及其效力检测[D]. 石河子：石河子大学, 2014.

苏俊强, 土建亮, 梁海锋, 等. 奶牛隐性乳房炎发生规律探讨[J]. 湛江海洋大学学报, 2004, 22（6）：65-68.

苏洋, 蒲万霞, 陈智华, 等. 牛源金黄色葡萄球菌的耐药性及耐甲氧西林金黄色葡萄球菌的检测[J]. 中国农业科学. 2012,（17）：3602-3607.

苏洋. 奶牛乳房炎耐甲氧西林金黄色葡萄球菌的检测及基因分型[D]. 兰州：甘肃农业大学, 2013.

孙怀昌, 于锋, 苏建华, 等. 人溶菌酶基因治疗奶牛乳腺炎的初步研究[J]. 畜牧兽医学报, 2004,（02）：227-232.

孙怀昌. 防治奶牛乳房炎的基因工程新药[J]. 北方牧业, 2005（19）：26.

孙凌志, 陈庆勋, 李凯, 等. 奶牛乳腺炎常见类型及治疗[J]. 中兽医医药杂志, 2007（03）：46-48.

孙淑霞. 我国部分地区奶牛乳腺炎流行病学调查及IRAK2基因与乳腺炎相关性分析[D]. 长春：吉林大学, 2013.

田海燕. 不同佐剂的奶牛乳房炎多联苗对小白鼠免疫效果影响研究[D]. 兰州：兰州畜牧与兽药研究所, 2009.

田萍, 蔡森, 姚武群. 温湿指数（THI）对荷斯坦生产奶量影响的研究[J]. 中国奶牛, 2002,（4）：13-15.

王丹敏, 吴胜吉, 吴玉秀, 等. 肺炎链球菌几种保存方法的比较[J]. 中华医学研究杂志, 2006, 6（1）：88.

王登峰, 李建军, 段新华, 等. 我国牛源金黄色葡萄球菌耐药现状及药敏检测方法探讨[J]. 中国动物传染病学报. 2011,（01）：31-38.

王芳, 胡松华. 体细胞含量与牛奶质量的关系[J]. 中国奶牛, 2005,（04）：51-52.

王凤, 宋立, 汤德元, 等. 奶牛乳房炎病原菌的分离鉴定、血清型及耐药性研究[J]. 动物医学进展. 2013,（06）：62-67.

参 考 文 献

王福慧, 杨帆. 预防奶牛隐性乳房炎的功能性奶牛饲料添加剂试验研究[J]. 中国乳业, 2011, (2): 42-45.
王福顺, 陈德, 王冬鸣, 等. 天津市某奶牛养殖场乳房炎的流行病学调查[J]. 中国动物检疫, 2013, (03): 30-32.
王桂琴, 杨萌萌, 邢燕, 等. 宁夏地区奶牛乳房炎金黄色葡萄球菌耐药性分析[J]. 动物医学进展. 2011, (10): 59-62.
王桂英. 中药复方制剂对奶牛临床型乳房炎病原菌的体外抑菌试验[J]. 黑龙江畜牧兽医, 2010, (3): 132-133.
王国建, 戴朴, 韩东一, 等. 基因芯片技术在非综合征性耳聋快速基因诊断中的应用研究[J]. 中华耳科学杂志, 2008, (1): 61-66.
王华, 王晓宁, 刘一鹤, 秦春华, 胡建宏. 奶牛隐性乳腺炎发病因素的研究[J]. 中国牛业科学, 2010, 36（5）: 30-32.
王会珍, 马玉华. 奶牛乳房炎治疗研究现状分析[J]. 山东省农业管理干部学院学报, 2006, (1): 163-164.
王会珍. 奶牛乳房炎的危害及分类[J]. 山东省农业管理干部学院学报, 2007, 23（04）: 151-152.
王加启, 赵圣国. 我国牛奶质量安全的现状、问题和对策[J]. 中国奶牛, 2009, （11）: 3-7.
王建军, 税丽. 奶牛乳房炎致病机理研究进展[J]. 中国畜牧兽医, 2009, 36（12）: 100-103.
王建忠. 郑州市奶牛乳房炎发病情况和病原菌调查及药物敏感性试验[D]. 郑州: 河南农业大学, 2007.
王娟. 致奶牛乳房炎大肠杆菌快速诊断及药敏试验方法的改进[D]. 扬州: 扬州大学, 2014.
王玲, 李宏胜, 陈炅然, 等. 细胞因子在奶牛乳房炎生物学防治中的应用[J]. 中国畜牧兽医, 2011, 38（12）: 173-177.
王娜, 高学军. 哈尔滨地区奶牛隐性乳房炎病原菌的分离鉴定[J]. 东北农业大学学报, 2011, 42（2）: 29-32.
王秋芳, 张淼涛, 效梅, 等. 中药对隐性乳房炎奶牛细胞免疫功能的影响[J]. 畜牧兽医学报, 2002, （04）: 408-411.
王思贤. 抗菌肽用于奶牛乳房炎的体细胞基因治疗[D]. 北京: 中国农业大学, 2006.
王天成. 甜菊苷对金黄色葡萄球菌性乳腺炎模型动物的保护作用及机制研究[D]. 长春: 吉林大学, 2015.
王相根, 胡书梅, 汪聪勇. 奶牛体细胞数（SCC）和隐性乳房炎变化规律的统计调查[J]. 中国奶牛, 2012（14）: 49-50.
王小龙. 奶牛乳腺炎应激条件下乳中脂肪酸变化及其调控[D]. 扬州: 扬州大学, 2014.
王雄清, 代敏, 罗英, 等. 10种中药对奶牛乳房炎病原菌的体外抑菌试验[J]. 绵阳师范学院学报, 2006, （02）: 52-55.
王英, 李文广, 周贞鉴, 等. 医学微生物实验室菌种的保存和保管[J]. 海南医学院学报, 2006, （01）: 57-58.
王雨玲. 中药材黄芩、双花、秦皮等对幽门螺旋杆菌体外抗菌活性的研究[J]. 实用心脑肺血管病杂志, 2010, （05）: 605.
王正兵, 严作廷, 罗金印, 等. 我国奶牛乳房炎常见病原菌的耐药性检测[J]. 贵州农业科学, 2011, 39（4）: 139-141.
王志刚, 于红花, 栾爽艳, 等. 中药在奶牛乳房炎防治中的应用[J]. 中国奶牛, 2006, （08）: 41-42.
王忠红, 张淑云, 李德竹, 等. 中草药在治疗奶牛乳房炎中的应用效果观察[J]. 河南畜牧兽医: 综合版, 2011, 25（5）: 7.
韦海飞, 郭锷锋. 奶牛肢蹄病的综合防治[J]. 广东畜牧兽医科技, 2004, 29（5）: 48-49.
魏成威. 哈尔滨部分奶牛场隐性乳房炎流行病学调查及中药治疗研究[D]. 哈尔滨: 东北农业大学, 2010.

魏伟, 苗永旺. 奶牛乳腺炎抗性候选基因多态性研究进展[J]. 中国牛业科学, 2011, 37（6）：49-51.

魏学良, 张家骅, 王豪举, 等. 高温环境对奶牛生理活动及生产性能的影响[J]. 中国农学通报, 2005, 21（5）：13-15.

温雅俐, 高民, 何钦, 等. 热应激对不同泌乳阶段奶牛瘤胃内环境发酵指标的影响[J]. 中国畜牧杂志, 2011, 47（19）：69-73.

文静, 孙建安, 周绪霞, 等. 屎肠球菌对仔猪生长性能、免疫和抗氧化功能的影响[J]. 浙江农业学报, 2011, 23（01）：70-73.

乌兰巴特尔. 呼市地区奶牛乳腺炎金黄色葡萄球菌的分离鉴定及基因分型[D]. 呼和浩特：内蒙古农业大学, 2007.

吴超, 邹全明. 新型疫苗佐剂的研究进展[J]. 中国生物工程杂志, 2005,（08）：10-15.

吴东桃, 周桂波, 陈明世, 等. 国产奶牛乳房炎疫苗应用效果探讨[J]. 中国奶牛, 2001,（6）：19-20.

吴国娟, 张中文, 李焕荣, 等. 中草药对奶牛乳房炎6种致病菌的抑菌效果观察[J]. 北京农学院学报, 2003,（3）：195-198.

吴继尧, 刘鹏龙, 张丹凤, 等. 琼脂双向免疫扩散法测定牛初乳中IgG[J]. 新疆农业科学, 2001,（5）：252-254.

吴静, 张彦明, 杨银萍. 中草药对奶牛乳房炎病原菌的体外抑菌作用[J]. 安徽农业科学, 2007,（28）：8896-8897.

吴利先, 黄文祥. 屎肠球菌耐万古霉素的分子生物学机制及毒力因子[J]. 国际流行病学传染病学杂志, 2006, 33（2）：132-135.

武泽轩. 抗金黄色葡萄球菌奶牛乳腺炎疫苗的研制[D]. 哈尔滨：东北农业大学, 2010.

夏祖和, 张淼. 关于中草药防治奶牛乳房炎[J]. 金陵科技学院学报, 2006,（02）：101-103.

项开合, 张乃生, 肖连明, 等. 奶牛乳房炎的生物学防治[J]. 动物医学进展, 2007, 28（3）：109-112.

肖定汉. 简明牛病防治手册[M]. 中国农业大学出版社, 2002：226-231.

肖峰. 应用金黄色葡萄球菌噬菌体裂解酶LysGH15治疗乳房炎的实验研究[D]. 长春：吉林大学, 2015.

肖颖, 谷维娜, 钱明明, 等. 奶牛隐性乳房炎多重PCR检测方法研究[J]. 安徽农业科学, 2010,（17）：9029-9031.

肖正中, 赖松家. 奶牛分子育种的研究进展[J]. 畜禽业, 2003,（6）：41-44.

谢怀根, 许世勇. 某奶牛场奶牛乳房炎调查结果及分析[J]. 当代畜牧, 2003,（5）：21-23.

邢慧敏, 云振宇, 李妍, 等. 牛乳体细胞数与乳中离子含量相关性的研究[J]. 食品科学, 2007, 28（05）：53-56.

邢玫, 王振雄, 李岩. 规模化奶牛场奶牛乳房炎感染情况调查报告[J]. 新疆畜牧业, 2007,（03）：33-34.

熊琪, 李晓峰, 索效军, 等. 改善奶牛热应激的研究进展[J]. 湖北农业科学, 2011, 50（11）：2161-2164.

徐继英, 刘俊林, 李先波, 等. 我国部分地区奶牛乳房炎源大肠杆菌生物学特性及耐药性分析[J]. 农业生物技术学报, 2012, 20（9）：1035-1041.

徐京平, 许文兵, 徐雷. 中草药防治奶牛乳房炎[J]. 中兽医学杂志, 2010,（1）：30.

徐君英. 金黄色葡萄球菌肠毒素I检测方法的建立[D]. 北京：北京中医药大学, 2009.

许金俊. 奶牛γ-干扰素基因的表达及其在乳房炎防治中的初步应用[D]. 扬州：扬州大学, 2004.

严作廷, 王东升, 张世栋, 等. 奶牛肢蹄病综合防治技术[J]. 兽医导刊, 2013,（1）：35-37.

燕霞, 陈俭, 张丽佳, 等. 浙江省主要奶牛养殖区乳房炎致病菌金黄色葡萄球菌的耐药性[J]. 福建农林大学学报（自然科学版）. 2012,（06）：633-636.

杨波, 张雪梅, 罗淑萍, 等. 奶牛乳房炎金黄色葡萄球菌毒力因子检测与基因分型[J]. 西北农业学报,

2009,（05）：1-6.

杨德英, 曹随忠, 刘长松, 等. 奶牛隐性乳房炎诊断新指标研究进展[J]. 安徽农业科学, 2008,（09）：3665-3666.

杨宏军. 奶牛乳腺炎金黄色葡萄球菌毒力因子的克隆表达及生物活性研究[D]. 济南: 山东农业大学, 2007.

杨洪森. 家兔乳房炎模型建立及对机体内抗氧化指标和急性期蛋白表达的影响[D]. 成都: 四川农业大学, 2011.

杨克礼, 潘玲, 章孝荣, 等. 合肥市某奶牛场隐性乳房炎发生规律的研究[J]. 中国奶牛, 2005,（4）：10-12.

杨锐, 李英伦, 李金良, 等. 四川雅安临床型奶牛乳房炎病原菌分离鉴定及耐药性分析[J]. 中国兽医杂志, 2009, 45（4）：41-42.

杨毅, 梁荣嵘, 刘庆华, 等. 轻微至中度热应激对荷斯坦奶牛生理指标及产奶性能的影响[J]. 中国农学通报, 2009, 25（24）：28-31.

叶纪梅, 王加启, 赵国琦. 体细胞数过高对乳制品品质的影响[J]. 中国奶牛, 2006,（05）：41-44.

伊岚, 靳亚平, 史薇, 等. 奶牛隐性乳房炎检测方法的研究进展[J]. 畜牧与兽医, 2006,（10）：52-55.

易本驰. 奶牛隐性乳房炎的流行病学及病原调查[J]. 中国畜牧兽医, 2007, 33（12）：37-38.

易明梅, 黄奕倩, 朱建国, 等. 上海地区奶牛乳房炎主要病原菌的分离鉴定及耐药性分析[J]. 中国兽医学报. 2009,（03）：360-363.

尹荣兰. 奶牛乳腺炎金黄色葡萄球菌基因工程疫苗构建及实验免疫[D]. 长春: 吉林大学, 2009.

于恩琪. 江苏部分地区奶牛乳房炎葡萄球菌调查及其部分耐药基因的分析[D]. 扬州: 扬州大学, 2014.

喻华英, 张苗, 陈伟, 等. 阿拉尔地区某奶牛场隐性乳房炎的病原分离鉴定[J]. 新疆农业科学, 2009, 46（04）：849-855.

袁文常. 金黄色葡萄球菌适应性耐药及MRSA耐药调控机制的研究[D]. 重庆: 第三军医大学, 2013.

寨鸿瑞, 田甜, 王开功, 等. 奶牛隐性乳房炎病原菌的分离鉴定及药敏试验[J]. 动物医学进展, 2008,（11）：40-43.

张慧林, 余文文, 刘小林, 等. 牛乳中体细胞数与产奶量和乳成分的相关分析[J]. 西北农业学报, 2010, 19（4）：1-4.

张静, 于三科, 王新, 等. 原料乳和临床乳房炎金黄色葡萄球菌毒力基因检测及药敏分析[J]. 中国兽医学报, 2012, 32（05）：759-764, 770.

张克春, 徐国忠, 吴显实, 等. 日粮添加富硒益生菌对奶牛乳房炎和乳汁体细胞数的影响[J]. 上海交通大学学报: 农业科学版, 2010, 28（01）：59-63.

张璐莹, 聂甜甜, 郑程远, 等. 牛乳溶菌酶概述及应用[J]. 吉林畜牧兽医, 2013, 33（12）：16-19.

张善瑞, 王长法, 高运东, 等. 检测奶牛乳房炎主要病原菌的多重PCR方法的建立与应用[J]. 中国兽医学报, 2008,（05）：573-575.

张少华, 孙丹丹, 史万玉, 等. 不同中药对奶牛乳房炎病原菌抑菌效果的观察[J]. 中国农学通报, 2009, 25（14）：11-14.

张雯. 硒对奶牛乳腺炎抗炎作用和炎症信号转导通路调节机制的研究[D]. 长春: 吉林大学, 2014.

张颖. 天津地区奶牛乳腺炎流行病学调查及耐药性分析[D]. 北京: 中国农业科学院, 2014.

张宇, 刘磊, 豆艳丽. 奶牛乳房炎葡萄球菌的分离鉴定及耐药性分析[J]. 甘肃农业大学学报. 2013,（06）：34-39.

张振国, 秦晓庆, 赵树臣, 等. 中药乳房灌注剂治疗奶牛临床型乳房炎[J]. 中国兽医杂志, 2010,（04）：92-94.

张振国. 中药乳头灌注剂治疗临床型奶牛乳腺炎的研究[D]. 哈尔滨：东北农业大学，2009.
赵红波，赵建宏，石磊. 质粒介导的耐药基因获得及转移研究进展[J]. 临床荟萃，2010，（11）：1005-1007.
赵建荣. 奶牛环境性乳腺炎灭活疫苗的研制及效果评价[D]. 呼和浩特：内蒙古农业大学，2009.
赵兴绪. 兽医产科学[M]. 第3版. 北京：中国农业出版社，2006.
赵彦杰. 金银花叶提取物的抑菌效果研究[J]. 食品科学，2007，（07）：63-65.
赵中利. MBL2基因与奶牛乳腺炎抗性研究[D]. 长春：吉林大学，2012.
郑国卫，潘鸿飞. 牛奶质量与体细胞数[J]. 中国奶牛，2006，（12）：43-45.
智宇. 奶牛乳腺炎乳汁中几种细胞因子与体细胞凋亡相关性的研究[D]. 呼和浩特：内蒙古农业大学，2014.
钟辉. 奶牛乳腺炎金黄色葡萄球菌油乳剂灭活疫苗的研制及免疫效果评价[D]. 南京：南京农业大学，2005.
周成青. 奶牛乳房炎的病因、诊断和综合防治[J]. 山东畜牧兽医，2013，（07）：29-30.
周传社，谭支良，赵陈锋等. 奶牛热应激的生理机制及其调控[J]. 家畜生态学报，2006，27（6）：173-177.
周二顺. 千金藤碱对LPS诱发的乳腺炎小鼠的保护效果及其机制的初步研究[D]. 长春：吉林大学，2015.
周亚平，刘琴，施开平，等. 乳体细胞数与产奶量、乳成分的关系研究[J]. 中国奶牛，2011，（04）：40-42.
朱立平. 免疫学常用实验方法[M]. 北京：人民军医出版社，2000.
朱贤龙，谢芝勋，刘加波，等. 广西奶牛隐性乳房炎的检测及病原分离、鉴定和药敏试验报告[J]. 广西农业科学，2004，（06）：488-490.
朱永平. 中药治疗奶牛乳腺炎的效果观察[J]. 北方牧业，2011，（01）：83-84.
朱志达，韩张兴. 中药"速效消炎膏"治疗奶牛临床型乳腺炎的疗效分析[J]. 乳业科学与技术，2001，（01）：47-48.
庄玉坚. 诃子炮制品鞣质类成分指纹图谱及药效学比较研究[D]. 广州：广州中医药大学，2009.
Abd Ellah M R，Okada K，Goryo M，et al. Superoxide dismutase activity as a measure of hepatic oxidative stress in cattle following ethionine administration[J]. The Veterinary Journal，2009，182：336-341.
Abd Ellah M R，Rushdi M，Keiji O，et al. Oxidative stress and bovine liver diseases：Role of glutathione peroxidase and glucose6-phosphate dehydrogenase[J]. Japanese Journal of Veterinary Research，2007，54：163-173.
Abd Ellah M R. Role of Free Radicals and Antioxidants in Mastitis[J]. Journal of Advanced Veterinary Research，2013，3：1-7.
Abdalla S F，Musa O A，Aradaib I E. Evaluation of Milk as a Source of Human DNA in Lactating Women Using Polymerase Chain Reaction[J]. International Journal of Molecular Medicine and Advance Sciences，2009，5：6-9.
Abera M，Elias B，Aragaw K，Denberga Y，Amenu K，Sheferaw D. Major causes of mastitis and associated risk factors in smallholder dairy cows in Shashemene，southern Ethiopia[J]. African Journal of Agricultural Research，2012b，7：3513-3518.
Abera M，Habte T，Aragaw K，Asmare K，Sheferaw D. Major causes of mastitis and associated risk factors in smallholder dairy farms in and around Hawassa，Southern Ethiopia[J]. Tropical Animal Health and Production，2012a，44：1175-1179.
Accarias S，Lugo-Villarino G，Foucras G，et al. Pyroptosis of resident macrophages differentially orchestrates inflammatory responses to Staphylococcus aureus in resistant and susceptible mice[J]. European Journal of Immunology，2015，45（3）：794-806.

Ahmadzadeh A, Frago F, Shafii B, et al. Effect of clinical mastitis and other diseases on reproductive performance of Holstein cows[J]. Animal Reproduction Science, 2009, 112: 273-282.

Akineden O, Annemuller C, Hassan A A, Lammler C, Wolter W, Zschock M. Toxin genes and other characteristics of Staphylococcus aureus isolates from milk of cows with mastitis[J]. Clinical and Diagnostic Laboratory Immunology, 2001, 8: 959-964.

Akira S, Uematsu S, Takeuchi O. Pathogen recognition and innate immunity[J]. Cell, 2006, 124 (4): 783-801.

Alain T. Extracellular ATP in the immune system: more than just a "danger signal"[J]. Science Signaling, 2009, 2 (56): pe6.

Alawneh J I, Laven R A, Stevenson M A. The effect of lameness on the fertility of dairy cattle in a seasonally breeding pasture-based system[J]. Journal of dairy science, 2011, 94 (11): 5487-5493.

Alberti C, Brun-Buisson C S, Guidici D, et al. Influence of systemic inflammatory response syndrome and sepsis on outcome of critically ill infected patients[J]. American Journal of Respiratory & Critical Care Medicine, 2003, 168 (1): 77-84.

Almaw G, Zerihun A, Asfaw Y. Bovine mastitis and its association with selected risk factors in smallholder dairy farms in and around Bahir Dar, Ethiopia[J]. Tropical Animal Health and Production, 2008, 40: 427-432.

Anonymous. 260/280 and 260/230 ratios. T009 technical bulletin. Thermo Scientific. 2008.

Archer S C, Green M J, Huxley J N. Association between milk yield and serial locomotion score assessments in UK dairy cows[J]. Journal of dairy science, 2010, 93 (9): 4045-4053.

Ata N, Zaki M S. New Approaches in Control of Mastitis in Dairy Animals[J]. Life Science Journal, 2014, 11: 275-277.

Atakisi O, Oral H, Atakisi E, Merhan O, Pancarci S M, Ozcana A, Marasli S, Polat B, Colak A, Kaya S. Subclinical mastitis causes alterations in nitric oxide, total oxidant and antioxidant capacity in cow milk[J]. Research in Veterinary Science, 2010, 89: 10-13.

Babaei H, Mansouri-Najand L, Molaei M M, et al. Assesscment of Lactate Dehydrogenase, Alkaline Phosphatase and Aspartate Aminotransferase Activities in Cow's Milk as an Indicator of Subclinical Mastitis[J]. Veterinary Research Communications, 2007, 31: 419-425.

Balaban N, Rasooly A. Staphylococcal enterotoxins[J]. International Journal of Food Microbiology, 2000, 61: 1-10.

Bannerman D D, Paape M J, Lee J W, et al. Escherichia coli and Staphylococcus aureus elicit differential innate immune responses following intramammary infection[J]. Clinical & Diagnostic Laboratory Immunology, 2004, 11 (3): 463-472.

Banu S, Serpil E, Dilek M, et al. Methicillin-resistant Staphylococcus aureus heterogeneously resistant to vancomycin in a Turkish university hospital[J]. Journal of Antimicrobial Chemotherapy, 2005, 56 (3): 519-523.

Barkema H W, Green M J, Bradley A J, et al. The role of contagious disease in udder health[J]. Journal of Dairy Science, 2009, 92 (10): 4717-4729.

Barkema H W, Schukken Y H, Zadoks R N. Invited Review: The Role of Cow, Pathogen, and Treatment Regimen in the Therapeutic Success of Bovine Staphylococcus aureus Mastitis[J]. Journal of Dairy Science, 2006, 89 (6): 1877-1895.

Batavani R A, Mortaz E, Falahian K, et al. Study on frequency, etiology and some enzymatic activities of subclinical ovine mastitis in Urmia, Iran[J]. Small Ruminant Research, 2003, 50: 45-50.

Berry D P, Lee J M, Macdonald K A, Stafford K, Matthews L, Roche J R. Associations among body condition score, body weight, somatic cell count, and clinical mastitis in seasonally calving dairy cattle[J]. Journal of Dairy Science, 2007, 90: 637-648.

Bicalho R C, Machado V S, Caixeta L S. Lameness in dairy cattle: A debilitating disease or a disease of debilitated cattle? A cross-sectional study of lameness prevalence and thickness of the digital cushion[J]. Journal of dairy science, 2009, 92 (7): 3175-3184.

Bicalho R C, Vokey F, Erb H N, et al. Visual locomotion scoring in the first seventy days in milk: Impact on pregnancy and survival[J]. Journal of dairy science, 2007, 90 (10): 4586-4591.

Biddle M K, Fox L K, Hancock D D, et al. Effects of storage time and thawing methods on the recovery of Mycoplasma species in milk samples from cows with intramammary infections[J]. Journal of Dairy Science, 2004, 87 (4): 933-936.

Biffa D, Debela E, Beyene F. Prevalence and risk factors of mastitis in lactating dairy cows in southern Ethiopia[J]. The International Journal of Applied Research in Veterinary Medicine, 2005, 3: 189-198.

Bjørg Heringstad, Gunnar Klemetsdal, John Ruane. Responses to Selection against Clinical Mastitis in the Norwegian Cattle Population[J]. Acta Agriculturae Scandinavica, 2001, 51 (2): 15-20.

Bohmanova J, Misztal I, Cole J B. Temperature-humidity indices as indicators of milk production losses due to heat stress[J]. Journal of dairy science, 2007, 90 (4): 1947-1956.

Booth C J, Warnick L D, Gröhn Y T, et al. Effect of lameness on culling in dairy cows[J]. Journal of dairy science, 2004, 87 (12): 4115-4122.

Boulanger V, Bouchard L, Zhao X, Lacasse P. Induction of nitric oxide production by bovine mammary epithelial cells and blood leukocytes[J]. Journal of Dairy Science, 2001, 84: 1430-1437.

Bourne N, Wathes D C, Lawrence K E, et al. The effect of parenteral supplementation of vitamin E with selenium on the health and productivity of dairy cattle in the UK[J]. Veterinary Journal, 2008, 177 (3): 381-387.

Bouwstra R J, Nielen M, Stegeman J A, et al. Vitamin E supplementation during the dry period in dairy cattle. Part 1: adverse effect on incidence of mastitis postpartum in a double-blind randomized field trial[J]. Journal of Dairy Science, 2010, 93 (12): 5684-5695.

Bowers S, Gandy S, Graves K, et al. Effects of prepartum milking on postpartum reproduction, udder health and production performance in first-calf dairy heifers[J]. Journal of Dairy Research, 2006, 73(3): 257-263.

Bradley A J. Bovine mastitis: an evolving disease[J]. The Veterinary Journal, 2002, 164 (2): 116-128.

Braem G, Vliegher S D, Supré K, et al. (GTG) 5-PCR fingerprinting for the classification and identification of coagulase[J]. Veterinary Microbiology, 2011, 147 (1-2): 67-74.

Breen J E, Green M J, Bradley A J. Quarter and cow risk factors associated with the occurrence of clinical mastitis in dairy cows in the United Kingdom[J]. Journal of Dairy Science, 2009, 92, 2551-2561.

Brinkmann V, Reichard U, Goosmann C, Fauler B, Uhlemann Y, Weiss D S, Weinrauch Y, Zychlinsky A. Neutrophil extracellular traps kill bacteria[J]. Science. 2004, 303: 1532-1535.

Brint E K, Xu D, Liu H, et al. ST2 is an inhibitor of interleukin 1 receptor and Toll-like receptor 4 signaling and maintains endotoxin tolerance[J]. Nature Immunology, 2004, 5 (4): 373-379.

Brozos C N, Kiossis E, Georgiadis M P, Piperelis S, Boscos C. The effect of chloride ammonium, vitamin E and Se supplementation throughout the dry period on the prevention of retained fetal membranes, reproductive performance and milk yield of dairy cows[J]. Livestock Science, 2009, 124: 210-215.

Bruno R G S, Rutigliano H, Cerri R L, et al. Effect of feeding yeast culture on reproduction and lameness in dairy cows under heat stress[J]. Animal reproduction science, 2009, 113 (1): 11-21.

Bubeck W J, Bae T, Otto M, et al. Poring over pores: alpha-hemolysin and Panton-Valentine leukocidin in Staphylococcus aureus pneumonia[J]. Nature Medicine, 2008, 13 (12): 1405-1406.

Burton J L, Erskine R J. Immunity and mastitis: Some new ideas for an old disease[J]. Veterinary Clinics of North America Food Animal Practice, 2003, 19 (1): 1-45.

Burvenich C, Van M V, Mehrzad J, et al. Severity of E. coli mastitis is mainly determined by cow factors[J]. Veterinary Research, 2003, 34 (5): 521-564.

Bytyqi H, Zaugg U, Sherifi K, Hamidi A, Gjonbalaj M, Muji S, Mehmeti H. Influence of management and physiological factors on somatic cell count in raw milk in Kosova[J]. Veterinarski Archiv, 2010, 80: 173-183.

Cavaillon J M, Adib-Conquy M, Fitting C, et al. Cytokine Cascade in Sepsis[J]. Scandinavian Journal of Infectious Diseases, 2003, 35 (9): 535-544.

Ceballos A, Kruze J, Barkema H W, et al. Barium selenate supplementation and its effect on intramammary infection in pasture-based dairy cows[J]. Journal of Dairy Science, 2010, 93 (4): 1468-1477.

Ceballos-Marquez A, Barkema H W, Stryhn H, et al. The effect of selenium supplementation before calving on early-lactation udder health in pastured dairy heifers[J]. Journal of Dairy Science, 2010, 93 (10): 4602-4612.

Čerňáková M, Košťálová D. Antimicrobial activity of berberine—a constituent of Mahonia aquifolium[J]. Folia Microbiologica, 2002, 47 (4): 375-378.

Cesa S J. Malondialdehyde contents in infant milk formulas[J]. Journal of Agricultural and Food Chemistry, 2004, 52: 2119-2122.

Chaiyotwittayakun A, Erskine R J, Bartlett P C, Herdt T H, Sears P M, Harmon R J. The effect of ascorbic acid and L-histidine therapy on acute mammary inflammation in dairy cattle[J]. Journal of Dairy Science, 2002, 85: 60-67.

Chandra G, Aggarwal A, Singh A K, Kumar1 M, Upadhyay R C. Effect of Vitamin E and Zinc Supplementation on Energy Metabolites, Lipid Peroxidation, and Milk Production in Peripartum Sahiwal Cows[J]. Asian-Australasian Journal of Animal Sciences, 2013, 26: 1569-1576.

Chang L, Karin M. Mammalian MAP kinase signaling cascades[J]. Nature, 2001, 410 (6824): 37-40.

Chapinal N, von Keyserlingk M A G, Cerri R L A, et al. Herd-level reproductive performance and its relationship with lameness and leg injuries in freestall dairy herds in the northeastern United States[J]. Journal of dairy science, 2013, 96 (11): 7066-7072.

Chawla R, Kaur H. Plasma antioxidant vitamin status of periparturient cows supplemented with α-tocopherol and β-carotene[J]. Animal Feed Science and Technology, 2004, 114: 279-285.

Cheung A L, Nishina K A, Trotonda M P, et al. The SarA protein family of Staphylococcus aureus[J]. International Journal of Biochemistry & Cell Biology, 2008, 40 (3): 355-361.

Cheung A L, Projan S J, Gresham H. The genomic aspect of virulence, sepsis, and resistance to killing mechanisms in Staphylococcus aureus [J]. Current Infectious Disease Reports, 2002, 4 (5): 400-410.

Cheung G Y C, Kevin R, Rong W, et al. Staphylococcus epidermidis Strategies to Avoid Killing by Human Neutrophils[J]. Plos Pathogens, 2010, 6 (10): 10943-10943.

Chiang Y C, Liao W W, Fan C M, Pai W Y, Chiou C S, Tsen H Y. PCR detection of Staphylococcal enterotoxins (SEs) N, O, P, Q, R, U, and survey of SE types in Staphylococcus aureus isolates from food-poisoning cases in Taiwan[J]. International Journal of Food Microbiology, 2008, 121: 66-73.

Cho J S, Pietras E M, Garcia N C, et al. IL-17 is essential for host defense against cutaneous Staphylococcus aureus infection in mice[J]. Journal of Clinical Investigation, 2010, 120 (5): 1762-1773.

Clyne L. A study investigating the prevalence and associated risk factors for heifer mastitis in the MID. In Countdown Symposium, Melbourne, Australia, 2013, page 16-19.

Colleen F, Qilin P, Mathison J C, et al. Triad3A Regulates Ubiquitination and Proteasomal Degradation of RIP1 following Disruption of Hsp90 Binding[J]. The Journal of Biological Chemistry, 2006, 281 (45): 34592-34600.

Compton C W R, Cursons R T M, Barnett C M E, et al. Expression of innate resistance factors in mammary secretion from periparturient dairy heifers and their association with subsequent infection status[J]. Veterinary Immunology & Immunopathology, 2009, 127 (3-4): 357-364.

Compton C W R, Heuer C, Parker K, et al. Epidemiology of mastitis in pasture-grazed peripartum dairy heifers and its effects on productivity[J]. Journal of Dairy Science, 2007, 90 (9): 4157-4170.

Compton C W R, Heuer C, Parker K, et al. Risk factors for peripartum mastitis in pasture-grazed dairy heifers[J]. Journal of Dairy Science, 2007, 90 (9): 4171-4180.

Cope C M, Mackenzie A M, Wilde D, Sinclair L A. Effects of level and form of dietary zinc on dairy cow performance and health[J]. Journal of Dairy Science, 2009, 92: 2128-2135.

Cortinhas C S, Botaro B G, Sucupira M C A, Renno F P, Santos M V. Antioxidant enzymes and somatic cell count in dairy cows fed with organic source of zinc, copper and selenium[J]. Livestock Science, 2010, 127: 84-87.

Cotter P D, Colin H R, Paul R. Bacteriocins: developing innate immunity for food[J]. Nature Reviews Microbiology, 2005, 3 (10): 777-788.

Cramer G, Lissemore K D, Guard C L, et al. The association between foot lesions and culling risk in Ontario Holstein cows[J]. Journal of dairy science, 2009, 92 (6): 2572-2579.

Cremonesi P, Luzzana M, Brasca M, Morandi S, Lodi R, Vimercati C, Agnellini D, Caramenti G, Moroni P, Castiglioni B. Development of a multiplex PCR assay for the identification of Staphylococcus aureus enterotoxigenic strains isolated from milk and dairy products[J]. Molecular and Cellular Probes, 2005, 19, 299-305.

Cuong V, Stanislava K, Voyich J M, et al. A crucial role for exopolysaccharide modification in bacterial biofilm formation, immune evasion, and virulence[J]. Journal of Biological Chemistry, 2004, 279 (52): 54881-54886.

Da Silva E R, Boechat J U D, Martins J C D, Ferreira W P B, Siqueira A P S, Da Silva N. Hemolysin production by Staphylococcus aureus species isolated from mastitic goat milk in Brazilian dairy herds[J]. Small Ruminant Research, 2005, 56, 271-275.

D'Angelo F, Santillo A, Sevi A, Albenzio M. Technical note: A simple salting-out method for DNA extraction from milk somatic cells: investigation into the goat CSN1S1 gene[J]. Journal of Dairy Science, 2007, 90: 3550-3552.

Daniels K J, Donkin S S, Eicher S D, et al. Prepartum milking of heifers influences future production and health[J]. Journal of Dairy Science, 2007, 90 (5): 2293-2301.

Daum R S, Spellberg B. Progress Toward a Staphylococcus aureus Vaccine[J]. Clinical Infectious Diseases An Official Publication of the Infectious Diseases Society of America, 2012, 54 (4): 560-567.

De S, Brahma B, Polley S, Mukherjee A, Banerjee D, Gohaina M, Singh K P, Singh R, Datta R K, Goswami

S L. Simplex and duplex PCR assays for species specific identification of cattle and buffalo milk and cheese[J]. Food Control, 2011, 22: 690-696.

De Vliegher S, Fox L K, Piepers S, McDougall S, Barkema H W. Invited review: Mastitis in dairy heifers: nature of the disease, potential impact, prevention, and control[J]. Journal of Dairy Science, 2012, 95: 1025-1040.

Dinges M M, Orwin P M, Schlievert P M. Exotoxins of Staphylococcus aureus [J]. Clinical Microbiology Reviews, 2000, 13 (1): 16-34.

Dippel S, Dolezal M, Brenninkmeyer C, et al. Risk factors for lameness in freestall-housed dairy cows across two breeds, farming systems, and countries[J]. Journal of dairy science, 2009, 92 (11): 5476-5486.

Dobson H, Smith R F. What is stress, and how does it affect reproduction?[J]. Animal reproduction science, 2000, 60: 743-752.

Dosogne H, Vangroenweghe F, Mehrzad J, et al. Differential Leukocyte Count Method for Bovine Low Somatic Cell Count Milk[J]. Journal of Dairy Science, 2003, 86 (3): 828-834.

Eland L E, Davenport R, Mota C R. Evaluation of DNA extraction methods for freshwater eukaryotic microalgae[J]. Water Research, 2012, 46 (16): 5355-5364.

El-Sayed A, Alber J, Lammler C, Bonner B, Huhn A, Kaleta E F, Zschock M. PCR-based detection of genes encoding virulence determinants in Staphylococcus aureus from birds[J]. Journal of veterinary medicine series B, Infectious diseases and veterinary public health, 2005, 52, 38-44.

Essmann F, Bantel H, Totzke G, et al. Staphylococcus aureus alpha-toxin-induced cell death: predominant necrosis despite apoptotic caspase activation[J]. Cell Death & Differentiation, 2003, 10 (11): 1260-1272.

Fattom A, Fuller S, Propst M, et al. Safety and immunogenicity of a booster dose of Staphylococcus aureus types 5 and 8 capsular polysaccharide conjugate vaccine (StaphVAX) in hemodialysis patients[J]. Vaccine, 2004, 23 (5): 656-663.

Ferens W A, Bohach G A. Persistence of Staphylococcus aureus on mucosal membranes: superantigens and internalization by host cells[J]. Journal of Laboratory & Clinical Medicine, 2000, 135 (3): 225-230.

Fischer G, Conceição F R, Leite F P L, et al. Immunomodulation produced by a green propolis extract on humoral and cellular responses of mice immunized with SuHV-1[J]. Vaccine, 2007, 25 (7): 1250-1256.

Fitzgerald J R, Hartigan P J, Meaney W J, Smyth C J. Molecular population and virulence factor analysis of Staphylococcus aureus from bovine intramammary infection[J]. Journal of Applied Microbiology, 2000, 88: 1028-1037.

Fluit A C, Schmitz F J, Verhoef J. Frequency of Isolation of Pathogens from Bloodstream, Nosocomial Pneumonia, Skin and Soft Tissue, and Urinary Tract Infections Occurring in European Patients[J]. European Journal of Clinical Microbiology & Infectious Diseases, 2001, 20 (3): 188-191.

Fowler T, Wann E R, Joh D, Johansson S, Foster T J, Hook M. Cellular invasion by Staphylococcus aureus involves a fibronectin bridge between the bacterial fibronectin-binding MSCRAMMs and host cell beta1 integrins[J]. European Journal of Cell Biology, 2000, 79, 672-679.

Fox L K, Hancock D D, Mickelson A, et al. Bulk tank milk analysis: Factors associated with appearance of Mycoplasma sp in milk[J]. Journal of Veterinary Medicine, 2003, 50 (5): 235-240.

Fox P F, Kelly A L. Indigenous enzymes in milk: Overview and historical aspects—Part 2[J]. International Dairy Journal, 2006, 16: 517-532.

Fuchs T A, Ulrike A, Christian G, et al. Novel cell death program leads to neutrophil extracellular traps[J].

Journal of Cell Biology, 2007, 176 (2): 231-241.

Gaafar H M A, Basiuoni M I, Ali M F E, Shitta A A, Shamas A S E. Effect of zinc methionine supplementation on somatic cell count in milk and mastitis in Friesian cows[J]. Archiva Zootechnica, 2010, 13: 36-46.

Gabriela T, Márcia D, Angela F, et al. Leukocyte populations and cytokine expression in the mammary gland in a mouse model of Streptococcus agalactiae mastitis[J]. Journal of Medical Microbiology, 2009, 58(7): 951-958.

Garbarino E J, Hernandez J A, Shearer J K, et al. Effect of lameness on ovarian activity in postpartum Holstein cows[J]. Journal of dairy science, 2004, 87 (12): 4123-4131.

Gaudreau M C, Lacasse P, Talbot B G. Protective immune responses to a multi-gene DNA vaccine against Staphylococcus aureus[J]. Vaccine, 2007, 25 (5): 814-824.

Gerhardt S, Abbott W M, Hargreaves D, et al. Structure of IL-17A in complex with a potent, fully human neutralizing antibody[J]. Journal of Molecular Biology, 2009, 394 (5): 905-921.

Getahun K, Kelay B, Bekana M, et al. Bovine mastitis and antibiotic resistance patterns in Selalle smallholder dairy farms, central Ethiopia[J]. Tropical Animal Health and Production, 2008, 40: 261-268.

Ghosh S, Karin M. Missing Pieces in the NF-κB Puzzle[J]. Cell, 2002, 109 (2Suppl1): S81-S96.

Gianneechini R, Concha C, Rivero R, et al. Occurrence of Clinical and Sub-Clinical Mastitis in Dairy Herds in the West Littoral Region in Uruguay[J]. Acta Veterinaria Scandinavica, 2002, 43: 221-230.

Gillespie B E, Oliver S P. Simultaneous detection of mastitis pathogens, Staphylococcus aureus, Streptococcus uberis, and Streptococcus agalactiae by multiplex real-time polymerase chain reaction[J]. Journal of Dairy Science, 2005, 88 (10): 3510-3518.

Girard C L, Matte J J. Effects of Intramuscular Injections of Vitamin B12 on Lactation Performance of Dairy Cows Fed Dietary Supplements of Folic Acid and Rumen-Protected Methionine[J]. Journal of Dairy Science, 2005, 88: 671-676.

Graulet B, Matte J J, Desrochers A, Doepel L, Palin M-F, Girard C L. Effects of Dietary Supplements of Folic Acid and Vitamin B12 on Metabolism of Dairy Cows in Early Lactation[J]. Journal of Dairy Science, 2007, 90: 3442-3455.

Green L E, Hedges V J, Schukken Y H, et al. The impact of clinical lameness on the milk yield of dairy cows[J]. Journal of Dairy Science, 2002, 85 (9): 2250-2256.

Green M J, Green L E, Medley G F, Schukken Y H, Bradley AJ. Influence of dry period bacterial intramammary infection on clinical mastitis in dairy cows[J]. Journal of Dairy Science, 2002, 85: 2589-2599.

Green M, Bradley A. The changing face of mastitis control[J]. Veterinary Record, 2013, 173 (21): 517-521.

Griffiths L M, Loeffler S H, Socha M T, Tomlinson D J, Johnson A B. Effects of supplementing complexed zinc, manganese, copper and cobalt on lactation and reproductive performance of intensively grazed lactating dairy cattle on the South Island of New Zealand[J]. Animal Feed Science and Technology, 2007, 137: 69-83.

Gropper S S, Smith J, Groff J. Advanced Nutrition and Human Metabolism: Copper transport and uptake. 4thed. Wadsworth. Belmont, CA; 2005, pp 449-451.

Gruet P, Maincent P, Berthelot X, et al. Bovine mastitis and intramammary drug delivery: review and perspectives[J]. Advanced Drug Delivery Reviews, 2001, 50: 245-259.

Gu B B, Zhu Y M, Zhu W, Miao J F, Deng Y E, Zou S X. Retinoid protects rats against neutrophil-induced oxidative stress in acute experimental mastitis[J]. International Immunopharmacology, 2009, 9: 223-229.

Güler L, Ok Ü, Gündüz K, et al. Antimicrobial susceptibility and coagulase gene typing of Staphylococcus aureus isolated from bovine clinical mastitis cases in Turkey[J]. Journal of Dairy Science, 2005, 88 (9): 3149-3154.

Gunawardana S, Thilakarathne D, Abegunawardana I S, et al. Risk factors for bovine mastitis in the Central Province of Sri Lanka[J]. Tropical Animal Health & Production, 2014, 46 (7): 1105-1112.

Gunay A, Gunay U. Effects of Clinical Mastitis on Reproductive Performance in Holstein Cows[J]. Acta Veterinaria Brno, 2008, 77: 555-560.

Gunaydin B, Aslantas O, Demir C. Detection of superantigenic toxin genes in Staphylococcus aureus strains from subclinical bovine mastitis[J]. Tropical Animal Health and Production, 2011, 43, 1633-1637.

Haas D M, Michael D, Todd S, et al. Human breast milk as a source of DNA for amplification.[J]. Journal of Clinical Pharmacology, 2011, 51 (4): 616-619.

Haftu R, Taddele H, Gugsa G, Kalayou S. Prevalence, bacterial causes, and antimicrobial susceptibility profile of mastitis isolates from cows in large-scale dairy farms of Northern Ethiopia[J]. Tropical Animal Health and Production, 2012, 44, 1765-1771.

Han H R, Pak S, Guidry A. Prevalence of capsular polysaccharide (CP) types of Staphylococcus aureus isolated from bovine mastitic milk and protection of S. aureus infection in mice with CP vaccine[J]. The Journal of Veterinary Medical Science, 2000, 62 (12): 1331-1333.

Hans-Curt F, Neu T R, Wozniak D J. The EPS matrix: the "house of biofilm cells"[J]. Journal of Bacteriology, 2007, 189 (22): 7945-7947.

Hansen P, Soto P, Natzke R. Mastitis and fertility in cattle—possible involvement of inflammation or immune activation in embryonic mortality[J]. American Journal of Reproductive Immunology, 2004, 51 (4): 294-301.

Hayajneh F M F. Antioxidants in dairy cattle health and disease[J]. Bulletin of University of Agricultural Sciences and Veterinary Medicine Cluj-Napoca. Veterinary Medicine, 2014, 71: 104-109.

He J, Wang Y, Xu L H, et al. Cucurbitacin IIa induces caspase-3-dependent apoptosis and enhances autophagy in lipopolysaccharide-stimulated RAW 264.7 macrophages[J]. International Immunopharmacology, 2013, 16 (1): 27-34.

Healy P J, Dennis J A, Windsor P A, et al. Genotyping cattle for inherited congenital myoclonus and maple syrup urine disease[J]. Australian Veterinary Journal, 2002, 80 (11): 695-697.

Heinrichs A J, Costello S S, Jones C M, et al. Control of heifer mastitis by nutrition[J]. Veterinary Microbiology, 2009, 134 (1-2): 172-176.

Heringstad B, Klemetsdal G, Ruane J. Responses to selection against clinical mastitis in the Norwegian cattle population[J]. Animal Science, 2001, 51: 155-165.

Heringstad B, Klemetsdal G, Ruane J. Selection for mastitis resistance in dairy cattle: a review with focus on the situation in the Nordic countries[J]. Livestock Production Science, 2000, 64: 95-106.

Hernandez J A, Garbarino E J, Shearer J K, et al. Comparison of the calving-to-conception interval in dairy cows with different degrees of lameness during the prebreeding postpartum period[J]. Journal of the American Veterinary Medical Association, 2005, 227 (8): 1284-1291.

Hernandez J, Shearer J K, Webb D W. Effect of lameness on the calving-to-conception interval in dairy cows[J]. Journal of the American veterinary medical association, 2001, 218 (10): 1611-1614.

Hernandez J, Shearer J K, Webb D W. Effect of papillomatous digital dermatitis and other lameness disorders on reproductive performance in a florida dairy herd[C]. International symposium on disorders of the ruminant digit. 2000, 11: 300-302.

Hertl J A, Grohn Y T, Leach J D, Bar D, Bennett G J, Gonzalez R N, Rauch B J, Welcome F L, Tauer L W, Schukken Y H. Effects of clinical mastitis caused by gram-positive and gram-negative bacteria and other organisms on the probability of conception in New York State Holstein dairy cows[J]. Journal of Dairy Science, 2010, 93: 1551-1560.

Hideyuki Takahashi, Masaharu Odai, Kenji Effect of intramammary injection of rboGM-CSF on milk levels of chemiluminescence activity, somatic cell count, and Staphylococcus aureus count in Holstein cows with S. aureus subclinical mastitis[J]. The Canadian Journal of Veterinary Research, 2004, 68 (3): 182-187.

Hiroaki M, Chiaki N, Naoki H, et al. Recombinant soluble forms of extracellular TLR4 domain and MD-2 inhibit lipopolysaccharide binding on cell surface and dampen lipopolysaccharide-induced pulmonary inflammation in mice[J]. Journal of Immunology, 2007, 177 (11): 8133-8139.

Hiss S, Mielenz M, Bruckmaier R M, et al. Haptoglobin concentrations in blood and milk after endotoxin challenge and quantification of mammary Hp mRNA expression[J]. Journal of Dairy Science, 2004, 87 (11): 3778-3784.

Hockett M E, Hopkins F M, Lewis M J, et al. Endocrine profiles of dairy cows following experimentally induced clinical mastitis during early lactation[J]. Animal Reproduction Science, 2000, 58: 241-251.

Hoedemaker M, Prange D, Gundelach Y. Body Condition Change Ante-and Postpartum, Health and Reproductive Performance in German Holstein Cows[J]. Reproduction in Domestic Animals, 2009, 44 (2): 167-173.

Hogan J, Larry S K. Coliform mastitis[J]. Veterinary Research, 2003, 34 (5): 507-519.

Huijps K, Lam T J G M, Hogeveen H. Costs of mastitis: facts and perception[J]. Journal of Dairy Research, 2008, 75: 113-120.

Hultgren O H, Stenson M, Tarkowski A. Role of IL-12 in Staphylococcus aureus-triggered arthritis and sepsis[J]. Arthritis Research, 2001, 3 (1): 41-47.

Huszenicza G, Janosi S, Kulcsar M, Korodi P, Reiczigel J, Katai L, Peters A R, De Rensis F. Effects of clinical mastitis on ovarian function in post-partum dairy cows[J]. Reproduction in Domestic Animals, 2005, 40: 199-204.

Israel A. The IKK complex, a central regulator of NF-kappaB activation[J]. Cold Spring Harbor Perspectives in Biology, 2010, 2 (3): a000158.

Iwasaki A, Medzhitov R. Toll-like receptor control of the adaptive immune responses[J]. Nature Immunology, 2004, 5 (10): 987-995.

Iyer A V, Ghosh S, Singh S N, et al. Evaluation of three 'ready to formulate' oil adjuvants for foot-and-mouth disease vaccine production[J]. Vaccine, 2000, 19 (9-10): 1097-1105.

Jaione V, Cristina L, Carmen G, et al. Bap, a biofilm matrix protein of Staphylococcus aureus prevents cellular internalization through binding to GP96 host receptor[J]. Plos Pathogens, 2012, 8 (8): e1002843.

Jerry W S, William P W. Role of antioxidants and trace elements in health and immunity of transition dairy cows[J]. The Veterinary Journal, 2008, 176: 70-76.

Jhambh R, Dimri U, Gupta V K, Rathore R. Blood antioxidant profile and lipid peroxides in dairy cows with clinical mastitis[J]. Veterinary World, 2013, 6: 271-273.

Jin G, Takayuki M, Jun-Ichiro I. Cutting edge: TNFR-associated factor (TRAF) 6 is essential for MyD88-dependent pathway but not toll/IL-1 receptor domain-containing adaptor-inducing IFN-beta

(TRIF) -dependent pathway in TLR signaling[J]. Journal of Immunology, 2004, 173 (2): 2913-2917.

Jin L, Yan S M, Shi B L, Bao H Y, Gong J, Guo X Y, Li J L. Effects of vitamin A on the milk performance, antioxidant functions and immune functions of dairy cows[J]. Animal Feed Science and Technology, 2014, 192: 15-23.

Johnson G L, Razvan L. Mitogen-activated protein kinase pathways mediated by ERK, JNK, and p38 protein kinases[J]. Science, 2002, 298 (5600): 1911-1912.

Jorgensen J H, Ferraro M J. Antimicrobial susceptibility testing: a review of general principles and contemporary practices[J]. Clinical Infectious Diseases, 2009, 49 (11): 1749-1755.

Juliane B W, Patel R J, Olaf S. Surface proteins and exotoxins are required for the pathogenesis of Staphylococcus aureus pneumonia[J]. Infection & Immunity, 2007, 75 (2): 1040-1044.

Juliarena M, Gutierrez S, Ceriani C. Chicken antibodies: a useful tool for antigen capture ELISA to detect bovine leukaemia virus without cross-reaction with other mammalian antibodies[J]. Veterinary Research Communications, 2007, 31 (1): 43-51.

Jungi T W. Research from the Division of Immunology: production of nitric oxide (NO) by macrophages in ruminants[J]. Schweizer Archiv Ffur Tierheilkunde, 2000, 142: 215-217.

Juniper D T, Phipps R H, Jones A K, et al. Selenium supplementation of lactating dairy cows: effect on selenium concentration in blood, milk, urine, and feces[J]. Journal of Dairy Science, 2006, 89: 3544-3551.

Kagan J C, Ruslan M. Phosphoinositide-mediated adaptor recruitment controls Toll-like receptor signaling[J]. Cell, 2006, 125 (5): 943-955.

Kalmus P, Viltrop A, Aasmae B, Kask K. Occurrence of clinical mastitis in primiparous Estonian dairy cows in different housing conditions[J]. Acta Veterinaria Scandinavica, 2006, 48, 21.

Kalorey D R, Shanmugam Y, Kurkure N V, et al. PCR-based detection of genes encoding virulence determinants in Staphylococcus aureus from bovine subclinical mastitis cases[J]. Journal of Veterinary Science, 2007, 8 (2): 151-154.

Kaper J B. Pathogenic Escherichia coli[J]. Nature Reviews Microbiology, 2004, 2 (2): 123-140.

Karin M, Greten F R. NF-kappaB: linking inflammation and immunity to cancer development and progression[J]. Nature Reviews Immunology, 2005, 5 (10): 749-759.

Katherine O, Lee J C. Staphylococcus aureus capsular polysaccharides[J]. Clinical Microbiology Reviews, 2004, 17 (1): 218-234.

Kellogg D W, Tomlinson D J, Socha M T, et al. Effects of zinc methionine complex on milk production and somatic cell count of dairy cows: twelve-trial summary[J]. The Professional Animal Scientist, 2004, 20: 295-301.

Kelly A L, Tiernan D, O'Sullivan C, Joyce P. Correlation between bovine milk somatic cell count and polymorphonuclear leukocyte level for samples of bulk milk and milk from individual cows[J]. J Dairy Sci, 2000, 83, 300-304.

Kerr D E, Plaut K, Bramley A J, et al. Lysostaphin expression in mammary glands confers protection against staphylococcal infection in transgenic mice[J]. Nature Biotechnology, 2001, 19 (1): 66-70.

Kerro D O, van Dijk J E, Nederbragt H. Factors involved in the early pathogenesis of bovine Staphylococcus aureus mastitis with emphasis on bacterial adhesion and invasion. A review[J]. Veterinary Quarterly, 2002, 24 (4): 182-198.

Kerro Dego O, Tareke F. Bovine Mastitis in Selected Areas of Southern Ethiopia[J]. Tropical Animal Health and

Production, 2003, 35: 197-205.

Keshari R S, Jyoti A, Dubey M, et al. Cytokines Induced Neutrophil Extracellular Traps Formation: Implication for the Inflammatory Disease Condition[J]. Plos One, 2012, 7 (10): 65-65.

Khan A Z, Muhammad G. 2005. Quarter-wise comparative prevalence of mastitis in buffaloes and crossbred cows[J]. Pakistan Veterinary Journal, 25, 9-12.

Khanal T, Pandit A. Assessment of sub-clinical mastitis and its associated risk factors in dairy livestock of Lamjung, Nepal[J]. International Journal of Infection and Microbiology, 2013, 2, 49-54.

Kimman T G, Vandebriel R J, Hoebee B. Genetic variation in the response to vaccination[J]. Community Genetics, 2007, 10 (4): 201-217.

Kivaria F M, Noordhuizen J P T M, Kapaga A M. Risk Indicators Associated with Subclinical Mastitis in Smallholder Dairy Cows in Tanzania[J]. Tropical Animal Health and Production, 2004, 36: 581-592.

Kivaria F M, Noordhuizen J P T M, Nielen M. Interpretation of California mastitis test scores using Staphylococcus aureus culture results for screening of subclinical mastitis in low yielding smallholder dairy cows in the Dar es Salaam region of Tanzania[J]. Preventive Veterinary Medicine, 2007, 78: 274-285.

Kivaria F M, Noordhuizen J P, Msami H M. Risk factors associated with the incidence rate of clinical mastitis in smallholder dairy cows in the Dar es Salaam region of Tanzania[J]. The Veterinary Journal, 2007, 173, 623-629.

Kiyoshi T, Shizuo A. Toll receptors and pathogen resistance[J]. Cellular Microbiology, 2003, 5 (3): 143-153.

Kleczkowski M, Kluciński W, Shaktur A, Sikora J. Concentration of ascorbic acid in the blood of cows with subclinical mastitis[J]. Polish Journal of Veterinary Sciences, 2005, 8: 121-125.

Klein E A. Selenium and Vitamin E: Interesting Biology and Dashed Hope[J]. Journal of the National Cancer Institute, 2009, 101 (5): 283-285.

Kolls J K, Lindén A. Interleukin-17 Family Members and Inflammation[J]. Immunity, 2004, 21 (4): 467-476.

Komine K, Kuroishi T, Komine Y, Watanabe K, Kobayashi J, Yamaguchi T, Kamata S, Kumagai K. Induction of nitric oxide production mediated by tumor necrosis factor alpha on staphylococcal enterotoxin C-stimulated bovine mammary gland cells[J]. Clinical and Diagnostic Laboratory Immunology, 2004, 11: 203-210.

Kromker V, Pfannenschmidt F, Helmke K, Andersson R, Grabowski N T. 2012. Risk factors for intramammary infections and subclinical mastitis in post-partum dairy heifers[J]. Journal of Dairy Research, 79, 304-309.

Lakew M, Tolosa T, Tigre W. Prevalence and major bacterial causes of bovine mastitis in Asella, South Eastern Ethiopia[J]. Tropical Animal Health and Production, 2009, 41 (7): 1525-1530.

Lalor S J, Dungan L S, Sutton C E, Basdeo S A, Fletcher J M, Mills K H. Caspase-1-Processed Cytokines IL-1β and IL-18 Promote IL-17 Production by γδ and CD4 T Cells That Mediate Autoimmunity[J]. Journal of Immunology, 2011, 186 (10): 5738-5748.

Lan L, Cheng A, Dunman P M, Missiakas D, He C. Golden pigment production and virulence gene expression are affected by metabolisms in Staphylococcus aureus[J]. Journal of Bacteriology, 2010, 192 (12): 3068-3077.

Larsen H D, Aarestrup F M, Jensen N E. Geographical variation in the presence of genes encoding superantigenic exotoxins and beta-hemolysin among Staphylococcus aureus isolated from bovine mastitis in Europe and USA[J]. Veterinary Microbiology, 2002, 85, 61-67.

Laura G Á, Holden M T G, Heather L, et al. Meticillin-resistant Staphylococcus aureus with a novel mecA homologue in human and bovine populations in the UK and Denmark: a descriptive study[J]. Lancet Infectious Diseases, 2011, 11 (8): 595-603.

LeBlanc SJ, Herdt TH, Seymour WM, Duffield TF, Leslie KE. Peripartum serum vitamin E, retinol, and beta-carotene in dairy cattle and their associations with disease[J]. Journal of Dairy Science, 2004, 87: 609-619.

Lee J H. Methicillin (Oxacillin) -resistant Staphylococcus aureus strains isolated from major food animals and their potential transmission to humans[J]. Applied & Environmental Microbiology, 2003, 69 (11): 6489-6494.

Lehtolainen T, Shwimmer A, Shpigel N Y, et al. In Vitro Antimicrobial Susceptibility of Escherichia coli Isolates from Clinical Bovine Mastitis in Finland and Israel[J]. Journal of Dairy Science, 2004, 86 (12): 3927-3932.

Leitner G, Lubashevsky E, Trainin Z. Staphylococcus aureus vaccine against mastitis in dairy cows, composition and evaluation of its immunogenicity in a mouse model[J]. Veterinary Immunology and Immunopathology, 2003, 93 (3-4): 159-167.

Leitner G, Shoshani E, Krifucks O, et al. Milk Leucocyte Population Patterns in Bovine Udder Infection of Different Aetiology[J]. Journal of Veterinary Medicine, 2000, 47 (8): 581-589.

Leyva A, Franco A, Gonzalez T, et al. A rapid and sensitive ELISA to quantify an HBsAg specific monoclonal antibody and a plant-derived antibody during their downstream purification process[J]. Biologicals, 2007, 35 (1): 19-25.

Li H, Wang Y, Ding L, et al. Staphylococcus sciuri Exfoliative Toxin C (ExhC) is a Necrosis-Inducer for Mammalian Cells[J]. Plos One, 2011, 6 (7): e23145-e23145.

Li M, Du X, Villaruz A E, et al. MRSA epidemic linked to a quickly spreading colonization and virulence determinant[J]. Nature Medicine, 2012, 18 (5): 816-819.

Li M, Lai Y, Villaruz A E, et al. Gram-positive three-component antimicrobial peptide-sensing system. Proc Natl Acad Sci USA[J]. Proceedings of the National Academy of Sciences, 2007, 104 (22): 9469-9474.

Li M, Rigby K, Lai Y P, et al. Staphylococcus aureus mutant screen reveals interaction of the human antimicrobial peptide dermcidin with membrane phospholipids[J]. Antimicrobial Agents & Chemotherapy, 2009, 53 (10): 4200-4210.

Liew F Y, Damo X, Brint E K, et al. Negative regulation of Toll-like receptor-mediated immune responses[J]. Nature Reviews Immunology, 2005, 5 (6): 446-458.

Lin A M, Rubin C J, Ritika K, et al. Mast cells and neutrophils release IL-17 through extracellular trap formation in psoriasis[J]. Journal of Immunology, 2011, 187 (1): 490-500.

Lindmark-Mansson H, Akesson B. Antioxidative factors in milk[J]. British Journal of Nutrition, 2000, 84 (Suppl. 1): 103-110.

Lindsay J A, Holden M T G. Staphylococcus aureus: superbug, super genome?[J]. Trends in Microbiology, 2004, 12 (8): 378-385.

Lippolis J D, Reinhardt T A, Goff J P, et al. Neutrophil extracellular trap formation by bovine neutrophils is not inhibited by milk[J]. Veterinary Immunology & Immunopathology, 2006, 113 (1-2): 248-255.

Liu G Y, Anthony E, Buchanan J T, et al. Staphylococcus aureus golden pigment impairs neutrophil killing and promotes virulence through its antioxidant activity[J]. Journal of Experimental Medicine, 2005, 202 (2): 209-215.

Lopez-Calleja I, Gonzalez I, Fajardo V, Rodriguez M A, Hernandez P E, Garcia T, Martin R. Rapid detection of cows' milk in sheeps' and goats' milk by a species-specific polymerase chain reaction technique[J]. Journal of Dairy Science, 2004, 87: 2839-2845.

Lovseth A, Loncarevic S, Berdal K G. Modified multiplex PCR method for detection of pyrogenic exotoxin genes in staphylococcal isolates[J]. Journal of Clinical Microbiology, 2004, 42 (8): 3869-3872.

Lucy M C. Reproductive loss in high-producing dairy cattle: where will it end?[J]. Journal of dairy science, 2001, 84 (6): 1277-1293.

Lykkesfeldt J, Svendsen O. Oxidants and antioxidants in disease: oxidative stress in farm animals[J]. The Veterinary Journal, 2007, 173: 502-511.

Macintosh R L, Brittan J L, Ritwika B, et al. The terminal A domain of the fibrillar accumulation-associated protein (Aap) of Staphylococcus epidermidis mediates adhesion to human corneocytes[J]. Journal of Bacteriology, 2009, 191 (22): 7007-7016.

Mack D, Davies A P, Harris L G, et al. Microbial interactions in Staphylococcus epidermidis biofilms[J]. Analytical & Bioanalytical Chemistry, 2007, 387 (2): 399-408.

Majee D N, Schwab E C, Bertics S J, Seymour W M, Shaver R D. Lactation Performance by Dairy Cows Fed Supplemental Biotin and a B-Vitamin Blend[J]. Journal of Dairy Science, 2003, 86: 2106-2112.

Makovec J A, Ruegg P L. Results of milk samples submitted for microbiological examination in Wisconsin from 1994 to 2001[J]. Journal of Dairy Science, 2003, 86 (11): 3466-3472.

Mancuso G, Gambuzza M, Midiri A, et al. Bacterial recognition by TLR7 in the lysosomes of conventional dendritic cells[J]. Nature Immunology, 2009, 10 (6): 587-594.

Mandron M, Ariès M F, Brehm R D, et al. Mandron, M. et al. Human dendritic cells conditioned with Staphylococcus aureus enterotoxin B promote TH2 cell polarization[J]. Journal of Allergy and Clinical Immunology, 2006, 117 (5): 1141-1147.

Martínez J L, Fernando B. Interactions among strategies associated with bacterial infection: pathogenicity, epidemicity, and antibiotic resistance[J]. Clinical Microbiology Reviews, 2002, 15 (4): 647-679.

McNamara P J, Syverson R E, Milligan-Myhre K, et al. Surfactants, aromatic and isoprenoid compounds, and fatty acid biosynthesis inhibitors suppress Staphylococcus aureus production of toxic shock syndrome toxin 1[J]. Antimicrobial Agents & Chemotherapy, 2009, 53 (5): 1898-1906.

Medzhitov R. Recognition of microorganisms and activation of the immune response[J]. Nature, 2007, 449 (7164): 819-826.

Mehrzad J, Duchateau L, Burvenich C. Viability of milk neutrophils and severity of bovine coliform mastitis[J]. Journal of Dairy Science, 2004, 87 (12): 4150-4162.

Mekibib B, Furgasa M, Abunna F, Megersa B, Regassa A. Bovine mastitis: Prevalence, risk factors and major pathogens in dairy farms of Holeta Town, Central Ethiopia[J]. Veterinary World, 2010, 3: 397-403.

Melchior M B, Vaarkamp H, Fink-Gremmels J. Biofilms: a role in recurrent mastitis infections?[J]. Veterinary Journal, 2006, 171 (3): 398-407.

Melendez P, Bartolome J, Archbald L F, et al. The association between lameness, ovarian cysts and fertility in lactating dairy cows[J]. Theriogenology, 2003, 59 (3): 927-937.

Meylan E, Burns K, Hofmann K, et al. RIP1 is an essential mediator of Toll-like receptor 3-induced NF-κB activation[J]. Nature Immunology, 2004, 5: 503-507.

Middleton J R, Timms L L, Bader G R, et al. Effect of prepartum intramammary treatment with pirlimycin hydrochloride on prevalence of early first-lactation mastitis in dairy heifers[J]. Journal of the American Veterinary Medical Association, 2005, 227 (12): 1969-1974.

Miranda M, Muriach M, Almansa I, Jareño E, Bosch-Morell F, Romero F J, Silvestre D. Oxidative status of human milk and its variations during cold storage[J]. BioFactors, 2004, 20: 129-137.

Moeini M M, Karami H, Mikaeili E. Effect of selenium and vitamin E supplementation during the late pregnancy on reproductive indices and milk production in heifers[J]. Animal Reproduction Science, 2009, 114: 109-114.

Momtaz H, Rahimi E, Tajbakhsh E. Detection of some virulence factors in Staphylococcus aureus isolated from clinical and subclinical bovine mastitis in Iran[J]. African Journal of Biotechnology, 2010, 25, 3753-3758.

Moon J S, Lee A R, Kang H M, et al. Phenotypic and genetic antibiogram of methicillin-resistant staphylococci isolated from bovine mastitis in Korea[J]. Journal of Dairy Science, 2007, 90 (3): 1176-1185.

Moore P C, Lindsay J A. Genetic variation among hospital isolates of methicillin-sensitive Staphylococcus aureus: evidence for horizontal transfer of virulence genes[J]. Journal of Clinical Microbiology, 2001, 39: 2760-2767.

Morris M J, Kaneko K, Walker S L, et al. Influence of lameness on follicular growth, ovulation, reproductive hormone concentrations and estrus behavior in dairy cows[J]. Theriogenology, 2011, 76 (4): 658-668.

Morris M J, Walker S L, Jones D N, et al. Influence of somatic cell count, body condition and lameness on follicular growth and ovulation in dairy cows[J]. Theriogenology, 2009, 71 (5): 801-806.

Mrode R A, Swanson G J T. Estimation of genetic parameters for somatic cell count in the first three lactations using random regression[J]. Livestock Production Science, 2003, 79 (s 2-3): 239-247.

Mukherjee R, De U K, Ram G C. Evaluation of mammary gland immunity and therapeutic potential of Tinospora cordifolia against bovine subclinical mastitis[J]. Tropical Animal Health & Production, 2010, 42 (4): 645-651.

Mukherjee R. Selenium and vitamin E increases polymorphonuclear cell phagocytosis and antioxidant levels during acute mastitis in riverine buffaloes[J]. Veterinary Research Communications, 2008, 32 (4): 305-313.

Mungube E O, Tenhagen B A, Kassa T, et al. Risk Factors for Dairy Cow Mastitis in the Central Highlands of Ethiopia[J]. Tropical Animal Health and Production, 2004, 36: 463-472.

Murphy M A, Shariflou M R, Moran C. High quality genomic DNA extraction from large milk samples[J]. Journal of Dairy Research, 2002, 69, 645-649.

Nakamura K, Miyazato A G, Hatta M, et al. Deoxynucleic Acids from Cryptococcus neoformans Activate Myeloid Dendritic Cells via a TLR9-Dependent Pathway[J]. Journal of Immunology, 2008, 180 (6): 4067-4074.

Naresh R, Dwivedi S K, Swarup D, Patra R C. Evaluation of ascorbic acid treatment in clinical and subclinical mastitis of Indian dairy cows[J]. Asian-Australasian Journal of Animal Sciences, 2002, 15: 905-911.

Nash D L, Rogers G W, Cooper J B, et al. Heritability of clinical mastitis incidence and relationships with sire transmitting abilities for somatic cell score, udder type traits, productive life, and protein yield[J]. Journal of Dairy Science, 2000, 83 (10): 2350-2360.

Nashev D, Toshkova K, Salasia S I, Hassan A A, Lammler C, Zschock M. Distribution of virulence genes of Staphylococcus aureus isolated from stable nasal carriers[J]. FEMS Microbiology Letters, 2004, 233, 45-52.

National Research Council (NRC). Subcommittee on Dairy Cattle Nutrition, Committee on Animal Nutrition,

National Research Council. Nutrient Requirements of Dairy Cattle, 7th rev. ed. Natl. Acad. Sci, 2001. Washington, DC.

Nava-Trujillo H, Soto-Belloso E, Hoet A E. Effects of clinical mastitis from calving to first service on reproductive performance in dual-purpose cows[J]. Animal Reproduction Science, 2010, 121: 12-16.

Nocek J E, Socha M T, Tomlinson D J. The effect of trace mineral fortification level and source on performance of dairy cattle[J]. Journal of Dairy Science, 2006, 89: 2679-2693.

Nonnecke B J, Kimura K, Goff J P, et al. Effects of the mammary gland on functional capacities of blood mononuclear leukocyte populations from periparturient cows[J]. Journal of Dairy Science, 2003, 86 (7): 2359-2368.

O'Flaherty C, Beorlegui N, Beconi M T. Participation of superoxide anion in the capacitation of cryopreserved bovine sperm[J]. International Journal of Andrology, 2003, 26: 109-114.

Olechnowicz J, Jaśkowski J M. Relation between clinical lameness and reproductive performance in dairy cows[J]. Medycyna Weterynaryjna, 2011, 67 (1): 5-9.

Ondiek J O, Ogore P B, Shakala E K, Kaburu GM. Prevalence of bovine mastitis, its therapeutics and control in Tatton Agriculture Park, Egerton University, Njoro District of Kenya[J]. Basic Research Journal of Agricultural Science and Review, 2013, 2 (1): 15-20.

O'Neill L A J, Bowie A G. The family of five: TIR-domain-containing adaptors in Toll-like receptor signalling[J]. Nature Reviews Immunology, 2007, 7 (5): 353-364.

O'Rourke D. Nutrition and udder health in dairy cows: a review[J]. Irish Veterinary Journal, 2009, 62: S15.

Orsatti C L, Missima F, Pagliarone A C, et al. Propolis immunomodulatory action in vivo on Toll-like receptors 2 and 4 expression and on pro-inflammatory cytokines production in mice [J]. Phytotherapy Research, 2010a, 24 (8): 1141-1146.

Orsatti C L, Missima F, Pagliarone A C, et al. Th1/Th2 cytokines' expression and production by propolis-treated mice[J]. Journal of Ethnopharmacology, 2010b, 129 (3): 314-318.

Osman K M, Hassan H M, Ibrahim I M, et al. The impact of staphylococcal mastitis on the level of milk IL-6, lysozyme and nitric oxide[J]. Comparative Immunology Microbiology & Infectious Diseases, 2010, 33 (1): 85-93.

Ote I, Taminiau B, Duprez J, Dizier I, Mainil J G. Genotypic characterization by polymerase chain reaction of Staphylococcus aureus isolates associated with bovine mastitis[J]. Veterinary Microbiology, 2011, 153: 285-292.

Otto M. Molecular basis of Staphylococcus epidermidis infections[J]. Seminars in Immunopathology, 2012, 34 (2): 201-214.

Otto M. Staphylococcus epidermidis pathogenesis[J]. Methods in Molecular Biology, 2014, 1106: 17-31.

Papp L V, Lu J, Holmgren A, Khanna K K. From selenium to selenoproteins: synthesis, identity, and their role in human health[J]. Antioxidants Redox Signaling, 2007, 9: 775-806.

Pareek R, Wellnitz O, Van D R, et al. Immunorelevant gene expression in LPS-challenged bovine mammary epithelial cells[J]. Journal of Applied Genetics, 2005, 46 (2): 171-177.

Park Y K, Fox L K, Hancock D D, et al. Prevalence and antibiotic resistance of mastitis pathogens isolated from dairy herds transitioning to organic management[J]. Journal of Veterinary Science, 2012, 13 (1): 103-105.

Parker K I, Compton C W R, Anniss F M, et al. Quarter-level analysis of subclinical and clinical mastitis in primiparous heifers following the use of a teat sealant or an injectable antibiotic, or both, precalving[J].

Journal of Dairy Science, 2008, 91 (1): 169-181.

Parker K I, Compton C W, Anniss F M, Weir A M, Heuer C, McDougall S. Subclinical and clinical mastitis in heifers following the use of a teat sealant precalving[J]. Journal of Dairy Science, 2007b, 90: 207-218.

Parker K I, Compton C W, Anniss F M, Weir A M, McDougal S. Management of dairy heifers and its relationships with the incidence of clinicalmastitis[J]. NewZealandVeterinary Journal, 2007a, 55: 208-216.

Passchyn P, Piepers S, Meulemeester L D, et al. Between-herd prevalence of Mycoplasma bovis in bulk milk in Flanders, Belgium[J]. Research in Veterinary Science, 2012, 92 (2): 219-220.

Paulrud C O. Basic Concepts of the bovine teat canal [J]. Veterinary Research Communications, 2005, 29 (3): 215-245.

Peacock, S J, Moore, C E, Justice, A, Kantzanou, M, Story, L, Mackie, K, O'Neill, G, Day, N P. Virulent combinations of adhesin and toxin genes in natural populations of Staphylococcus aureus[J]. Infection and Immunity, 2002, 70: 4987-4996.

Peake K A, Biggs A M, Argo C M, et al. Effects of lameness, subclinical mastitis and loss of body condition on the reproductive performance of dairy cows[J]. Veterinary Record, 2011, 168 (11): 301-301.

Philippa M, Mckee A S, Munks M W. Towards an understanding of the adjuvant action of aluminium[J]. Nature Reviews Immunology, 2009, 9 (4): 287-293.

Phuektes P, Mansell P D, Browning G F. Multiplex polymerase chain reaction assay for simultaneous detection of Staphylococcus aureus and streptococcal causes of bovine mastitis[J]. Journal of Dairy Science, 2001, 84 (5): 1140-1148.

Piepers S, De M L, De K A, et al. Prevalence and distribution of mastitis pathogens in subclinically infected dairy cows in Flanders, Belgium[J]. Journal of Dairy Research, 2007, 74 (4): 478-483.

Piepers S, Opsomer G, Barkema H W, et al. Heifers infected with coagulase-negative staphylococci in early lactation have fewer cases of clinical mastitis and higher milk production in their first lactation than noninfected heifers[J]. Journal of Dairy Science, 2010, 93 (5): 2014-2024.

Piessens V, Coillie E V, Verbist B, et al. Distribution of coagulase-negative Staphylococcus species from milk and environment of dairy cows differs between herds[J]. Journal of Dairy Science, 2011, 94 (6): 2933-2944.

Piessens V, Supré K, Heyndrickx M, et al. Validation of amplified fragment length polymorphism genotyping for species identification of bovine associated coagulase-negative staphylococci[J]. Journal of Microbiological Methods, 2010, 80 (3): 287-294.

Piette A, Verschraegen G. Role of coagulase-negative staphylococci in human disease[J]. Veterinary Microbiology, 2009, 134 (1-2): 45-54.

Pinedo P J, Melendez P, Viiiagomez-Cortes J A, et al. Effect of high somatic cell counts on reproductive performance of Chilean dairy cattle[J]. Journal of Dairy Science, 2009, 92 (4): 1575-1580.

Pirondini A, Bonas U, Maestri E, Visioli G, Marmiroli M, Marmiroli N. Yield and amplificability of different DNA extraction procedures for traceability in the dairy food chain[J]. Food Control, 2010, 21: 663-668.

Pitkälä A, Haveri M, Pyörälä S, et al. Bovine mastitis in Finland 2001--prevalence, distribution of bacteria, and antimicrobial resistance[J]. Journal of Dairy Science, 2004, 87 (8): 2433-2441.

Popovic Z. Performance and udder health status of dairy cows influenced by organically bound zinc and chromium[D]. Ph. D. Thesis, University of Belgrade. 2004.

Przybylska J, Albera E, Kankofer M. Antioxidants in Bovine Colostrum[J]. Reproduction in Domestic Animals, 2007, 42: 402-409.

Puvogel G, Baumrucker C R, Sauerwein H, Rühl R, Ontsouka E, Hammon H M, Blum J W. Effects of an Enhanced Vitamin A Intake During the Dry Period on Retinoids, Lactoferrin, IGF System, Mammary Gland Epithelial Cell Apoptosis, and Subsequent Lactation in Dairy Cows[J]. Journal of Dairy Science, 2005, 88: 1785-1800.

Pyörälä S. Mastitis in Post-Partum Dairy Cows[J]. Reproduction in Domestic Animals, 2008, 43(s2): 252-259.

Pyörlä S. Indicators of inflammation in the diagnosis of mastitis[J]. Veterinary Research, 2003, 34: 565-578.

Rainard P, Fromageau A, Cunha P, et al. Staphylococcus aureus lipoteichoic acid triggers inflammation in the lactating bovine mammary gland[J]. Veterinary Research, 2008, 39 (5): 498-498.

Rainard P, Riollet C. Innate immunity of the bovine mammary gland[J]. Veterinary Research, 2006, 37 (3): 369-400.

Rainard P, Riollet C. Mobilization of neutrophils and defense of the bovine mammary gland[J]. Reproduction Nutrition Development, 2003, 43 (5): 439-457.

Rainard P. The complement in milk and defense of the bovine mammary gland against infections[J]. Veterinary Research, 2003, 34 (5): 647-670.

Ranjan R, Swarup D, Naresh R, Patra RC. Enhanced erythrocytic lipid peroxides and reduced plasma ascorbic acid, and alteration in blood trace elements level in dairy cows with mastitis[J]. Veterinary Research Communications, 2005, 29: 27-34.

Raulo S M, Sorsa T, Tervahartiala T, et al. Increase in milk metalloproteinase activity and vascular permeability in bovine endotoxin-induced and naturally occurring Escherichia coli mastitis[J]. Veterinary Immunology and Immunopathology, 2002, 85 (3-4): 137-145.

Reinoso E B, El-Sayed A, Lammler C, Bogni C, Zschock M. Genotyping of Staphylococcus aureus isolated from humans, bovine subclinical mastitis and food samples in Argentina[J]. Research in Microbiology, 2008, 163: 314-322.

Renzoni A, Francois P, Li D, et al. Modulation of Fibronectin Adhesins and Other Virulence Factors in a Teicoplanin-Resistant Derivative of Methicillin-Resistant Staphylococcus aureus[J]. Antimicrobial Agents and Chemotherapy, 2004, 48 (8): 2958-2965.

Rich M. Staphylococci in animals: prevalence, identification and antimicrobial susceptibility, with an emphasis on methicillin-resistant Staphylococcus aureus[J]. British Journal of Biomedical Science, 2005, 62: 98-105.

Riffon R, Sayasith K, Khalil H, et al. Development of a rapid and sensitive test for identification of major pathogens in bovine mastitis by PCR[J]. Journal of Clinical Microbiology, 2001, 39 (7): 2584-2589.

Riollet C, Rainard P, Poutrel B. Cell subpopulations and cytokine expression in cow milk in response to chronic Staphylococcus aureus infection[J]. Journal of Dairy Science, 2001, 84 (5): 1077-1084.

Roark C L, Simonian P L, Fontenot A P, Born W K and O'Brien R L. γδ T cells: an important source of IL-17[J]. Current Opinion in Immunology, 2008, 20 (3): 353-357.

Rogers K L, Fey P D, Rupp M E. Coagulase-Negative Staphylococcal Infections[J]. Infectious Disease Clinics of North America, 2009, 23 (1): 73-98.

Roghmann M, Taylor K L, Gupte A, Zhan M, Johnson J A, Cross A, Edelman R, Fattom A I. Epidemiology of capsular and surface polysaccharide in Staphylococcus aureus infections complicated by bacteraemia[J]. The Journal of Hospital Infection, 2005, 59: 27-32.

Rosendo O, Staples C R, McDowell L R, McMahon R, Badinga L, Martin F G, Shearer J F, Seymour W M, Wilkinson N S. Effects of biotin supplementation on peripartum performance and metabolites of

Holstein cows[J]. Journal of Dairy Science, 2004, 87: 2535-2545.

Rowe D C, Mcgettrick A F, Eicke L, et al. The myristoylation of TRIF-related adaptor molecule is essential for Toll-like receptor 4 signal transduction[J]. Proceedings of the National Academy of Science, 2006, 103 (16): 6299-6304.

Roy J P, Francoz D, Labrecque O. Mastitis in a 7-week old calf caused by Mycoplasma bovigenitalium[J]. Veterinary Journal, 2008, 176 (3): 403-404.

Rubinstein E, Kollef M H, Nathwani D. Pneumonia caused by methicillin-resistant Staphylococcus aureus[J]. Clinical Infectious Diseases, 2008, 46 Suppl 5 (Suppl 5): S378-S385.

Sabini L, Torres C, Demo M, et al. Effect of Staphylococcus toxins isolated from dairy cow milk on Vero cell monolayers[J]. Revista Latinoamericana De Microbiología, 2001, 43 (1): 13-18.

Sabour P M, Gill J J, Lepp D, et al. Molecular typing and distribution of Staphylococcus aureus isolates in Eastern Canadian dairy herds[J]. Journal of Clinical Microbiology, 2004, 42 (8): 3449-3455.

Sacadura FC, Robinson PH, Evans E, Lordelo M. Effects of a ruminally protected B-vitamin supplement on milk yield and composition of lactating dairy cows[J]. Animal Feed Science and Technology, 2008, 144: 111-124.

Sahebekhtiari N, Nochi Z, Eslampour M A, et al. Characterization of Staphylococcus aureus strains isolated from raw milk of bovine subclinical mastitis in Tehran and Mashhad[J]. Acta microbiologica et immunologica Hungarica, 2011, 58 (2): 113-121.

Salasia S I O, Khusnan Z, Lämmler C, et al. Comparative studies on pheno-and genotypic properties of Staphylococcus aureus isolated from bovine subclinical mastitis in central Java in Indonesia and Hesse in Germany[J]. Journal of Veterinary Science, 2004, 5 (2): 103-109.

Sampimon O C, Zadoks R N. Performance of API Staph ID 32 and Staph-Zym for identification of coagulase-negative staphylococci isolated from bovine milk samples[J]. Veterinary Microbiology, 2009, 136 (3-4): 300-305.

Sampimon O, van den Borne B H, Santman-Berends I, Barkema H W, Lam T. Effect of coagulase-negative staphylococci on somatic cell count in Dutch dairy herds[J]. Journal of Dairy Research, 2010, 77: 318-324.

Santman-Berends I M, Olde R R, Sampimon O C, van Schaik G, Lam T J. Incidence of subclinicalmastitis inDutch dairy heifers in the first 100 days in lactation and associated risk factors[J]. Journal of Dairy Science, 2012, 95: 2476-2484.

Santos J, Cerri R, Ballou M, et al. Effect of timing of first clinical mastitis occurrence on lactational and reproductive performance on Holstein dairy cows[J]. Animal Reproduction Science, 2004, 80: 31-45.

Sargeant J M, Leslie K E, Shirley J E, et al. Sensitivity and specificity of somatic cell count and California mastitis test for identifying intramammary infection in early lactation[J]. Journal of Dairy Science, 2001, 84: 2018-2024.

Sarker S C, Parvin M S, Rahman A K, Islam M T. Prevalence and risk factors of subclinical mastitis in lactating dairy cows in north and south regions of Bangladesh[J]. Tropical Animal Health and Production, 2013, 45: 1171-1176.

Satorres S E, Alcaráz L E, Cargnelutti E, et al. IFN-gamma plays a detrimental role in murine defense against nasal colonization of Staphylococcus aureus[J]. Immunology Letters, 2009, 123 (2): 185-188.

Scaletti R W, Harmon R J. Effect of dietary copper source on response to coliform mastitis in dairy cows[J]. Journal of Dairy Science, 2012, 95 (2): 654-662.

Scaletti R W, Trammell D S, Smith B A. Role of Dietary Copper in Enhancing Resistance to Escherichia coli Mastitis[J]. Journal of Dairy Science, 2003, 86: 1240-1249.

Schmaler M, Jann N, Ferracin F, Landmann R. T and B cells are not required for clearing Staphylococcus aureus in systemic infection despite a strong TLR2-MyD88-dependent T cell activation[J]. Journal of Immunology, 2011, 186 (1): 443-452.

Schrick F, Hockett M, Saxton A, et al. Influence of subclinical mastitis during early lactation on reproductive parameters[J]. Journal of Dairy Science, 2001, 84 (6): 1407-1412.

Schukken Y H, González R N, Tikofsky L L, et al. CNS mastitis: nothing to worry about?[J]. Veterinary Microbiology, 2008, 134 (1-2): 9-14.

Seeley E J, Matthay M A, Wolters P J. Inflection points in sepsis biology: From local defense to systemic organ injury[J]. American Journal of Physiology Lung Cellular & Molecular Physiology, 2012, 303 (5): L355-63.

Shamay A, Shapiro F, Mabjeesh S J, et al. Casein-derived phosphopeptides disrupt tight junction integrity, and precipitously dry up milk secretion in goats[J]. Life Sciences, 2002, 70 (23): 2707-2719.

Sharma N K, Rees C E, Dodd C E. Development of a single-reaction multiplex PCR toxin typing assay for Staphylococcus aureus strains[J]. Applied and Environmental Microbiology, 2000, 66, 1347-1353.

Sharma N, Kang T Y, Lee S J, Kim J N, Hur C H, Ha J C, Vohra V, Jeong D K. Status of bovine mastitis and associated risk factors in subtropical Jeju Island, South Korea[J]. Tropical Animal Health and Production, 2013, 45: 1829-1832.

Sharma N, Singh N K, Bhadwal M S. Relationship of somatic cell count and mastitis: An overview[J]. Asian-Australasian Journal of Animal Sciences, 2011, 24: 429-438.

Shaver RD, Bal MA. Effect of Dietary Thiamin Supplementation on Milk Production by Dairy Cows[J]. Journal of Dairy Science, 2000, 83: 2335-2340.

Sheppard S C, Sanipelli B. Trace elements in feed, manure, and manured soils[J]. Journal of Environmental Quality, 2012, 41 (6): 1846-1856.

Shinefield H R. Use of a conjugate polysaccharide vaccine in the prevention of invasive staphylococcal disease: is an additional vaccine needed or possible?[J]. Vaccine, 2006, 24 (Suppl 2): 65-69.

Shittu A, Abdullahi J, Jibril A, Mohammed A A, Fasina F O. Sub-clinical mastitis and associated risk factors on lactating cows in the Savannah Region of Nigeria[J]. BMC Veterinary Research, 2012, 8: 134.

Shkreta L, Talbot B G, Diarra M S, et al. Immune responses to a DNA/protein vaccination strategy against Staphylococcus aureus induced mastitis in dairy cows[J]. Vaccine, 2004, 23 (1): 114-126.

Shpigel N Y, Elazar S, Rosenshine I. Mammary pathogenic Escherichia coli[J]. Current Opinion in Microbiology, 2008, 11 (1): 60-65.

Sieradzki K, Tomasz A. Alterations of cell wall structure and metabolism accompany reduced susceptibility to vancomycin in an isogenic series of clinical isolates of Staphylococcus aureus[J]. Journal of Bacteriology, 2003, 185 (24): 7103-7110.

Sierra D, Sánchez A, Luengo C, et al. Temperature effects on Fossomatic cell counts in goats milk[J]. International Dairy Journal, 2006, 16: 385-387.

Silva E R, Boechat J U D, Martins J C D, et al. Hemolysin production by Staphylococcus aureus species isolated from mastitic goat milk in Brazilian dairy herds[J]. Small Ruminant Research, 2005, 56: 271-275.

Song H R, Li Z G, Luo R Q, Guang H, Gu X L, Zhang F Zhou, X. Isolation and Idlemirlaltion on Maslilis Pathogen of Cow in Shihezi Area and Bacteriostatic Test with Chinese Herbal Medicine[J]. Journal of

Anhui agriculture Science, 2009, 37: 4044-4047.

Sood P, Nanda A S. Effect of lameness on estrous behavior in crossbred cows[J]. Theriogenology, 2006, 66 (5): 1375-1380.

Sordillo L M, O'Boyle N, Gandy J C, Corl C M, Hamilton E. Shifts in thioredoxin reductase activity and oxidant status in mononuclear cells obtained from transition dairy cattle[J]. Journal of Dairy Science, 2007, 90: 1186-1192.

Sordillo L M, Streicher K L. Mammary gland immunity and mastitis susceptibility[J]. Journal of Mammary Gland Biology & Neoplasia, 2002, 7 (2): 135-146.

Sordillo L M. Factors affecting mammary gland immunity and mastitis susceptibility [J]. Livestock Production Science, 2005, 98 (1): 89-99.

Sordillo L M. New concepts in the causes and control of mastitis[J]. Journal of Mammary Gland Biology & Neoplasia, 2011, 16 (4): 271-273.

Sori H, Zerihun A, Abdicho S. Dairy Cattle Mastitis In and Around Sebeta, Ethiopia[J]. Journal of Applied Research in Veterinary Medicine, 2005, 4: 332-338.

Spears J W, Weiss W P. Role of antioxidants and trace elements in health and immunity of transition dairy cows[J]. Veterinary Journal, 2008, 176 (1): 70-76.

Spears J W. Micronutrients and immune function in cattle[J]. Proceedings of the Nutrition Society, 2000, 59: 587-594.

Spears J W. Trace mineral bioavailability in ruminants[J]. Journal of Nutrition, 2003, 133: 1506S-1509S.

Statistical Analyses Systems SAS, 2004, . SAS/STAT User's Guide, 9.1 edition. SAS Institute Inc, Cary, North Carolina, USA.

Stephan R, Annemüller C, Hassan A A, et al. Characterization of enterotoxigenic Staphylococcus aureus st rains isolated from bovine mastitis in northeast Switzerland[J]. Veterinary Microbiology, 2000, 78(4): 373-382.

Su W J, Chang C J, Peh H C, et al. Apoptosis and oxidative stress of infiltrated neutrophils obtained from mammary glands of goats during various stages of lactation[J]. American Journal of Veterinary Research, 2002, 63: 241-246.

Suojala L, Orro T, Järvinen H, et al. Acute phase response in two consecutive experimentally induced E. coli intramammary infections in dairy cows[J]. Acta Veterinaria Scandinavica, 2008, 50 (1): 18.

Supré K, Haesebrouck F, Zadoks R N, et al. Some coagulase-negative Staphylococcus species affect udder health more than others[J]. Journal of Dairyence, 2011, 94 (5): 2329-2340.

Supré K, Vliegher S D. Use of transfer RNA-intergenic spacer PCR combined with capillary electrophoresis to identify coagulase-negative Staphylococcus species originating from bovine milk and teat apices[J]. Journal of Dairyence, 2009, 92 (7): 3204-3210.

Suriyasathaporn W, Vinitketkumnuen U, Chewonarin T, et al. Higher somatic cell counts resulted in higher malondialdehyde concentrations in raw cows' milk[J]. International Dairy Journal, 2006, 16: 1088-1091.

Suzuki C, Yoshioka K, Iwamura S, et al. Endotoxin induces delayed ovulation following endocrine aberration during the proestrous phase in Holstein heifers[J]. Domestic Animal Endocrinology, 2001, 20 (4): 267-278.

Svensson C, Nyman A K, Waller K P, et al. Effects of housing, management, and health of dairy heifers on first-lactation udder health in southwest Sweden[J]. Journal of Dairy Science, 2006, 89 (6): 1990-1999.

Swanson K M, Stelwagen K, Dobson J, et al. Transcriptome profiling of Streptococcus uberis-induced mastitis reveals fundamental differences between immune gene expression in the mammary gland and in a primary cell culture model[J]. Journal of Dairy Science, 2009, 92 (1): 117-129.

Takahashi H, Odai M, Mitani K, et al. Effect of intramammary injection of rboGM-CSF on milk levels of chemiluminescence activity, somatic cell count, and Staphylococcus aureus count in Holstein cows with S. aureus subclinical mastitis[J]. Canadian journal of veterinary research = Revue canadienne de recherche vétérinaire, 2004, 68 (3): 182-187.

Talbot B G, Lacasse P. Progress in the development of mastitis vaccines[J]. Livestock Production Science, 2005, 98 (s1-2): 101-113.

Tamarapu, S, McKillip, J L, Drake, M. Development of a multiplex polymerase chain reaction assay for detection and differentiation of Staphylococcus aureus in dairy products[J]. Journal of Food Protection, 2001, 64: 664-668.

Tammy K, Phulwani N K, Nilufer E, et al. MyD88-dependent signals are essential for the host immune response in experimental brain abscess[J]. Journal of Immunology, 2007, 178 (7): 4528-4537.

Taponen S, Koort J, Björkroth J, et al. Bovine intramammary infections caused by coagulase-negative staphylococci may persist throughout lactation according to amplified fragment length polymorphism-based analysis[J]. Journal of Dairy Science, 2007, 90 (7): 3301-3307.

Téllez-Pérez A D, Alva-Murillo N, Ochoa-Zarzosa A, et al. Cholecalciferol (vitamin D) differentially regulates antimicrobial peptide expression in bovine mammary epithelial cells: implications during Staphylococcus aureus internalization[J]. Veterinary Microbiology, 2012, 160 (1-2): 91-98.

Tenhagen B A, Hansen I, Reinecke A, Heuwieser W. Prevalence of pathogens in milk samples of dairy cows with clinical mastitis and in heifers at first parturition[J]. Journal of Dairy Research, 2009, 76: 179-187.

Terhune T D, Deth R C. How aluminum adjuvants could promote and enhance non-target IgE synthesis in a genetically-vulnerable sub[J]. Journal of Immunotoxicology, 2013, 10 (2): 210-222.

Thorberg B M, Danielsson-Tham M L, Emanuelson U, et al. Bovine subclinical mastitis caused by different types of coagulase-negative staphylococci[J]. Journal of Dairy Science, 2009, 92 (10): 4962-4970.

Thorberg B M. Coagulase-negative staphylococci in bovine sub-clinical mastitis[D]. Swedish University of Agricultural Sciences, 2008.

Tollersrud T, Norstebo P E, Engvik J P, et al. Antibody responses in sheep vaccinated against Staphylococcus aureus mastitis: a comparison of two experimental vaccines containing different adjuvants[J]. Veterinary Research Communications, 2002, 26 (8): 587-600.

Tollersrud T, Zernichow L, Andersen S R, et al. Staphylococcus aureus capsular polysaccharide type 5 conjugate and whole cell vaccines stimulate antibody responses in cattle[J]. Vaccine, 2001, 19: 3896-3903.

Tuchscherr L P, Buzzola F R, Alvarez L P, Caccuri R L, Lee J C, Sordelli D O. Capsule-negative Staphylococcus aureus induces chronic experimental mastitis in mice[J]. Infection and Immunity, 2005, 73: 7932-7937.

Turashvili G, Yang W, McKinney S, Kalloger S, Gale N, Ng Y, Chow K, Bell L, Lorette J, Carrier M, Luk M, Aparicio S, Huntsman D, Yip S. Nucleic acid quantity and quality from paraffin blocks: defining optimal fixation, processing and DNA/RNA extraction techniques[J]. Experimental and Molecular Pathology, 2012, 92: 33-43.

Uchida K, Mandebvu P, Ballard C S, Sniffen CJ, Carter MP. Effect of feeding a combination of zinc, manganese and copper amino acid complexes, and cobalt glucoheptonate on performance of early lactation high producing dairy cows[J]. Anim. Feed Sci. Tech. 2001, 93: 193-203.

Uddin M A, Kamal M M, Haque M E. Epidemiological study of udder and teat diseases in dairy cows[J]. Bangladesh Journal of Veterinary Medicine, 2009, 7: 332-340.

Ugall S, Parker K I, Heuer C, et al. A review of prevention and control of heifer mastitis via non-antibiotic strategies[J]. Veterinary Microbiology, 2009, 134 (s 1-2): 177-185.

Ugall S, Parker K I, Weir A M, et al. Effect of application of an external teat sealant and/or oral treatment with a monensin capsule pre-calving on the prevalence and incidence of subclinical and clinical mastitis in dairy heifers[J]. New Zealand Veterinary Journal, 2008, 56 (3): 120-129.

Urban C F, Reichard U, Brinkmann V, et al. Neutrophil extracellular traps capture and kill Candida albicans yeast and hyphal forms [J]. Cellular microbiology, 2006, 8 (4): 668-676.

Vaarst M, T. W. Bennedsgaard, I. Klaas, et al. Development and Daily Management of an Explicit Strategy of Nonuse of Antimicrobial Drugs in Twelve Danish Organic Dairy Herds[J]. Journal of Dairy Science, 2006, 89: 1842-1853.

van Beelen A J, Zelinkova Z, Taanman-Kueter E W, Muller F J, Hommes D W, Zaat S A, Kapsenberg M L, de Jong E C. Stimulation of the Intracellular Bacterial Sensor NOD2 Programs Dendritic Cells to Promote Interleukin-17 Production in Human Memory T Cells[J]. Immunity, 2007, 27 (4): 660-669.

Vanderhaeghen W, Cerpentier T. Methicillin-resistant Staphylococcus aureus (MRSA) ST398 associated with clinical and subclinical mastitis in Belgian cows[J]. Veterinary Microbiology, 2010, 144 (1-2): 166-171.

Verdier I, Durand G, Bes M, Taylor K L, Lina G, Vandenesch F, Fattom A I, Etienne J. Identification of the capsular polysaccharides in Staphylococcus aureus clinical isolates by PCR and agglutination tests[J]. Journal of Clinical Microbiology, 2007, 45: 725-729.

Verma G, Bhatia H, Datta M. Gene expression profiling and pathway analysis identify the integrin signaling pathway to be altered by IL-1β in human pancreatic cancer cells: role of JNK[J]. Cancer letters, 2012, 320 (1): 86-95.

Viguier C, Arora S, Gilmartin N, et al. Mastitis detection: current trends and future perspectives[J]. Trends in Biotechnology, 2009, 27 (8): 486-493.

Vijaya Reddy L, Choudhuri P C, Hamza P A. Sensitivity, specificity and predictive values of various indirect tests in the diagnosis of subclinical mastitis[J]. Indian Veterinary Journal, 1998, 75: 1004-1005.

Virgilio F D, Ceruti S, Bramanti P, et al. Purinergic signalling in inflammation of the central nervous system[J]. Trends in Neurosciences, 2009, 32 (2): 79-87.

Visai L, Yanagisawa N, Josefsson E, et al. Immune evasion by Staphylococcus aureus conferred by iron-regulated surface determinant protein IsdH[J]. Microbiology, 2009, 155 (Part 3): 667-679.

Vliegher S D, Fox L K, Piepers S, et al. Invited review: Mastitis in dairy heifers: nature of the disease, potential impact, prevention, and control[J]. Journal of Dairy Science, 2012, 95 (3): 1025-1040.

von Eiff, C, Friedrich, A W, Peters, G, Becker, K. Prevalence of genes encoding for members of the staphylococcal leukotoxin family among clinical isolates of Staphylococcus aureus[J]. Diagnostic Microbiology and Infectious Disease, 2004, 49: 157-162.

Waage S, Skei H R, Rise J, Rogdo T, Sviland S, Odegaard S A. Outcome of clinical mastitis in dairy heifers assessed by reexamination of cases one month after treatment[J]. Journal of Dairy Science, 2000, 83: 70-76.

Walker S L, Smith R F, Jones D N, et al. Chronic stress, hormone profiles and estrus intensity in dairy cattle[J]. Hormones and behavior, 2008a, 53 (3): 493-501.

Walker S L, Smith R F, Routly J E, et al. Lameness, activity time-budgets, and estrus expression in dairy cattle[J]. Journal of dairy science, 2008b, 91 (12): 4552-4559.

Waller K P, Bengtsson B, Lindberg A, et al. Incidence of mastitis and bacterial findings at clinical mastitis in Swedish primiparous cows-influence of breed and stage of lactation[J]. Veterinary Microbiology, 2009, 134 (1-2): 89-94.

Wang C, Liu Q, Yang WZ, Dong Q, Yang XM, He DC, Zhang P, Dong KH, Huang YX. Effects of selenium yeast on rumen fermentation, lactation performance and feed digestibilities in lactating dairy cows[J]. Livestock Science, 2009, 126: 239-244.

Wang R N, Wang Y B, Geng J W, et al. Enhancing immune responses to inactivated porcine parvovirus oil emulsion vaccine by co-inoculating porcine transfer factor in mice[J]. Vaccine, 2012, 30 (35): 5246-5252.

Wang S C, Wu C M, Xia S C, Qi Y H, Xia L N, Shen J Z. Distribution of superantigenic toxin genes in Staphylococcus aureus isolates from milk samples of bovine subclinical mastitis cases in two major diary production regions of China[J]. Veterinary Microbiology, 2009, 137: 276-281.

Wang Z, Huang J, Xia P, et al. Alternative splicing and expression of cattle clusterin gene in healthy and mastitis-infected mammary tissues[J]. Journal of Animal Science Advances, 2012, 2 (4): 425-428.

Wartha F, Beiter K, Normark S, et al. Neutrophil extracellular traps: casting the NET over pathogenesis [J]. Current opinion in microbiology, 2007, 10 (1): 52-56.

Weese J S, Dick H, Willey B M, et al. Suspected transmission of methicillin-resistant Staphylococcus aureus between domestic pets and humans in veterinary clinics and in the household[J]. Veterinary Microbiology, 2006, 115 (s 1-3): 148-155.

Weigel L M, Clewell D B, Gill S R, et al. Genetic analysis of a high-level vancomycin-resistant isolate of Staphylococcus aureus[J]. Science, 2003, 302 (5650): 1569-1571.

Weiss W P, Hogan J S, Smith K L. Changes in vitamin C concentrations in plasma and milk from dairy cows after an intramammary infusion of Escherichia coli[J]. Journal of Dairy Science, 2004, 87: 32-37.

Weiss W P, Spears J W. Vitamin and trace mineral effects on immune function of ruminants. In: Sejrsen K, Hvelplund T, Nielsen M O (Eds.), Ruminant Physiology. Wageningen Academic Publishers, Utrecht, The Netherlands, 2006, pp. 473-496.

Werling D, Piercy J, Coffey T J. Expression of TOLL-like receptors (TLR) by bovine antigen-presenting cells-Potential role in pathogen discrimination?[J]. Veterinary Immunology & Immunopathology, 2006, 112 (s1-2): 2-11.

West J W. Effects of heat-stress on production in dairy cattle[J]. Journal of Dairy Science, 2003, 86 (6): 2131-2144.

Wieliczko R J, Williamson J H, Cursons R T, et al. Molecular typing of Streptococcus uberis strains isolated from cases of bovine mastitis[J]. Journal of Dairy Science, 2002, 85 (9): 2149-2154.

Williams R J, Henderson BSharp L J, Nair S P. Identification of a fibronectin-binding protein from Staphylococcus epidermidis[J]. Infection & Immunity, 2003, 70 (12): 6805-6810.

Winn W, Allen S, Janda W, Koneman E, Procop G, Schreckenberger P, Woods G. Koneman's Color Atlas and Textbook of Diagnostic Microbiology. Sixth Edition, Lippincott Williams & Wilkins, Philadelphia, 2006, pp.623-671.

Workineh S, Bayleyegn M, Mekonnen H, et al. Prevalence and Aetiology of Mastitis in Cows from Two Major Ethiopian Dairies[J]. Tropical Animal Health and Production, 2002, 34: 19-25.

Workineh S, Bayleyegn M, Mekonnen H, Potgieter L N D. Prevalence and Aetiology of Mastitis in Cows from Two Major Ethiopian Dairies[J]. Tropical Animal Health and Production, 2002, 34: 19-25.

Wu B Zhang W, Huang J. et al. Effect of recombinant Panton-Valentine leukocidin in vitro on apoptosis and cytokine production of human alveolar macrophages[J]. Canadian Journal of Microbiology, 2010, 56 (3): 229-235.

Xi C W, Wu J F. dATP/ATP, a Multifunctional Nucleotide, Stimulates Bacterial Cell Lysis, Extracellular DNA Release and Biofilm Development[J]. Plos One, 2010, 5 (10): e13355.

Yang B, Zhang X M, Luo S P, Liu M J. Investigation on Virulence Determinants and Genetic Typing of Staphylococcus aureus Isolated from Mastitis Milk[J]. Acta Agriculturae Boreali-Occidentalis Sinica, 2009, 18, 1-6.

Yang F L, Li X S, He B X, et al. Bovine Mastitis in Subtropical Dairy Farms, 2005-2009[J]. Journal of Animal and Veterinary Advances, 2011, 10 (1): 68-72.

Yang F L, Li X S, He B X, Yang X L, Li G H, Liu P, Huang Q H, Pan X M, Li J. Malondialdehyde level and some enzymatic activities in subclinical mastitis milk[J]. African Journal of Biotechnology, 2011b, 10: 5534-5538.

Yang F L, Li X S, He B X. Effects of vitamins and trace-elements supplementation on milk production in dairy cows: A review[J]. African Journal of Biotechnology, 2011a, 10: 2574-2578.

Yang F L, Li X S, Liang X W, et al. Detection of virulence-associated genes in *Staphylococcus aureus* isolated from bovine clinical mastitis milk samples in Guangxi[J]. Tropical Animal Health & Production, 2012, 44 (8): 1821-1826.

Yohannis M, Molla W. Prevalence, risk factors and major bacterial causes of bovine mastitis in and around Wolaita Sodo, Southern Ethiopia[J]. African Journal of Microbiology Research, 2013, 7: 5400-5405.

Zadoks R N, Watts J L. Species identification of coagulase-negative staphylococci: genotyping is superior to phenotyping[J]. Veterinary Microbiology, 2009, 134 (1-2): 20-28.

Zecconi A, Cesaris L, Liandris E, Dapra V, Piccinini R. Role of several Staphylococcus aureus virulence factors on the inflammatory response in bovine mammary gland[J]. Microbial Pathogenesis, 2006, 40: 177-183.

Zhang H, Wei H W, Cui Y, Zhao G Q, Feng F Q. Antibacterial interactions of monolaurin with commonly used antimicrobials and food components[J]. Journal of Food Science, 2009, 74: 418-421.

Zhang R, Brennan M L, Shen Z, et al. Myeloperoxidase functions as a major enzymatic catalyst for initiation of lipid peroxidation at sites of inflammation[J]. The Journal of Biological Chemistry, 2002, 277 (48): 46116-46122.

Zheng Y, Danilenko D M, Valdez P, et al. Interleukin-22, a TH17 cytokine, mediates IL-23-induced dermal inflammation and acanthosis[J]. Nature, 2007, 445 (7128): 648-651.

Zheng Y, Valdez P A, Danilenko D M, Hu Y, Sa S M, Gong Q, Abbas A R, Modrusan Z, Ghilardi N, de Sauvage F J, Ouyang W. Interleukin-22 mediates early host defense against attaching and effacing bacterial pathogens[J]. Nature Medicine, 2008, 14 (3): 282-289.

Zhou H, Xiong Z Y, Li H P, et al. An immunogenicity study of a newly fusion protein Cna-FnBP vaccinated against Staphylococcus aureus infections in a mice model[J]. Vaccine, 2006, 24 (22): 4830-4837.

Zhou X, Bailey-Bucktrout S, Jeker L T, Bluestone J A. Plasticity of CD4+ FoxP3+ T cells[J]. Current Opinion in Immunology, 2009, 21 (3): 281-285.

Zschöck M, Botzler D, Blöcher S, et al. Detection of genes for enterotoxins (ent) and toxic shock syndrome toxin-1 (tst) in mammary isolates of Staphylococcus aureus by polymerase-chain-reaction [J]. International Dairy Journal, 2000, 10 (8): 569-574.